Molecular Plant Breeding and Genome Editing Tools for Crop Improvement

Pradip Chandra Deka
Sir Padampat Singhania University, India

IGI Global
PUBLISHER of TIMELY KNOWLEDGE

A volume in the Advances in Environmental Engineering and Green Technologies (AEEGT) Book Series

Published in the United States of America by
IGI Global
Engineering Science Reference (an imprint of IGI Global)
701 E. Chocolate Avenue
Hershey PA, USA 17033
Tel: 717-533-8845
Fax: 717-533-8661
E-mail: cust@igi-global.com
Web site: http://www.igi-global.com

Library of Congress Cataloging-in-Publication Data

Names: Deka, Pradip Chandra, 1947- author.
Title: Molecular plant breeding and genome editing tools for crop improvement / Pradip Chandra Deka.
Description: Hershey, PA : Engineering Science Reference, [2020] | Includes bibliographical references and index. | Summary: "This book examines the application of various molecular breeding techniques for crop improvements, identification of cultivars, and germplasm preservation"-- Provided by publisher.
Identifiers: LCCN 2020001215 (print) | LCCN 2020001216 (ebook) | ISBN 9781799843122 (hardcover) | ISBN 9781799852995 (paperback) | ISBN 9781799843139 (ebook)
Subjects: LCSH: Plant breeding. | Plant biotechnology. | Plant genomes.
Classification: LCC SB123 .D453 2020 (print) | LCC SB123 (ebook) | DDC 631.5/2--dc23
LC record available at https://lccn.loc.gov/2020001215
LC ebook record available at https://lccn.loc.gov/2020001216

This book is published in the IGI Global book series Advances in Environmental Engineering and Green Technologies (AEEGT) (ISSN: 2326-9162; eISSN: 2326-9170)

British Cataloguing in Publication Data
A Cataloguing in Publication record for this book is available from the British Library.

Advances in Environmental Engineering and Green Technologies (AEEGT) Book Series

ISSN:2326-9162
EISSN:2326-9170

Editor-in-Chief: Sang-Bing Tsai, Zhongshan Institute, University of Electronic Science and Technology of China, China & Wuyi University, China & Ming-Lang Tseng, Lunghwa University of Science and Technology, Taiwan & Yuchi Wang, University of Electronic Science and Technology of China Zhongshan Institute, China

MISSION

Growing awareness and an increased focus on environmental issues such as climate change, energy use, and loss of non-renewable resources have brought about a greater need for research that provides potential solutions to these problems. Research in environmental science and engineering continues to play a vital role in uncovering new opportunities for a "green" future.

The **Advances in Environmental Engineering and Green Technologies (AEEGT)** book series is a mouthpiece for research in all aspects of environmental science, earth science, and green initiatives. This series supports the ongoing research in this field through publishing books that discuss topics within environmental engineering or that deal with the interdisciplinary field of green technologies.

COVERAGE

- Alternative Power Sources
- Cleantech
- Radioactive Waste Treatment
- Contaminated Site Remediation
- Water Supply and Treatment
- Policies Involving Green Technologies and Environmental Engineering
- Electric Vehicles
- Waste Management
- Industrial Waste Management and Minimization
- Green Technology

IGI Global is currently accepting manuscripts for publication within this series. To submit a proposal for a volume in this series, please contact our Acquisition Editors at Acquisitions@igi-global.com or visit: http://www.igi-global.com/publish/.

The Advances in Environmental Engineering and Green Technologies (AEEGT) Book Series (ISSN 2326-9162) is published by IGI Global, 701 E. Chocolate Avenue, Hershey, PA 17033-1240, USA, www.igi-global.com. This series is composed of titles available for purchase individually; each title is edited to be contextually exclusive from any other title within the series. For pricing and ordering information please visit http://www.igi-global.com/book-series/advances-environmental-engineering-green-technologies/73679. Postmaster: Send all address changes to above address.

Titles in this Series

For a list of additional titles in this series, please visit:
http://www.igi-global.com/book-series/advances-environmental-engineering-green-technologies/73679

Solar Concentrating Modules With Louvered Heliostats Emerging Research and Opportunities
Dmitry Strebkov (Federal State Budget Scientific Institution, Russia & Federal Scientific Agroengineering Center VIM, Russia) Natalya Filippchenkova (Federal State Budget Scientific Institution, Russia & Federal Scientific Agroengineering Center VIM, Russia) and Anatoly Irodionov (Federal State Budget Scientific Institution, Russia & Federal Scientific Agroengineering Center VIM, Russia)
Engineering Science Reference • © 2021 • 200pp • H/C (ISBN: 9781799842767) • US $190.00

Recent Advancements in Bioremediation of Metal Contaminants
Satarupa Dey (Shyampur Siddheswari Mahavidyalaya, India) and Biswaranjan Acharya (School of Computer Engineering, KIIT University (Deemed), India)
Engineering Science Reference • © 2021 • 363pp • H/C (ISBN: 9781799848882) • US $195.00

Global Environmental Impacts on Food Security, Nutrition, and Human Health
Hicham Chatoui (Private University of Marrakech, (UPM), Morocco & Cady Ayyad University, Morocco) Mohamed Merzouki (Faculty of Science and Technology, Sultan Moulay Slimane University, Beni Mellal, Morocco) Hanane Moummou (Private University of Marrakech (UPM), Marrakech, Morroco) Mounir Tilaoui (Private University of Marrakech (UPM), Sultan Moulay Slimane University, Beni Mellal, Morocco) and Nabila Saadaoui (Private University of Marrakech (UPM), Marrakech, Morroco)
Engineering Science Reference • © 2020 • 300pp • H/C (ISBN: 9781799837480) • US $195.00

Spatial Information Science for Natural Resource Management
Suraj Kumar Singh (Suresh Gyan Vihar University, Jaipur, India) Shruti Kanga (Suresh Gyan Vihar University, Jaipur, India) and Varun Narayan Mishra (Suresh Gyan Vihar University, Jaipur, India)
Engineering Science Reference • © 2020 • 355pp • H/C (ISBN: 9781799850274) • US $195.00

For an entire list of titles in this series, please visit: http://www.igi-global.com/book-series/advances-environmental-engineering-green-technologies/73679

IGI Global
PUBLISHER of TIMELY KNOWLEDGE

701 East Chocolate Avenue, Hershey, PA 17033, USA
Tel: 717-533-8845 x100 • Fax: 717-533-8661
E-Mail: cust@igi-global.com • www.igi-global.com

Table of Contents

Preface

Plant Breeding has played a very significant role in increasing global food production during last several decades. However, after exploitation of all the traditional approaches it has become necessary to seek other tools to increase food production.

Two disciplines that have revolutionized crop improvement in the recent past are molecular breeding and plant genomics. Application of molecular markers in plant genetics started in late 1980s and has accelerated the pace and precision of plant breeding process. Genetic variation is usually detected by identifying the polymorphisms exhibited at restriction site, as fragment lengths, or at single nucleotide levels either in genic or intergenic regions of the genome. Traditionally, the development of markers such as microsatellites, RFLPs and ALFPs was a costly iterative process that involved time-consuming, cloning and primer design steps that could not easily be parallelized.

Discovery of markers based on simple sequence repeats (SSRs), along with single nucleotide polymorphisms (SNPs) and the availability of high-throughput (HTP) genotyping platforms have accelerated the generation of dense genetic linkage maps, which in turn contributed towards routine application of marker-assisted breeding in several crops. In recent years, SNPs have been the markers of choice for the researchers, due to their high abundance and amenability for automation and high-throughput (HTP) genotyping capabilities. However, use of marker-assisted selection (MAS) remained a challenge for traits which are complex in their inheritance pattern such as yield, improved nutrition value and resistance to several biotic and abiotic stresses. Therefore, there was a dire need for the molecular dissection of these traits in the context of the whole genome. Introduction of plant genomics has contributed towards resolving this issue. Availability of a multitude of "omics" technologies and computational tools has provided unprecedented ability to dissect the molecular and genetic basis of traits as well as characterization of whole genome.

The traditional approach to species or variety identification involves observation and recording of morphological characters or descriptors. This approach is not precise, less informative and time consuming. The legal right to market a newly-bred cultivar depends on the results of statutory testing, which provides information regarding distinctness and uniformity and stability (DUS). However, it is less suitable when results are required rapidly and more plant samples need to be separated. Furthermore, morphological characters are often multigenic, not available at all growth stages and influenced by environmental factors. Use of molecular markers serves as modern and suitable approach to cultivar and variety identification, as it is more accurate, rapid and cost effective. Different molecular marker techniques have been used in plant population genetics, phylogenetics and biodiversity studies, analysis of recombination frequencies between genotypes, and identification of genes for important agricultural traits. Molecular markers are currently used to protect plant breeders' right (PBR) accurately, and thereby encourage breeders for continuous development of new varieties.

Similarly, other molecular biological methods, especially DNA fingerprinting techniques, have promising applications in the identification of plant genotypes including varieties and cultivars. Recent years have witnessed increasingly rapid development of molecular phylogenetics and systematics. This is caused by the development of new diverse methods of analysis of molecular DNA markers. These methods allow researchers to assess genetic relationships among taxa at a new, more advanced level and obtain new evidence concerning their phylogeny and biodiversity.

A major task for any plant systematics, field ecologist, evolutionary biologist, conservationist, or applied forensic specialist is to determine the correct identification of a plant sample in a rapid, repeatable, and reliable fashion. DNA barcoding i.e. standardized short sequences of DNA between 400 and 800 base pairs long that in theory can be easily isolated and characterized for all species of plant on the planet, was originally conceived to facilitate this task. By combining the strength of molecular genetics, sequencing technologies, and bioinformatics, DNA barcodes offer a quick and accurate means to recognize previously known, described, and named species and to retrieve information about them. This tool also has the potential to speed the discovery of the thousands of plant species yet to be named, especially in tropical biomes.

Functional genomics utilizes the vast wealth of data generated through genome sequencing projects. It involves understanding the gene functions and their interactions. Functional genomics helps to understand the mechanism of a biological function and usually involves combination of both transcriptions and proteomics.

Genome-wide expression analysis is rapidly becoming an essential tool for identifying and analyzing genes involved in, or controlling, various biological processes ranging from development to responses to environmental cues. The advent of genomics tools and technologies has provided unprecedented capabilities for understanding the molecular basis of plant growth, development and key traits towards improving crop productivity in the 21st century.

The advent of NSG technologies has changed the dynamics and the pace and genomic research in human, plants, animals and microorganisms because of their rapid, inexpensive and highly accurate sequencing capabilities. NSG technologies offer a wide variety of applications such as whole genome de novo and re-sequencing, transcriptome sequencing (RNA-seq), microRNA sequencing, amplicon sequencing, targeted sequencing, chromatin immunoprecipated DNA sequencing (ChlP-seq), methylome sequencing and many more. Analysis of NSG data from genome-wide association studies, transcriptomics and epigenomics in combination with data from proteomics, metabolomics and other "omics" can provide an integrative systems biology approach to understand the regulation of complex traits in plants.

Since the advent of recombinant DNA technology in 1972, genetic engineering has come a long way and achieved enormous success. Investigations on molecular genetics and biochemistry of bacteria and viruses have resulted into development of new methods of manipulating DNA through creation of various vector systems and tools for their delivery into the cell. All these led to successful creation of genetically modified higher plants. However, conventional genetic engineering strategies have several issues and limitations. The complexity associated with the manipulation of large genome of higher plants and utilization of bacterial origin marker genes are the two major concerns. Currently, several tools that help to solve the problems of precise genome editing of plants are at scientists' disposal. Important among them are: zinc finger nucleases (ZNFs), transcription activator-like effector nucleases (TALENs), and clustered regularly interspaced short palindromic repeats (CRISPER).

Genome editing is defined as a collection of advanced molecular biological techniques that facilitates precise, efficient, and targeted modifications at genomic loci. Genomic editing using zinc-finger nucleases (ZNFs) and transcription activator-like effector nucleases (TALENs) have been around for two decades. Recently, a new technology called the clustered regularly interspersed short repeats (CRISPER/Cas) system has been introduced, which provide simplicity and ease of targeted gene editing. Many gene knockout mutants and some gene replacement and insertion mutants have been produced through the use of genome-editing technologies in a wide variety of plants, and many of these mutants have shown to be useful for crop improvement. The risk involved in altering genomes through the use of genome-editing technology are significantly lower than those associated with GM crops because most edits alter only a few nucleotides, producing changes that are not unlike those found throughout naturally occurring populations. Very recently, a modified version of CRISPER/Cas genome editing system called "Prime Editing" was developed, which could overcome all the limitations posed by CRISPER/Cas system. Thus introduction of genome editing into modern plant breeding programs should facilitate rapid and precise crop improvement.

Several recent studies have demonstrated the potential of CRISPER/Cas to generate a broad range of allelic diversity at specific locus, thereby help to create large genetic diversity. Creation of novel genetic diversity shall help in resolving diverse plant breeding objectives.

During the process of domestication, different crops have been selected on the basis of analogous traits such as favorable plant architecture, simultaneous flowering, ease of harvesting, and larger fruits size for higher yield, etc. With the increase in our understanding of the genetic basis for these domestication traits, it has become easier to identify several so-called domestication genes. By targeting these genes with CRISPER, the domestication process has been shown to be accelerated in several crop plants.

ORGANIZATION OF THE BOOK

Exciting opportunities have been generated with the development of molecular plant breeding and genome editing systems for addressing various crop improvement related issues. Apart from improving crop plant, these techniques have the potential to resolve issues related to germplasm conservation, cultivar identification and protection, genetic diversity creation and de-novo domestication of plants. Accordingly, the book is organized into twelve chapters. A brief description of each of the chapters follows:

Chapter 1 deals with the concepts of molecular plant breeding and genome editing systems and prospects of their utilization for crop improvement.

Chapter 2 describes the various types of genetic markers with special emphasis on molecular markers, their occurrence, basic principles and applications.

Chapter 3 describes various applications of molecular markers in plant breeding programs.

Chapter 4 describes the molecular basis and advantages of using molecular markers for plant variety identification and protection.

Chapter 5 deals with historical developments on molecular phylogeny, different molecular markers used in phylogenetic studies, and evolution of phylogenetic tree building methods. Utilization of molecular markers for germplasm conservation is also discussed.

Chapter 6 deals with the basic requirements for selection of markers, available databases, advantages and limitations of utilizing DNA barcoding for phylogenetic studies in plants.

Chapter 7 describes the basic principle and methodology of gene cloning technology which include cutting, modifying, rejoining and replication of genomic DNA, to produce transgenic plants with the desired traits.

Chapter 8 describes the mechanism of induction of hairy roots in various plant species, and their various applications.

Chapter 9 describes the various molecular genetic tools available for plant genome editing, their mechanism of operations, and merits and demerits.

Chapter 10 provides an overview of the key software support tools which can assist the plant breeders in their decision making process while conducting various breeding programs.

Chapter 11 describes the basic concepts and methods to analyze the genomic, proteomic and metabolomics data for their utilization in crop improvement.

Chapter 12 describes the basic principles and procedures of some of most important molecular biological techniques invented and applied in molecular plant breeding.

An attempt has been made to provide up-to-date information on all the topics with illustrations, wherever necessary. Comparative assessment of various techniques, their relevance, and future prospects has also been discussed.

The approach made in this book should be suitable for the students, researchers and teachers of Plant Breeding, Genetics, Molecular Biology, Molecular Taxonomy and Biotechnology, pursuing their studies and/or involved in research on crop improvement programs and other related activities.

Pradip Chandra Deka
Sir Padampat Singhania University, India

Acknowledgment

I extend my gratitude to Dr. Pallavi Dwivedi and Mr. Lalitesh Sharma for their help in the preparation of some of the figures of the book. I would like to thank Dr. Monika Anand and Mr. Sanjay Gupta for their technical support during the preparation of the manuscript. The facilities receive from Sir Padampat Singhania University, Udaipur, Rajasthan, India during preparation of the manuscript is thankfully acknowledged.

It is difficult to imagine writing a book without the full support and understanding of one's family. My greatest thanks go to my wife, Nanda, who has given me her wholehearted and unwavering support, and my son, Gautam and daughter Namrata, who stood by me in all my academic pursuit. Finally, to my parents for their love, affection and encouragement which helped me to do what I love to do.

I feel immense pleasure in thanking IGI Global, Pennsylvania, USA for bringing out the book in the present form.

Pradip Chandra Deka
Sir Padampat Singhania University, India

Chapter 1
Concepts of Molecular Plant Breeding and Genome Editing

ABSTRACT

Traditional plant breeding depends on spontaneous and induced mutations available in the crop plants. Such mutations are rare and occur randomly. By contrast, molecular breeding and genome editing are advanced breeding techniques that can enhance the selection process and produce precisely targeted modifications in any crop. Identification of molecular markers, based on SSRs and SNPs, and the availability of high-throughput (HTP) genotyping platforms have accelerated the process of generating dense genetic linkage maps and thereby enhanced application of marker-assisted breeding for crop improvement. Advanced molecular biology techniques that facilitate precise, efficient, and targeted modifications at genomic loci are termed as "genome editing." The genome editing tools include "zinc-finger nucleases (ZNFs)," "transcription activator-like effector nucleases (TALENs)," oligonucleotide-directed mutagenesis (ODM), and "clustered regularly interspersed short palindromic repeats (CRISPER/Cas) system," which can be used for targeted gene editing. Concepts of molecular plant breeding and genome editing systems are presented in this chapter.

DOI: 10.4018/978-1-7998-4312-2.ch001

INTRODUCTION

Molecular Plant Breeding is an interdisciplinary science that combines molecular genetic tools and methodologies with conventional approaches for crop improvement. Several modern breeding strategies have been included under molecular plant breeding. These include: marker-assisted selection (MAS), marker assisted backcrossing (MABC), marker-assisted recurrent selection (MARS), Genome wide selection (GWS) or genome selection (GS). The methods of molecular plant breeding continue to evolve and generated great interest among plant breeders involved in various crop improvement projects.

Genetic markers are basically determined by allelic forms of genes (loci), which can transmit from generation to generation. Accordingly they can be used as experimental probes or tags to monitor its presence in an individual, a tissue, a cell, a nucleus, a chromosome or a gene. In plant breeding, genetic markers are classified into two categories: classical markers and DNA markers. The morphological markers, cytological markers and biochemical markers are classical markers. On the other hand, DNA markers have been developed into many systems based on polymorphic-detection techniques or methods. The polymorphic detecting techniques include: Southern blotting- nuclear hybridization, Polymorphic chain reaction (PCR), and DNA sequencing which are used in RFLP, AFLP, RAPD, SSR, SNP etc. (Eathington et al., 2007)

Plant breeders have used mutagenic agents to create variability for their use in crop improvement. However, application of mutagenic agents has its own drawbacks, such as non-specificity and random nature, simultaneous effect on large numbers of genes, induction of chromosomal aberrations etc. To overcome these limitations, several genome editing systems have been developed. Important among them are: zinc finger nucleases (ZNFs), transcription activator-like effector nucleases (TALENs), oligonucleotide-directed mutagenesis (ODM), and clustered regularly interspersed short palindromic repeats (CRISPER) systems. These techniques are much simpler and efficient. Therefore, plant breeders are progressively adopting these techniques for crop improvement (Mao et al., 2019).

HISTORICAL DEVELOPMENT OF MOLECULAR PLANT BREEDING

Plant breeding encompasses methods applied for the creation, identification and selection of superior plant types in the development of improved cultivars to meet the requirements of the farmers and consumers. Primary objective of any plant breeding program is to improve yield, nutritional quality, and other commercial traits. There has been enormously successful plant breeding programs on a global scale, both in the agricultural and horticultural crops. Thus many products of plant breeding have contributed sustainable supply of carbon that has been harvested as food, feed, fiber, forest, and fuel (Bliss, 2007).

Selection and harvesting of phenotypically superior plant type/products have led to increased production, which in turn motivated to domesticate the first crop, during prehistoric time. Darwin laid down the scientific principles of hybridization and selection, and Mendel enunciated the relationship between phenotype and genotype. Scientific approach to plant breeding was initiated at the beginning of 20th century. Although importance of Mendelian genetics was realized by the plant breeders, full integration of genetics into plant breeding was seen, only when quantitative genetics reconciled Mendelian principles with continuous variation observed in most traits, having importance from the plant breeding point of view. Subsequently advancement in the understanding of plant biology, identification and analysis of genetic variations, cytogenetics, quantitative genetics, molecular biology, genetic engineering, and genomics have been successively applied in various crop improvement process.

The era of plant biotechnology began with the landmark achievement of producing transgenic plants using *Agrobacterium,* in the early 1980s. Thereafter, molecular marker systems for crop plants were developed, wherein high-resolution genetic maps were created and genetic linkage between DNA markers and important phenotypic traits of crop plants were established. In the late 1990s commercialization of transgenic crops has become a reality, which implied successful integration of biotechnology into plant breeding and crop improvement strategies. Over the years, application of plant biotechnological tools, molecular markers and genomics has made remarkable progress on utilization of genetic variations and development of new improved cultivars in many crop plants. Molecular breeding has now become a standard practice in many agricultural and horticultural plants (Winzel 2006, Nadeem et al., 2018).

GENETIC GAIN THROUGH MOLECULAR PLANT BREEDING

In simple terms, plant breeding is a cross between the best plants and to recover progeny that outperform the parents. However, in practice, three steps are involved in plant breeding: collection of germplasm with useful genetic variations, identification of superior phenotypes, and development of improved cultivars from the selected individuals. Accordingly, various breeding approaches were developed, to meet the breeding objectives, of diverse crop plants. In general, three basic breeding methods are employed for crop improvement. When the objective is to upgrade an elite genotype with trait(s) controlled by one or few genes, backcross method is used. In the case of genetically complex characters, reshuffling of the genome to produce new favorable gene combinations is required, for germplasm improvement. Through pedigree breeding method, it is possible to produce new genotype through crossing and recombination among superior genotypes. The novel genotypes thus produced may show improved performance. Adoption of recurrent selection can increase the frequency of favorable alleles at multiple loci (Jiang 2013, Grover & Sharma 2016).

One of the important concepts of quantitative genetics is genetic gain, which refers to the change in the mean value of a trait within a population, after selection. Thus genetic gain is a simple universal expression for expected genetic improvement of any crop. There are four core factors that influence genetic gain: degree of phenotypic variation (Standard deviation, σ^2), heritability (h^2, probability that a phenotypic trait will be transmitted from parent to offspring), selection intensity (i, proportion of the population selected as parents for the next generation), and length of time necessary to complete a cycle of selection (L). The genetic gain can be enhanced by increasing the σ^2, h^2, or i, and by decreasing L. Accordingly, it is possible to predict the effectiveness of a particular breeding objective through analysis of genetic gain, and can be used as a guide for judicious allocation of resources to meet the breeding objectives. Molecular plant breeding offers powerful approaches which can overcome the limitations in maximizing the genetic gain (Lorz & Winzel 2005, Zargar et al., 2015).

GENETIC DIVERSITY THROUGH MOLECULAR PLANT BREEDING

Genetic diversity is directly related to phenotypic variations present in the population and in the subsequent cycles of selection process of any breeding program. Interactions between genotype and environmental factors determine the phenotypic manifestation of any trait. Diversity in the gene pool may be obtained from breeding population, segregating progeny, exotic materials, wide crosses, natural or induced mutations, and transgenic plants.

Phenotypic variations that can be observed under different situations and circumstances are not equal. In certain cases, use of exotic germplasm may be successful for improving the many crop plants, but introduction of undesirable alleles and lack of adaptation of some traits may create problem in other plant species. It is important, therefore, to maintain a balance genetic diversity by choosing the best parents which have the potentiality to maximize the improvement of the crop successively, which is the real challenge for the plant breeders.

Advances in high-throughput genome sequencing and molecular markers analysis have made characterization of the genetic diversity in the germplasm much easier and faster. Information generated from different plant species have enriched our knowledge about evolution of plants, population structure, genetic response to selection, comparative genomics and to identify and maintain reservoirs of genetic variability for future use. It has also provided knowledge about the genetic relationships among different germplasm sources, which helps in selecting the parents for including into different crop improvement programs.

Plant biotechnology has opened new horizon wherein new genetic diversity can now be created which extends beyond species boundaries. In other words, genes hitherto not available through crossing have been created as an essential infinite pool of novel genetic variation. Genes can now be acquired from any genome spanning all kingdoms of life, or can be designed and assembled in the laboratory.

GENE ACTION THROUGH MOLECULAR PLANT BREEDING

Heritability is normally influenced by the genetic architecture of the trait, which includes number of genes involved, magnitude of effect of each one of them, and the type of gene action. Thus knowledge about the genetic architecture and favorable gene action can help to determine the impact on improving the genetic gain.

It has now been possible to simultaneously define gene action and breeding value at hundredths and thousands of loci distributed across the genome of many plant species. Such mapping studies have provided improved estimates of loci number, effect of alleles, and gene action controlling various traits of interest. Genomic segments that show statistically significant associations with quantitative traits (QTLs) can now be easily identified. Information on QTLs can effectively be used in several ways to enhance the heritability and favorable gene action.

Effective utilization of QTLs to increase genetic gain dependent on: magnitude of QTL effects, precise position of the QTLs, stability of QTL effects across different environmental conditions, and robustness across breeding germplasm. With genetic fine mapping it is possible to precisely locate the position of the QTLs, which facilitates to determine QTL effect and breeding values in additional populations. With molecular isolation of QTLs it is possible to increase the specificity and precision of estimating genetic effects in breeding program (Zagar et al., 2015).

Transgenes exert strong genetic effects at single loci, which may exhibit dominant gene action where only one copy of the gene is required for maximum expression of the phenotypic trait, in a hybrid cultivar. Such features of transgenes may bring about radical change in the complex quantitatively inherited trait to a much simpler straightforward solution. For example, partial resistance in corn germplasm to European corn borer and corn rootworm beetle had been characterized as quantitatively inherited trait with low heritability. But the expression of insecticidal toxin proteins from *Bacillus thuringenesis* (*Bt*) to reduce feeding damage by larvae of the above mentioned insect pest in transgenic corn hybrids has been found to be a simply inherited trait, which can be utilized to produce resistant cultivars with high efficiency.

Transgenes may also be used to disrupt allelic interactions between factors controlling the trait of interest and other factors controlling important characters. For example, a single locus *Bt* transgene may facilitate selection of favorable alleles for yield, that are otherwise linked with genes for resistance to the same class of insects.

Transgenes can be incorporated to intervene at key regulatory steps of a metabolic pathway, such that gene actions for the corresponding traits are inherited as dominant loci that are less sensitive to environmental factors.

Molecular cloning of QTLs has provided vital information about the biology of quantitative traits that were not likely to be discovered through conventional breeding procedures. Molecular markers, genomics, and biotechnology are now being applied simultaneously to exploit genetic diversity for crop improvement.

SELECTION THROUGH MOLECULAR PLANT BREEDING

In conventional plant breeding, phenotypic selection has been used effective for most traits. But for certain traits, phenotypic selection is not effective due to difficulties in measuring phenotypes or identifying individual having highest breeding value. Environmental factors and interactions between genotype and environment also effect phenotypic selection. Replicated trials in multiple locations (environment) allow better estimation of breeding values, but require additional resources and time. For certain traits, it may be required to sacrifice the individual to measure the phenotype. Molecular plant breeding can be applied to increase the efficiency of selection.

Molecular marker that are present within a gene or linked tightly to QTLs that influence the traits under consideration, can be employed as a supplement to phenotypic selection process. An effective strategy could be to use selection indices that consider multiple factors in while choosing the final genotype.

In situations where phenotype is difficult to evaluate due to lack of resources or influence of environment, the genetic gain for traits can be significantly enhanced through marker assisted selection. Molecular markers have also been used to increase the possibility of identifying superior genotypes by early elimination of inferior genotypes, thereby decreasing the number of progeny to be screened to recover a given level of gain. Application of such strategies has helped to develop resistant plants against diseases (cereals), nematodes (soybean) and draught (maize) (Gozal et al., 2016, Bishit et al., 2019).

Inclusion of physiological and biochemical parameters, as secondary traits, can increase the efficiency of phenotypic selection for some complex traits. However, there must be strong correlation with the target trait and should be highly heritable. Advances in functional genomics have now permit the population scale profiling of RNA abundance, levels and activities of proteins, and metabolites associated with targeted traits. In addition to molecular markers, such genomic approaches may provide additional phenotypic selection targets.

Marker-assisted selection can also be used to accelerate the transgenes deployment in commercial cultivars. This can be achieved through marker-assisted backcross breeding. Marker-assisted selection could be used in forward breeding as well.

ADOPTION OF MOLECULAR PLANT BREEDING

There have been different rates of adoption of molecular plant breeding approaches among crop species and institutes involved in crop improvement. The reasons being: scientific development and capability, infrastructure availability, economic and sociological factors. One of the earlier scientific barriers was non-applicability of *Agrobacterium*-mediated transformation in cereals crops. However, technological developments and continued research could overcome all the obstacles, and now nearly all important agricultural and horticultural species are amenable for *Agrobacterium*-mediated transformation.

Research on genomics in plants species has generated huge information about gene structure and function, and large number of molecular markers, providing opportunity for use in plant breeding. However, application of these resources remained elusive, without restructuring and integrating knowledge of pedigrees, phenotypes and genotypes of the markers, for providing optimum response to selection. Modifications of regulatory functions have remained a challenge to the molecular biologists, as determination of the sequence basis for regulatory changes and prediction of their phenotypic effect, remained difficult.

Once the scientific challenges are overcome, economic factors determine the possibility of integrating these innovations into the existing plant breeding programs. The costs associate with different operations of molecular plant breeding is much greater than the conventional plant breeding. Currently there has been a growing recognition about the potential of recent advances in molecular biology, biotechnology, and genomics and the need for increased investments for adoption of molecular plant breeding for crop improvement.

GENOME EDITING TOOOLS

Genetic engineering has achieved enormous success in crop improvement after its introduction by Paul Berg in 1972. After several decades of investigations, many new molecular biological phenomenon and mechanisms, having relevance to genetic engineering, have been discovered which helped the researchers to accomplish their objectives. Long and arduous experiments on bacteria and viruses have resulted in understanding and developing new methods of manipulating DNA through creation of various vector systems and their delivery into the plant cells. These in turn allowed successful creation of genetically modified higher crop plants. However, conventional strategies for genetic engineering have several limitations. The complexity involved with the manipulation of large genome of higher plants is one of these limitations. Application of foreign genes as markers during the selection process of transgenic plants is another factor which played adversely in releasing genetically modified plants for cultivation.

Several tools are now available to resolve this issue. In 1996, the first such tool called ZFN (zinc finger nucleases) was developed to address this issue. In this technique the protein domains such as 'zinc finger' coupled with *Fok*I endonuclease domains acts as site-specific nucleases and cleave the DNA in strictly defined site(s) *in vitro*. The protein has a modular structure and each of the domains of zinc finger recognizes one triplet of nucleotides. The method has been used for editing cultured plant cells (Kamburova et al., 2017).

Thereafter the genome editing tool TALEN (transcription activator-like effectors nucleases) was developed. TALEN requires designing of a new protein for each target site. This process has been streamlined by making the modules of repeat combinations available to reduce the requirement of cloning for the design (Satheesh et al., 2019).

CRISPER (clustered regularly interspersed short palindromic repeats) is a comparatively recent addition to the genome editing tools, which is much simpler and efficient. In this method the adaptive bacterial and archaeal immune system is utilized, the mechanism of which relies on the presence of a site called CRISPER loci in the bacterial chromosome. CRISPER loci are composed of operons which encodes the Cas9 protein and array of repeated spacer sequences. The spacers are short fragments and derived from foreign DNA (plasmids and viruses) that become integrated into bacterial genome following recombination. Recognition of the target site by CRISPER system is accomplished by complimentary sequence based interactions between DNA of the target site and noncoding guide RNA (gRNA) (Makarova et al., 2011, Molla et al., 2019). Due to its efficiency, simplicity and wide capabilities CRISPER has become a useful genome editing tool.

CONCLUSION

Molecular plant breeding became a new member in the family of plant breeding as various types of molecular markers in crop plants were developed over the years. Application of molecular markers has now become a powerful and reliable tool not only for crop improvement but also for variety identification, germplasm evaluation, genetic mapping, map-based gene discovery, and characterization of traits. Compared to conventional breeding methods, molecular breeding methods have several advantages. These include: selection of all kinds of traits at seedling stage, not being affected by environmental factors, co-dominance in nature, ease in identification of QTLs, and reliable, fast and cheap to execute.

However, marker-assisted plant breeding technology is facing following challenges: not all markers are breeder friendly, not all markers can be applicable across populations due to lack of marker polymorphism or reliable market-trait association, false selection may occur due to recombination between the markers and the genes/QTLs of interest, and imprecise estimate of QTL locations. Accordingly breeder has to ascertain the feasibility of MAS before executing the breeding programme.

Development of genome editing tools have been able to overcome some of the problems faced by the breeders and expected to revolutionized the plant breeding methodologies applied for crop improvement.

REFERENCES

Bisht, D. S., Bhatia, V., & Bhattacharya, R. (2019). Improving plant-resistance to insect-pest and pathogens: The new opportunities through targeted genome editing. *Seminars in Cell & Developmental Biology*, *96*, 65–76. doi:10.1016/j.semcdb.2019.04.008 PMID:31039395

Eathington, S. R., Crosbie, T. M., Edwards, M. D., Reiter, R. S., & Bull, J. K. (2007). Molecular markers in a commercial breeding program. *Crop Science*, *47*, S154–S163. doi:10.2135/cropsci2007.04.0015IPBS

Gazal, A., Dar, Z. A., Wani, S. H., Loan, A., Shikari, A. B., Ali, G., & Abidi, I. A. (2016). Molecular breeding for enhancing resilience against biotic and abiotic stress in major cereals. *SABRAO Journal of Breeding and Genetics, 48*, 1–32.

Grover, A., & Sharma, P. (2016). Development and use of molecular markers: Past and present. *Critical Reviews in Biotechnology, 36*(2), 290–302. doi:10.3109/07388551.2014.959891 PMID:25430893

Jiang, G.L. (2013). *Molecular markers and marker-assisted breeding in plants.* doi:10.5772/52583

Kamburova, V. S., Nikitina, E. V., Shermatov, S. E., Buriev, Z. T., Kumpatla, S. P., Emani, C., & Abdurakhmonov, I. Y. (2017). Genome editing in plants: An overview of tools and applications. *International Journal of Agronomy. Article ID, 7315351*, 1–15.

Lorz, H., & Winzel, G. (2005). *Molecular marker system in plant breeding and crop improvement.* Springer. doi:10.1007/b137756

Makarova, K. S., Haft, D. H., Barrangou, R., Brouns, S. J., Charpentier, E., Horvath, P., & Koonin, E. V. (2011). Evolution and classification of the CRISPER-Cas system. *Nature Reviews. Microbiology, 9*(6), 467–477. doi:10.1038/nrmicro2577 PMID:21552286

Mao, Y., Botella, J. R., Lin, Y., & Zhu, J. K. (2019). Gene editing in plants: Progress and challenges. *National Science Review, 6*(3), 421–437. doi:10.1093/nsr/nwz005

Molla, K. A., & Yang, Y. (2019). CRISPER/Cas-mediated base editing: Technical considerations and practical applications. *Trends in Biotechnology, 37*(10), 1121–1142. doi:10.1016/j.tibtech.2019.03.008 PMID:30995964

Nadeem, M. A., Nawaz, M. A., Shahid, M. Q., Dogan, Y., Comprtpay, G., Yildiz, M., ... Baloch, F. S. (2018). DNA molecular markers in plant breeding: Current status and recent advancements in genomic selection and genome editing. *Biotechnology, Biotechnological Equipment, 32*(2), 261–285. doi:10.1080/13102818.2017.1400401

Satheesh, V., Zhang, H., Wang, X., & Lei, M. (2019). Precise editing of plant genome- prospects and challenges. *Seminars in Cell & Developmental Biology, 96*, 115–123. doi:10.1016/j.semcdb.2019.04.010 PMID:31002868

Winzel, G. (2006). Molecular plant breeding: Achievements in green biotechnology and future perspectives. *Applied Microbiology and Biotechnology*, *24*(6), 490–499. doi:10.100700253-006-0375-9

Zargar, S. M., Raatz, B., Nazir, M., Bhat, J. A., Dar, Z. A., Agrarwal, G. K., & Rakwal, R. (2015). Recent advances in molecular marker techniques: Insight into QTL mapping, GWAS and genomic selection in plants. *Journal of Crop Science and Biotechnology*, *18*(5), 293–308. doi:10.100712892-015-0037-5

ADDITIONAL READING

Abduvakhmonov, J. V. (2016). *Microsatellite markers*. InTech. doi:10.5772/62560

Arebncibia, V., D'Afonseca, V., Chakravarthi, M., & Castiglione, S. (2019). Learning from transgenics: Advanced gene editing technologies should also bridge the gap with traditional genetic selection. *Electronic Journal of Biotechnology*, *41*, 22–29. doi:10.1016/j.ejbt.2019.06.001

Bharat, S.S., Li, S., Li, J., Yan, L., & Xia, L. (2019). Base editing in plants: current status and challenges. *The Crop Journal*, Retrieved from: . 2019.10.002 doi:10.1016/ j.cj

Bliss, F. A. (2007). Education and preparation of plant breeders for careers in global crop improvement. *Crop Science*, *47*, S250–S261. doi:10.2135/cropsci2007.04.0017IPBS

Breseghello, F., & Coelho, A. S. G. (2013). Traditional and modern plant breeding methods with examples in rice (*Oryza sativa* L.). *Journal of Agricultural and Food Chemistry*, *61*(35), 8277–8286. doi:10.1021/jf305531j PMID:23551250

Delannay, X., McLaren, G., & Ribaut, J. M. (2012). Fostering molecular breeding in developing countries. *Molecular Breeding*, *29*(4), 857–873. Advance online publication. doi:10.100711032-011-9611-9

Henry, R. J. (Ed.). (2012). *Molecular markers in plants*. John Wiley & Sons, Inc., doi:10.1002/9781118473023

Kleter, G. A., Kuiper, H. A., & Kok, E. J. (2019). Gene-edited crops: Towards a harmonized safety assessment. *Trends in Biotechnology*, *37*(5), 443–447. doi:10.1016/j.tibtech.2018.11.014 PMID:30616999

Kumpatia, S. P., Buyyarapu, R., Abdurakhmonov, I. Y., & Mammadov, J. A. (2012). *Genomics-assisted plant breeding in the 21st century: technological advances and progress*. InTech. Retrieved from., doi:10.5772/37458

Madhumati, B. (2014). Potential and application of molecular marker techniques for plant genome analysis. *International Journal of Pure Applied Biosciences*, *2*, 169–188.

Metje-Sprink, J., Menz, J., Modrzejewski, D., & Sprink, T. (2019). DNA-free genome editing: Past, present and future. *Frontiers in Plant Science*, *9*, 1–9. doi:10.3389/fpls.2018.01957 PMID:30693009

Mishra, R., Joshi, R. K., & Zhao, K. (2020). Base editing in crops: Current advances, limitations and future implications. *Plant Biotechnology Journal*, *18*(1), 20–31. doi:10.1111/pbi.13225 PMID:31365173

Moose, S. P., & Mumm, R. T. (2008). Molecular plant breeding as a foundation for 21st century crop improvement. *Plant Physiology*, *147*(3), 969–977. doi:10.1104/pp.108.118232 PMID:18612074

Pandey, P. K., Quilichini, T. D., Vaid, N., Gao, P., Xiang, D., & Datla, R. (2019). Versatile and multifaceted CRISPER/Cas gene editing tool for plant research. *Seminars in Cell & Developmental Biology*, *96*, 107–114. doi:10.1016/j.semcdb.2019.04.012 PMID:31022459

Razzaq, A., Saleem, F., Kanwal, M., Mustafa, G., Yousaf, S., Arshad, H. M. I., Hameed, M. K., & Khan, M. S. (2019). Modern trends in plant genome editing: An inclusive review of the CRISPR/Cas9 Toolbox. *International Journal of Molecular Sciences*, *20*(16), 4045–4098. doi:10.3390/ijms20164045 PMID:31430902

Semagn, K., Bjornstad, A., & Ndjiondjop, M. N. (2014). An overview of molecular marker methods for plants. *African Journal of Biotechnology*, *2450*, 25–68.

Shenoy, V., & Sharma, N. P. (2012). New facets of 21st century plant breeding. *Journal of Rice Research*, *5*, 1–16.

Singh, B. D., & Singh, A. K. (2015). *Marker-assisted plant breeding: principles and practices*. Springer. doi:10.1007/978-81-322-2316-0

Singh, B. P., & Gupta, V. K. (2017). *Molecular markers in mycology: diagnostics and marker developments*. Springer. doi:10.1007/978-3-319-34106-4

Xu, Y., Li, Z. K., & Thomson, M. J. (2012). Molecular breeding in plants: Moving into the mainstream. *Molecular Breeding*, *29*(4), 831–832. doi:10.100711032-012-9717-8

APPENDIX

1. How molecular plant breeding has changed the approach in plant breeding methodologies?
2. How genetic gain can be achieved through molecular plant breeding?
3. How genetic diversity can be utilized in more efficient ways through molecular plant breeding?
4. How gene action can be increased through molecular plant breeding?
5. How the efficiency of selection process can be enhanced through molecular plant breeding?

Chapter 2
Molecular Markers

ABSTRACT

Conventionally, establishment of relationship between the genotype and phenotype through genetic analysis was considered as key to success in plant breeding. The discovery of molecular markers has changed the entire scenario of genome analysis. Coinheritance of a gene of interest and a marker suggests that they are physically close on the chromosome. A marker must be polymorphic in nature for their identification and utilization. Such polymorphism can be detected at three levels: phenotype (morphological), difference in biomolecules (biochemical), or differences in the nucleotide sequence of DNA (molecular). These markers act as a versatile tool and find their importance in taxonomy, plant breeding, gene mapping, cultivar identification, and forensic science. They have several advantages over the conventional methods of plant breeding for developing new varieties with higher rate of success. This chapter covers the basic principles and applications of various types of markers with special emphasis on molecular markers.

INTRODUCTION

Genes and their alleles have been used as genetic markers on the basis of their phenotypic expressions. Information about their inheritance pattern, linkage and segregation during their movement from generation to generation, and their exact location on chromosomes is essential in order to carry out genetic studies. Conventionally, all this is done by evaluating the relationship between genotype and governed phenotype. Such types of markers are

DOI: 10.4018/978-1-7998-4312-2.ch002

called morphological marker or phenotypic marker. Later, the discovery of biochemical and molecular markers has changed the entire scenario of genome analysis and has supplemented the existing knowledge of genetics. In the classical genetics, location of a gene in the chromosome and its inheritance pattern can be studied through linkage analysis. In molecular genetics, clues to the location of a gene can come from comparing the inheritance of a gene with the inheritance of a molecular marker. Coinheritance or genetic linkage of a gene of interest and a molecular marker suggests that they are physically close together on the chromosome. However, the molecular marker must be polymorphic in nature, meaning thereby that it must be found in variable forms so that chromosome with the mutant gene can be distinguished from the chromosome with normal gene through marker which it carries. This polymorphic nature of marker can be analyzed at three levels: phenotype (morphological), differences in biomolecules (biochemical) or differences in the nucleotide sequence of DNA (molecular). These markers act as a versatile tool and find their own importance in various fields like taxonomy, plant breeding, gene mapping, genome and cultivar identification, forensic science etc. They have several advantages over conventional methods and assist to reduce the overall time span of developing new varieties with higher rate of success.

CLASSICAL MARKERS

During early history of plant breeding, only visible markers were used as selectable markers. They may be classified as morphological markers and cytological markers.

Morphological markers

The morphological features or phenotypes which are governed by genes are considered as morphological markers or phenotypic markers. They are generally qualitative traits like color of the flower, albinism, and altered leaf morphology that can be scored visually. Morphological markers are usually dominant or recessive. Most of the identified morphological markers are alleles of the wild type phenotypes which have accumulated in the population through mutations. Since the rate of spontaneous mutation is slow, number of such mutations found in natural population is also low. The problem is more acute in the case of forestry species and animal systems (Avise, 1994, Grover & Sharma, 2016).

Further, most of the quantitative characters are governed by polygenes, each contributing a small portion. Therefore, they are difficult to identify. Therefore, the genes controlling quantitative traits cannot be used as morphological markers. Availability of good number of genetic marker is the key to success for any plant or animal improvement program. Discovery of biochemical and molecular markers have been able to solve this problem to a large extent.

Cytological markers

Karyotypes and banding patterns of the chromosomes can be used as cytological markers. The physical structure of the chromosomes observed at the metaphase stage of mitosis has been used to construct the karyotypes of the organisms. The banding patters of the chromosomes derived through e.g. G-banding, Q-banding, R-banding etc. can also be used as markers. The color, width, order and position of the bands can be considered as important features of the markers. Different landmarks of the chromosomes are used for characterization and detection of chromosome mutation, linkage group identification and physical mapping. The physical maps developed through cytological and morphological markers are used for construction of linkage maps. However, in plant breeding cytological markers have limited use (Avise, 1994, Grover & Sharma, 2016).

BIOCHEMICAL MARKERS

Biochemical markers are biomolecules which are produced by gene expression. Two biomolecules primarily used as genetic markers are, monoterpenes and allozymes.

Monoterpenes

Monoterpenes are found in the resins and essential oils of plants. They play an important role in the defense mechanism of plants against diseases and pests. Several types of monoterpenes such as α-pinene, β-pinene, myrcene, 3-carene, and limolene have been identified. Concentration of each of these monoterpenes can be determined by gas-chromatography and used as genetic markers.

During 1960s and early 1970s monoterpenes were the best available genetic markers for forest species. They were primarily used for taxonomic and evolutionary studies and to limited extent for estimation of genetic patterns of geographic variation within species. The simple Mendelian inheritance of these monoterpenes emphasizes their suitability as a marker for clonal identification and for the study of population genetics mainly in tree species. Application of monoterpenes as genetic marker has several disadvantages, such as, availability of relatively few marker loci, lack of expression of co-dominant phenotype, unable to distinguish between heterozygotes from homozygotes genotype, and requirement of specialized and expensive equipment (Avise, 1994, Grover & Sharma, 2016). Monoterpenes genetic markers were gradually replaced by isozyme and allozyme genetic markers, because of several advantages.

Isozymes and Allozymes

Isozymes and Allozymes were the most frequently used biochemical markers because of the polymorphic nature of the enzymes. Isozymes refer to multiple form of an enzyme, sharing a catalytic activity derived from a tissue of single organism. It also refers to a detectably different enzyme, which catalyze the same reaction. Allozyme are slightly different from Isozymes and should not treat as interchangeable. Isozymes are enzymes that convert the same substrate, but are not necessarily products of the same gene. Allozymes are isozymes but are encoded by different alleles of the same gene (due to allelic polymorphism). Thus allozymes may differ from one another by one or more amino acids. Isozymes may be active at different life stages or in different cell compartments.

Isozymes are differently charged protein molecules that can be separated by gel electrophoresis (on the basis of molecular size, shape and electrical charge), because they have the same substrate specificity but different electrophoretic mobility. Due to specificity of enzyme, the location of a particular enzyme on a gel can be visualized by supplying the appropriate substrate and cofactor in a color producing reaction. In the simplest protocol, starch gels are used. Similarly, two alleles in an allozyme locus in a heterozygous individual can be detected. Allozymes are thus codominant markers. They can also be multiallelic. The overall procedure of allozyme analysis is shown in Figure 1.

Varying number of bands (one to many) are visualized depending on the number of loci, their state of homo- or heterozygosity, and the enzyme configuration (*i.e* the number of separate subunits). These bands may polymorphic and thereby provide information regarding the status of the marker gene(s).

Figure 1. Procedure of allozyme analysis

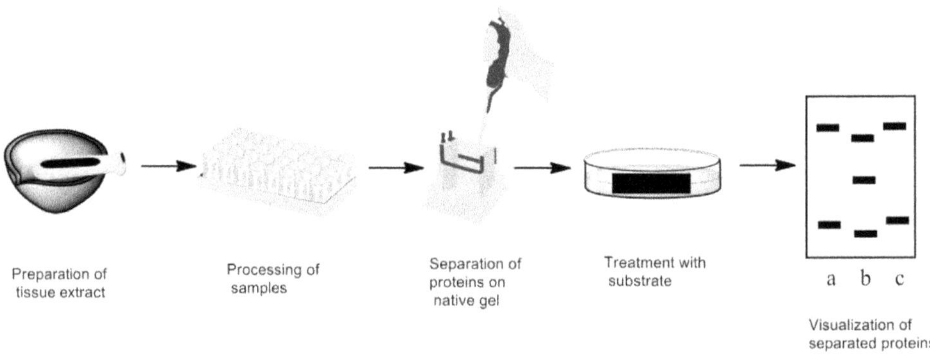

The advantages of allozyme markers are their codominant inheritance, technical simplicity and low cost. Presence of limited number of suitable allozyme loci in the genome, requirement of fresh tissue, sometimes limited variations, requirement of similar type of material for all experiment, tissue specific gene expression and environmental influence on gene expression are major drawbacks of allozyme markers (Avise, 1994).

Figure 2. The electrophoretic separation of enzyme β-glucosidase from tissue extracts of fifteen different varieties of rat

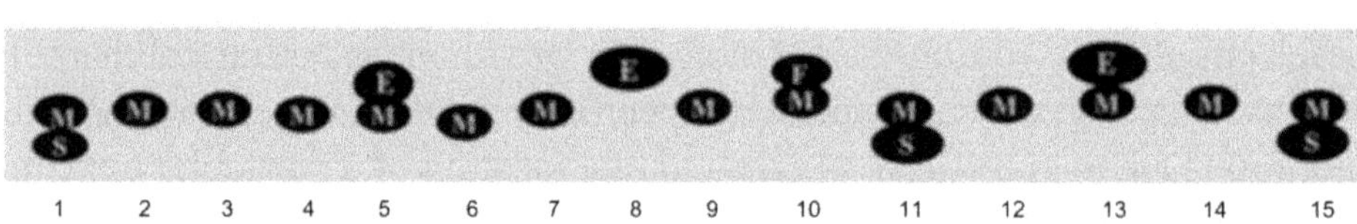

The procedure adopted for analysis of allozymes is explained through an example. A liver tissue extract is prepared and electrophoresed from 15 different strains of rat (strain 1- 15) for the enzyme β-glucosidase. The pattern of protein bands obtained is shown in Figure 2. The 4 alleles of the enzyme β-glucosidase can be distinguished by the different electrophoretic mobility of their protein products. They are named as fast (F), moderately fast (E), medium (M) and slow (S). Following conclusions can be drawn from the results shown in Figure 2.

1. Eight strains (2, 3, 4, 6, 7, 9, 12, and 14) were homozygous for allele M.
2. Strain 8 was homozygous for allele E.
3. Three strains (1, 11, 15) are heterozygous for the M and S alleles.
4. Two strains (5, 13) were heterozygous for M and E.
5. Strain (10) was heterozygous for M and F.
6. Allozyme markers have successfully been applied to several organisms ranging from microbes to plants to animals and significantly contributed to enrich the information in various fields like physiology, biochemistry, systematics, genetics, and breeding.

MOLECULAR MARKERS

Molecular marker is defined as a fragment of DNA which is quickly detectable and whose migration from generation to generation is monitored easily. Application of molecular markers is based on the availability of naturally occurring DNA polymorphism in the genome of the individual. Genetic polymorphism is a phenomenon of occurrence of multiple forms of a gene, within the same population. The different forms may represent two or more discontinuous variants or genotypes, which arise due to mutation. Genetic polymorphisms can be revealed by molecular markers at DNA level. Various types of DNA markers are implemented to examine DNA polymorphism. There are two types of molecular markers: hybridization based markers, and PCR amplification based markers. In hybridization based markers, genetic polymorphism of individual genes or sequences can be analyzed by using a small piece of DNA as a probe. The property of complimentary base pairing in DNA, will allow the probe to hybridize with the complimentary segment of the genomic DNA and thereby establish polymorphism. Restriction fragment length polymorphism (RFLP) is an example of this technique. The principle of PCR based markers is the *in-vitro* amplification of a specific fragment of DNA by using specific or arbitrary primers. Primers are oligonucleotides of known sequences (Caetano-Anolles & Gresshoff, 1997, Henry, 2012, Boopathi,

2013). Examples of PCR based markers included RAPD (random amplified polymorphic DNA), AFLP (amplified fragment length polymorphism), SSR (simple sequence repeat), ESTPs (expressed sequence tagged polymorphism), SNPs (single nucleotide polymorphism), and SSCP (single strand conformation polymorphism).

Properties Desirable for Ideal DNA Markers

Ideal DNA markers should have the following properties: (1) Highly polymorphic nature, (2) Co-dominant inheritance (3) Frequent occurrence in genome, (4) The DNA sequences should remain unaffected to environmental conditions, (5) Easy access (availability), (6) Easy and fast assay, (6) High reproducibility, and (7) Exchangeability of data among laboratories.

A single molecular marker might not possess all the desirable features at a time. On the basis of the aim of study to be undertaken an appropriate marker has to be selected that will fulfill some of the desirable characteristics. Molecular markers are described under four major groups as follows: (1) Single or low copy number markers, (2) Multi locus markers, (3) Arbitrary sequence markers, and (4) Organelle genome markers

SINGLE OR LOW COPY NUMBER MARKERS

Single and low copy number markers include: restriction fragment length polymorphism (RFLP), random amplified polymorphic DNA (RAPD), amplified fragment length polymorphism (AFLP), simple sequence repeats (SSR), single nucleotide polymorphism (SNP), expressed sequence tag polymorphism (ESTP) and, single strand conformation polymorphism (SSCP).

Restriction Fragment Length Polymorphism (RFLP)

RFLPs are heritable co-dominant molecular markers. The acronym is pronounced "riflip". Restriction fragment length polymorphism is defined as the identification of specific restriction enzymes that reveals a pattern difference between the DNA fragment sizes in individual organisms. This phenomenon was first described by Grodzicker for mutant strains of adeno-virus. Restriction enzymes are endonucleases produced by a variety of prokaryotes. Their natural functions and properties of restriction endonucleases are described in Chapter-6. A list of all known restriction enzymes, their recognition sequences, methylation sensitivity, commercial availability and other useful information is compiled in the REBASE database which is available at http://rebase.neb.com/rebase/rebase.html.

Genome of all organisms contains recognition sites for various restriction enzymes. Therefore, when the DNA is incubated with these enzymes, a mixture of DNA fragments of varying size will be obtained. These fragments are known as restriction fragments. The recognition site for restriction enzymes may change because of any kind of mutation like base substitution, insertion or deletion of bases which will result into altered pattern of fragments. By comparing the obtained fragment pattern, after digesting the DNA of different individuals with a particular restriction enzyme, polymorphism among them is analyzed.

The detection of an RFLP requires extraction and purification of the DNA from an individual followed by digestion with a restriction endonuclease, to form a mixture of restriction fragments, varying in length when electrophoresed. Length of the fragments varies according to the distribution of restriction sites on the genome for that particular enzyme. The DNA fragments thus obtained are then fixed by transferring them onto a membrane via southern blotting (see Chapter-12). Labeled probes are used to visualize the bands on the basis of complementary base pairing (hybridization). The overall procedure of RFLP analysis is shown in Figure 3. Probes are oligonucleotide of about 0.5-2.0 Kb in size, obtained either from cDNA library or genomic library which are mostly species specific. For preparation of probes, DNA is isolated from the species of interest, digested with a restriction enzyme to generate relatively small fragments. Individual restriction fragments are ligated into a bacterial plasmid and the plasmid is transformed into a bacterial cell. By growing these transformed bacteria, it is possible to obtain large quantity of DNA restriction fragments, which are suitable for use as a hybridization probe. Allelic variations are, therefore, identified as differences in the size of the restriction fragments to which the probe hybridizes. For example, the genome of an individual "A" contains three restriction sites as shown in Figure 4, which are indicated by arrows. In another individual "a" second restriction site is lost because of mutation. When their genomes are digested with restriction enzyme four and three restriction fragments are produced by "A" and "a", respectively. Among these only one fragment is detected by DNA probe which is indicated by double headed arrow. Figure 5 shows how this fragment size variation would look on a southern blot, and how each allele (two per individual) might be inherited in members of a family. The chances of finding a useful polymorphism can be increased by using enzymes that cut more frequently or simply by using a greater variety of enzymes. For the proper interpretation of RFLP bands, it is absolutely imperative that complete digestion occurs. Therefore, after digestion, the DNA must be run on an electrophoresis gel to be visually checked for complete digestion and concentration, and prepared for southern blotting. Good mixing of the reaction mixture is also very important.

Figure 3. Generalized procedure of conventional RFLP analysis

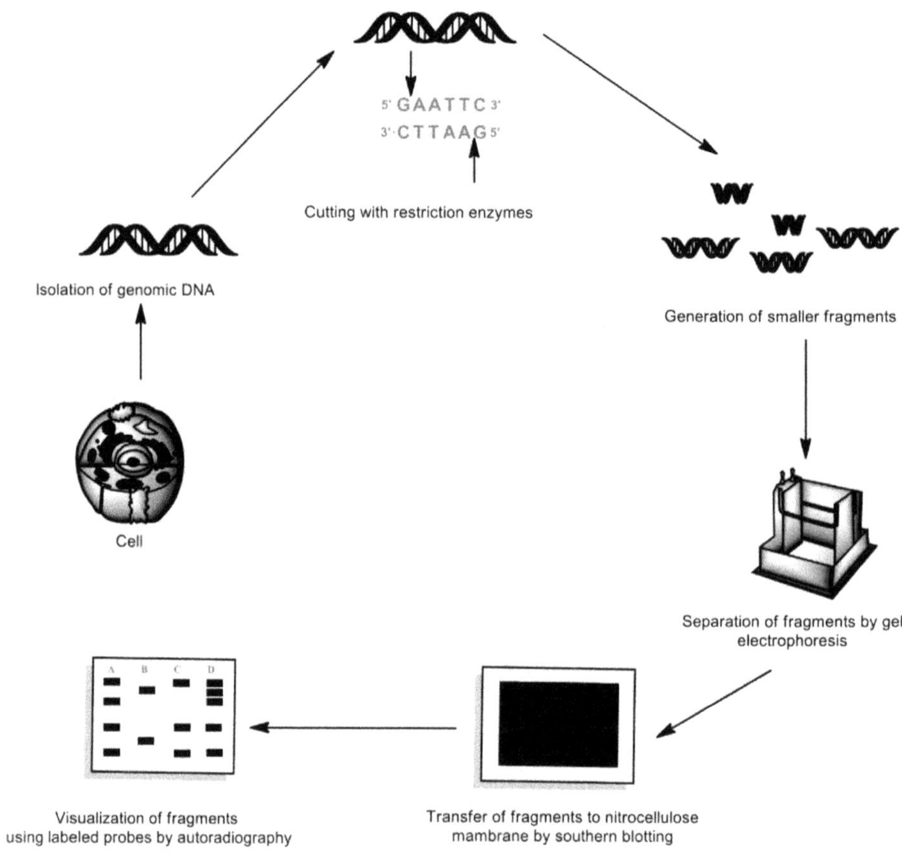

RFLP markers are most suited to study polymorphism at the intra specific level or among closely related taxa. The genome of two individuals who belong to the same species is similar but not identical. It varies at few nucleotides. These variations are high among the less related individuals, and there are more chances of finding a RFLP. The similarities and differences can be used to infer phylogenetic relationships. These markers are also used for constructing genetic maps, solving paternity case and disease diagnosis. They have been used to prepare chromosome maps in humans, mice, fruit fly, maize, tomato, lettuce and rice. Previously, construction of genetic maps was based on linkage and recombination relationship during gamete formation. The only way to find the movement of chromosomal fragments and their recombination during gamete formation is the analysis of phenotype which is caused by the action of genes. Later, RFLPs proved itself as potent tool for gene mapping. These RFLP markers show linkage with important genes, therefore, instead of looking at the phenotype caused by the presence of genes on a chromosome segment, direct look for associated RFLP marker is sufficient to trace the movement of chromosomal segment. RFLP markers segregate exactly in the same way as do conventional gene markers and follow strict Mendelian rules. Therefore, gene maps using RFLP markers can be constructed in the same way as maps of conventional markers. They are considered as reliable marker for linkage analysis and breeding exercises. The proposed reason is that homozygous or heterozygous state of an individual can be very easily determined by using RFLP markers as they show co-dominant inheritance. These RFLP markers can supplement the plant breeding exercises when used in conjugation with conventional markers. They assist in quick indirect selection of desirable gene for breeding experiments. Direct selection of that gene would be expensive, difficult or time taking. With indirect selection, one does not directly select for the gene of interest, but rather for one or more closely linked RFLP markers. If the RFLP markers are indeed closely linked, they will remain associated with the gene of interest during segregation (Nadeem et al., 2018). For example, conventionally, selection of a recessive gene in a cultivar is done by backcross breeding program. It involves alternate backcross and selection phases. It is necessary to select progeny bearing the desired gene at several points in the backcross cycle. This is difficult because recessive gene will not be expressed in any of the backcrossed plants and it would be necessary to carry out progeny testing by selfing the backcrossed plants to test for the presence of the recessive gene. However, a recessive allele could be indirectly selected, by looking for a linked RFLP marker.

When their genomes are digested with restriction enzyme three and two restriction fragments are produced by “A” and “a” respectively.

One more important application of RFLP is in selection of quantitative trait loci (QTL). Quantitative traits are those which are governed by multiple genes (polygenic traits) and inherited quantitatively. Each of the individual

Figure 4. RFLP analysis of genomic DNA of two different individuals to study the allelic variation. Individual "A" contains three restriction sites while one restriction site is lost in another individual "a" because of any mutation

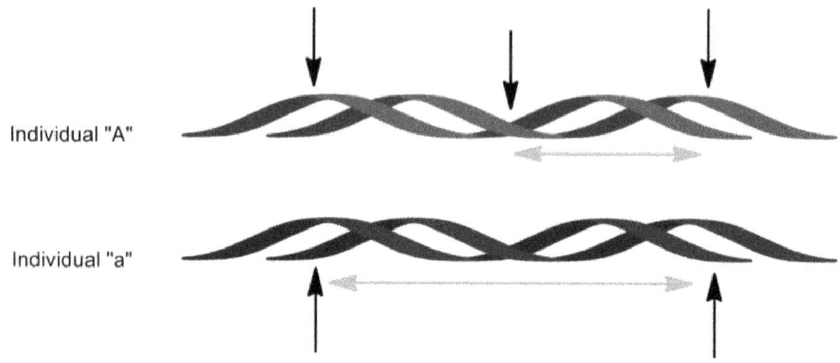

genes of such polygenic system contributes a small positive or negative effect to the trait of interest. Analysis of inheritance of such quantitative traits is very difficult by conventional Mendelian method because clear dominance is not exhibited and the phenotype has a large environmental component. The study of such quantitative trait is possible by the use of RFLP markers, if a correlation is established between the quantitative trait of interest and specific chromosomal segment marked by RFLPs. Sometimes association of RFLPs with important disease causing gene (or mutant gene) makes them suitable as diagnostic marker, for calculating the probability of developing a particular disease and its diagnosis. Several genetic disorders have been reported to be tightly linked with an RFLP marker. Presence or absence of that RFLP gives a direct indication about the possibility of developing and transmitting the disease. In 1978, Kan and Dozy used this concept first time for the diagnosis of sickle cell anemia. They found a 13.0 kb RFLP using a restriction enzyme *Hpa*I which is highly associated with mutant gene for human β-globulin gene but not with the normal allele. β–thalassemia, phenylketonuria, cystic fibrosis are other genetic diseases whose diagnosis is possible by using RFLP markers.

However, their utility has been restricted due to the requirement of large amount of highly pure DNA for restriction digestion. Besides, it requires expensive and hazardous southern blotting which makes it time consuming and non-amenable to automation. Constant good supplies of probes are needed. These probes are generally radioactively labeled hence expertise in autoradiography is desirable.

Figure 5. Analysis and inheritance of allelic RFLP fragments

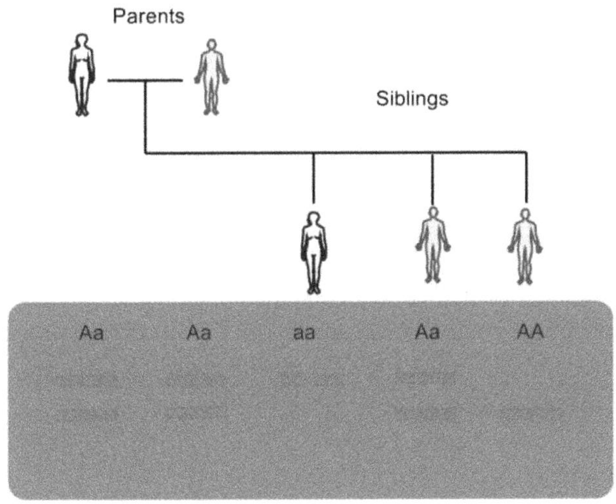

Cleaved Amplified Polymorphic Sequences (PCR-RFLP)

Significant quantity of DNA is required for RFLP analysis. DNA isolation is a laborious and time taking exercise. However, the required quantity of DNA for RFLP analysis can be amplified through PCR from a very small amount of DNA within 2-3 hours. Thus, in less time more samples can be analyzed. This technique is known by several acronyms like cleaved amplified polymorphic sequences (CAPS) or PCR-RFLP. These CAPS markers are generated in two steps. In the first step a defined DNA fragment is amplified using specific primer pairs. This may already result in differently sized and hence informative PCR fragments. In the next step, the PCR product is digested with a restriction enzyme. The digested amplification products may or may not reveal polymorphisms after separation on agarose gels. As opposed to conventional RFLP analysis, the CAPS approach does not require radioactivity or blotting steps, but instead exhibits all the attractive attributes of PCR based technique. CAPS markers are also co-dominant (Nadeem et al. 2018).

Single Nucleotide Polymorphism (SNP)

Since late 1990s, single nucleotide polymorphisms (SNPs) have become increasingly popular as a molecular marker system. SNP (pronounced as "snips") is characterized as a single base substitution at a particular position in the genome of two (or more) individuals, at which different sequence alternatives (alleles) exist in populations. For example a SNP might change the DNA sequence ATCATGCT to AACATGCT. As per definition, the least frequent allele should have abundance of at least 1% SNPs, originated as a result of either transition or transversion events. They differ from several type of naturally occurring mutations by their sheer numbers per genome, relatively low mutation rates, even distribution across the genomes and relative ease of detection. In principle, a SNP locus can have two, three, or four alleles in a population, but biallelic SNPs are most common. Therefore, they are considered as biallelic markers. Based on their exact location in the genome and on their effect on the encoded protein, SNPs are classified into the following types (Gupta et al., 2001).

1. Noncoding SNPs (ncSNPs): They are found in the noncoding DNA. A subset of these ncSNPs resides in introns.
2. Coding SNPs, exonic SNPs, cDNA SNPs: They reside in exon and the corresponding cDNA.
3. Synonymous SNPs: They are exonic SNPs which do not change the amino acid sequence of encoded domain or protein.
4. Nonsynonymous SNPs: They are exonic SNPs which change the amino acid sequence of encoded domain or protein and affect the functioning of that particular protein or cause altered phenotype. These SNPs are also known as diagnostic SNPs because of their association with certain diseases in human and with certain agronomic traits in plants. The detection of diagnostic SNPs is a major aim of many SNP discovery projects.
5. Promoter SNPs (pSNPs) or Regulatory SNPs: They are located in the promoters or regulatory regions of genes, respectively. Such pSNPs can strongly affect the activity of the associated gene. In contrast, intronic SNPs are more or less inert.
6. Reference SNPs (refSNP): Any SNP at a specific site of a genome that serves as a reference point for the detection of other SNP in its neighborhood is called a reference SNP. A reference SNP number which is also known as rsID is assigned to each SNP at the time of its submission to the databases (e.g., the public dbSNP at the National Centre for Biotechnology Information [NCBI]; https://www.ncbi.nlm.nih.gov/SNP). As more and more SNPs are accumulating in the databases, they are labeled with the organism from which they originate (e.g., yeast SNP, human SNP, wheat SNP).

In general, SNPs are highly abundant in the genome, but their density differs substantially in different regions of a genome. It varies from genome to genome in any species, and more so from species to species. For example, the average density of SNPs in the human genome is about 1 in 1000 bp, while it is relatively high in plant species as 1 in 200 to 500 bp. As may be expected, SNP density is generally higher in intergenic and intronic regions compared to that in exons. The implementation of SNPs as molecular marker involves two major stages: SNP detection, and SNP genotyping.

SNP Detection Technology

Through detection of SNP (also known as SNP discovery) it is possible to know the number of SNPs and their precise location in the genome. Presence of SNPs can be determined through one of the two approaches. In the first approach called database approach, SNPs are identified by mining sequence databases and are then coined *in silico* or electronic SNPs (isSNPs, eSNPs). For model organisms and major crops SNP maps are already established and available in public databases (NCBI database). The second approach called experimental approach is used for uncharacterized individuals, desired genes or genomic regions. In this approach screening for SNPs is carried out by a series of techniques such as microchip hybridization, direct sequencing, and electrophoresis of PCR fragments containing candidate sequence through single strand conformation polymorphism (SSCP) or denaturing gradient gels (DGGE). Single strand conformation polymorphism (SSCP) is based on the variation in mobility of small polymorphic single-strand DNA fragments in non-denaturing acrylamide gels. Sequence analysis is the most direct way of identifying SNPs but is also the most time consuming and costly. Another problem with this strategy is sequencing errors. A sequencing error rate of just one base pair per 100 would make a huge difference in the rate at which SNPs occur in nature. If these sequencing errors are not detected at an early stage, they would result in a considerable waste of resources in both designing allele-specific oligonucleotide (ASOs) and carrying out the SNP assay (Gupta et al., 2001).

SNP GENOTYPING

SNP genotyping describes the confirmation about the presence of selected SNPs and their allelic variation among different individuals of a species or population. For model organisms, including animals as well as plants, SNP maps are available in the database (dbSNP). First few SNP markers are selected through mining of databases and then genotyping is performed. The major SNP genotyping techniques fall into six groups:

1. **Direct Sequencing:** Fluorescent based sequencing by automated slab gel or capillary electrophoresis is the standard method for SNP genotyping, but it is one of the slowest techniques.
2. **Restriction Enzyme Digestion:** This is the most convenient method for SNP genotyping, also known as SNP-RFLP. If one allele contains a recognition site for a restriction enzyme while the other does not, digestion of the two alleles will give rise to fragments of different length.
3. **Allele Specific PCR:** In this, SNP specific primer is used to amplify DNA. DNA containing SNP will be amplified while the absence of that particular SNP is indicated by no amplification. In order to study polymorphism, primer is used to amplify one SNP at a time but not the other.
4. **Allele Specific Primer Extension**: In the primer extension technique only a single fluorophore labeled dideoxynucleotide is either incorporated at the SNP position or not, depending on the allelic state.
5. **Allele Specific Oligonucleotide Hybridization:** In this technique, fluorescence labeled PCR fragments are hybridized to immobilized oligonucleotides, each representing a particular SNP allele. After stringent hybridization and washing, fluorescence intensity is measured for each SNP nucleotide separately.
6. **Allele Specific Oligonucleotide Ligation:** For this technique, the genomic target sequence is first PCR amplified. Then allele specific oligonucleotides complementary to the target sequence are ligated to the DNA adjacent to the polymorphic site, at the 3'- or 5'-ends. The ligation is possible only in the case of a complete match.

SNP markers are more applicable for animal and human genomics. Therefore, major advances in SNPology occurred in mammalian system. SNP markers find applications in forensic science, comparative and evolutionary genetics and disease diagnosis. The association of SNPs with disease susceptibility genes for human disorders such as type-II diabetes, hypertension, and cancer has been identified which is expected to help for their early diagnosis and treatment. Once the culprit genes are identified, the encoded proteins can be targeted by novel therapeutic drugs. Despite of their increasing use in genotyping and gene mapping, use of SNPs in plants was rather delayed. Initially, SNPs have been rigorously searched for only in a few plant species. These include several major crops like barley, rice, maize, wheat, sugar beet, *Arabidopsis thaliana* and some forest trees. With the reduction in cost for next generation sequencing (NGS), application of SNPs is getting extended to carryout phylogenic analysis, marker-assisted selection, genome selection, genome-wide association studies (GWAS) and genetic mapping of quantitative character (QTLs) in most of the important crop plants and forestry species (Kumar et al., 2012, Zargar et al., 2015).

Expressed Sequence Tag Polymorphism (ESTPs)

Expressed sequence tags are short, single stranded, randomly selected, unedited sequences which are obtained from cDNA libraries. These EST markers provide a low cost alternative for whole genome sequencing (also known as "poor" man's genome) and especially relevant to the transcriptome of an organism at various stages of development and under different experimental conditions. EST markers find application in gene discovery, whole genome annotation, gene structure identification, cloning of specific gene of interest, SNP characterization and proteomic exploration. EST markers are utilized mainly for rice and *Arabidopsis*. The established EST markers for model organisms are compiled and available in public database (http://www.ncbi.nlm.nih.gov). Implementation of EST markers for gene mapping and analysis of polymorphism is analogous to SNP marker.

Single Strand Conformation Polymorphism (SSCP)

The principle of single strand conformation polymorphism (SSCP) relies on the fact that the mobility of a single stranded DNA fragment (ssDNA) on a non-denaturing poly acrylamide gel is not only governed by its size but also by its three dimensional conformation. The conformation of ssDNA is affected by its nucleotide sequence. Two DNA fragments of identical size which vary in their base sequence even by a single nucleotide have different conformation and migrate differently on gel. Therefore, SSCP is considered as a powerful technique to analyze the DNA polymorphism and can detect the heterozygosity of DNA fragments of similar size on the basis of their conformation. This technique is applicable only for ssDNA because a single nucleotide change in a particular sequence cannot be distinguished by electrophoresis in a double-stranded DNA. The possible reason is that the physical properties of both strands are almost identical in dsDNA. A single-stranded DNA undergoes a 3-dimensional folding and may assume a unique conformational state based on its DNA sequence.

A standard SSCP experiment involves only few steps which are diagrammatically shown in Figure 6. First, the desired DNA fragments from different individuals are obtained either from restriction digestion of genomic DNA or by PCR amplification by using specific primers. These dsDNA fragments are denatured into ssDNA fragments by heating, and immediately resolved by poly acrylamide gel electrophoresis. SSCPs are detected as mobility shifts of individual ssDNA fragments relative to each other. Detection on gel is facilitated by labeling either with radioisotopes or fluorescent dyes by autoradiography or fluorimetry, respectively. Alternatively, unlabeled fragments are visualized by silver staining. Ethidium bromide can also be used, but is less efficient for ssDNA.

Figure 6. Single strand conformational polymorphism analysis

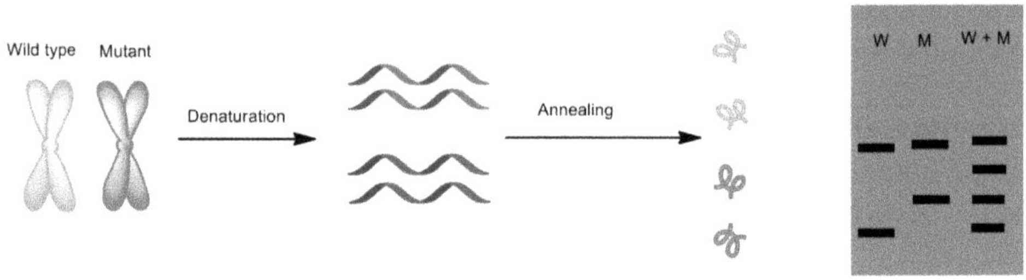

Single point mutations can cause major differences in the folded form of single stranded DNA. These differences can be detected as differences in electrophoretic mobility

A sample SSCP gel is shown in Figure 7. It describes the inheritance of multiple haplotypes or sets of alleles as a unit. The difference in the position of bands in adjacent lanes is related to the difference in the number of nucleotides (in parentheses) in separating fragments: lanes 1-2 (2), lanes 2-3 (0), lanes

3-4 (3), lanes 4-5 (1), lanes 5-6 (3), lanes 6-7 (1), lanes 7-8 (1), and lanes 8-9 (4). The lanes 2 and 3 represent identical haplotypes from two individuals.

Figure 7. A SSCP gel showing the inheritance of multiple haplotypes or sets of alleles as a unit

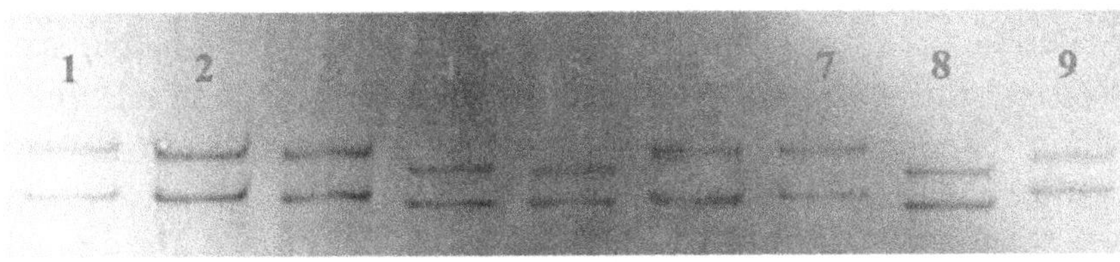

The difference in the position of bands in adjacent lanes is related to the difference in the number of nucleotides in separating fragments

Advantages of SSCP analysis include its technical simplicity, high sensitivity even for a single nucleotide difference, rapid and inexpensive. The precise electrophoresis conditions used may determine whether a given PCR fragment runs as a monomorphic or polymorphic band, or even produces a complicated banding pattern on the gel. Multiple bands may result from, e. g., the existence of multiple stable ssDNA conformations, partial unfolding and formation of heteroduplex. Because the exact migration behavior of ssDNA fragment is sensitive to a number of parameters (temperature, solvent, pH etc.), development of consistent experimental protocol is essential to ensure reproducibility, and the precise running conditions of each SSCP may have to be decided one by one. SSCP has been most extensively applied in diagnosis of heritable human diseases. Other application areas include linkage analysis, genetic mapping of cDNAs, developing nuclear markers and population genetics. In plants, the potential of SSCP analysis still seems to be underexploited. Besides SSCP, there are other techniques like heteroduplex analysis, denaturing gradient gel electrophoresis (DGGE), temperature gradient gel electrophoresis (TSGE) to assess DNA sequence variation on the basis of 3-D conformation through electrophoresis.

MULTI LOCUS MARKERS

In eukaryotes, approximate 40% of genomic DNA is noncoding which is also known as "junk DNA". This does not contain any protein coding genes, therefore, considered as nonfunctional DNA. This is highly polymorphic in nature and characterized by repetitive sequences. They may either be present in discontinuous manner (known as interspersed repeated DNA) or in continuous array (known as tandem repetitive DNA). Interspersed repeats exemplified by transposable elements, are present at multiple sites throughout the genome. In contrast, tandem repeats are restricted to fewer sites, arranged in head-to-tail fashion. This kind of organization is also exhibited by some genes, such as genes for histone protein and rRNA. The repetitive DNA in general is characterized by highly variable copy number in different individuals, therefore, gained the term Variable Number of Tandem Repeats (VNTRs). Each variant acts as an inherited allele, allowing them to be used for personal or parental identification. This repetitive DNA is further classified into different categories depending upon the length, copy number of basic repeat unit and genomic localization (Caetano-Anolles & Gresshoff, 1997). The overview of this classification is represented in Table 1.

Table 1. Categories of tandem repetitive DNA

Name	Copy number	Repeat unit (bp)	Location in the genome	Features
Satellite DNA	1000-100,000	100-300	Telomeres and centromeres	Rarely used as molecular marker
Minisatellite	-	10-60	Many loci in the genome, unevenly distributed	Used as molecular marker
Microsatelite	100-500	1-6	Many loci in the genome, evenly distributed	Used as molecular marker

Minisatellite Markers

The abundance of polymorphic minisatellite in eukaryotic genome allows their use as ideal molecular marker. They were initially discovered in the human during early eighties and later detected from several organisms including cattle, mouse, birds, and plants. They are not only present in nuclear genome but also reported in mitochondrial and chloroplast DNA. They have been exploited as molecular marker in several ways, but two techniques are more common. In the first method, the minisatellite, complementary probes are hybridized to restriction enzyme digested genomic DNA to produce a complicated banding pattern upon electrophoresis and southern blotting. In the second method, the minisatellites are amplified by PCR using single specific primer. This technique is known as direct amplification of minisatellite DNA (DAMD-PCR). The amplification products are then separated by electrophoresis to generate a specific banding pattern which varies according to genotype. An individual's minisatellite pattern is based either on father's or mother's minisatellite pattern or it is a combination of both. He or she never possesses a minisatellite which either of his or her parents do not have. Because these markers follow genetic inherited pattern, they are individual specific.

These minisatellite markers find tremendous applications both for animals and plants. They form the basis of DNA fingerprinting technique for personal identification. DNA fingerprinting is a very common technique for identification of suspect at crime scenes, solving paternity and maternity cases, diagnosis of genetic diseases etc. In plants the use of minisatellite marker is comparatively restricted, still they find application in plant systematics, species and cultivar identification, protecting plant varieties, gene mapping, and study of genetic diversity in a particular population. The uneven distribution of minisatellite is a major drawback in its use as an effective marker. Like rRNA genes, minisatellite loci are also concentrated to a particular location on chromosome like subtelomeric region and thus may not be able to provide the desired density of markers.

Microsatellites Markers

Litt & Lutty (1989) coined the term microsatellite. The tandem repetitive DNA of short length (1-6 bp), which are more frequent in occurrence and distributed evenly are defined as microsatellites analogous to minisatellites. Several synonyms exist for microsatellites like short tandem repeats (STRs), and simple sequence repeats (SSRs). Allele size difference of a single base pair can be revealed by this technique, therefore, they are also called simple sequence length polymorphisms (SSLPs), or sequence-tagged microsatellite sites (STMS). These microsatellites vary in terms of length, copy number and localization among genotypes. The most abundant motifs found in mammalian genomes are $(A)_n$ and $(CA)_n$ as well as their complements; whereas $(A)_n$,

$(AT)_n$, $(GA)_n$ and $(GAA)_n$ repeats are the most frequent motifs in plants. Microsatellites composed of tri, tetra and pentanucleotide motifs are generally less common than mono and dinucleotide repeats. Another way to categorize microsatellites relates to the degree of perfectness of the arrays and they are of three types (1) Perfect repeat, which consist of a single, uninterrupted array of a particular motif; (2) imperfect repeats, in which the array is interrupted by one or several out of frame bases; and (3) compound repeats, with intermingled perfect or imperfect arrays of several motifs. Examples of these different categories are given in Figure 8. Like minisatellites, occurrence of microsatellite is also detected in nuclear as well as organelle genome but they are evenly distributed throughout the genome. Numerous methods have been developed that exploit microsatellite as molecular markers, but the most important one is the locus specific PCR amplification of microsatellites with flanking sequence specific primers. The flanking ends of these microsatellites are represented by unique sequence DNA and locus specific primers are designed according to this region, in order to facilitate the amplification of desired microsatellite (Figure 9). If a microsatellite is flanked by a stretch of an unordered DNA sequence of 30-50 nucleotides long then the probability of occurrence of that particular sequence more than once in the genome is very small. (if all the four nucleotides have equal chances of occurrence, then the probability of a given 50 bp fragment is 0.25^{50}). In contrast a repeating unit with less number of nucleotides (say GC16) may present at thousands of places in the genome. The amplified products are than separated either by polyacrylamide gel electrophoresis or capillary electrophoresis. The size of each PCR product depends on the copy number of basic repeat units. In the case of poly acrylamide gel electrophoresis detection of bands is done by silver staining, but capillary electrophoresis is preferred. In a single capillary, markers of multiple base sizes can be separated simultaneously. Up to four fluorescent dyes can be used in the same PCR reaction, enabling several microsatellites to be analyzed in a single run. Suppose five different plant samples i. e. 06, 07, 08. 09 and 10 are amplified using a microsatellite marker RM-224. The amplified products are separated through capillary electrophoresis (Figure 10). Two peaks are observed in sample 06 and 09 which indicates heterozygous state of plant. Single peak is obtained in sample 07 and 08, which points towards homozygosity. Absence of any kind of amplification in sample 10 denotes the unavailability of particular microsatellite in that plant.

Other methods used for microsatellite detection are single primer PCR, PCR primers in combination with other primer types and hybridization with complementary probes.

Figure 8. Examples of different types of microsatellites (a) On the basis of length of repeat units (b) On the basis of degree of perfectness of the arrays

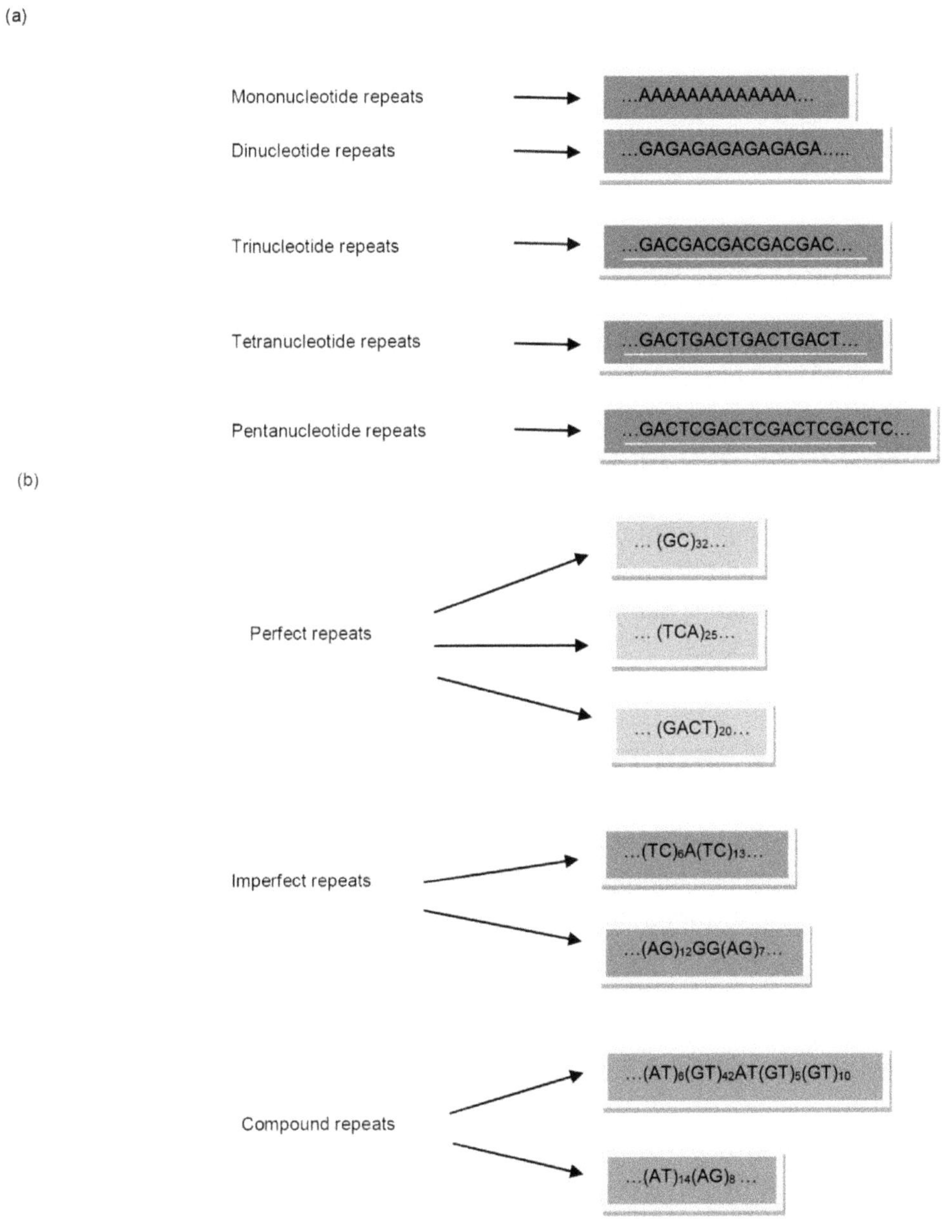

Microsatellites are proved to be useful for a range of applications covering mapping genomes, diagnosis of certain diseases, personal identification, parentage analysis, in population genetics, genetic diversity analysis etc.

Figure 9. Detecting microsatellites from genomic DNA by PCR amplification

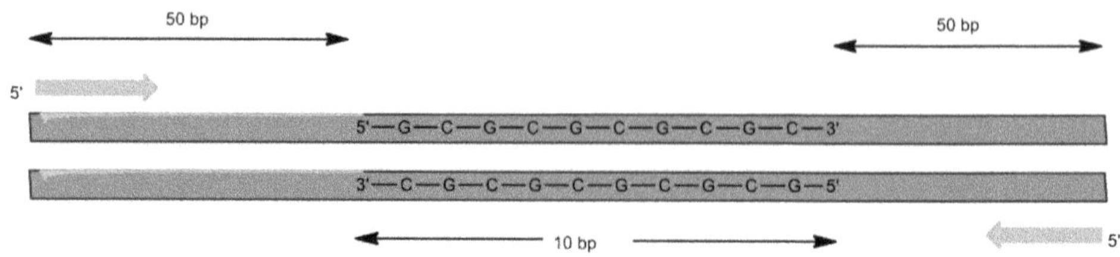

These markers are superior to minisatellite markers because of their locus specificity, co-dominant, highly polymorphic nature and ease of amplification due to their small size. Besides having several advantages microsatellite is a rarely used marker for higher-level systematics, because of high mutation rate.

Inter Simple Sequence Repeat Markers (ISSR)

In this technique primers are designed on the basis of microsatellite sequence and used for the amplification of inter SSR DNA sequence through PCR. These are mostly dominant markers, although few of them exhibit co-dominance (Ng & Tan, 2015).

ARBITRARY SEQUENCE MARKERS

All those markers which make use of primers designed on the basis of arbitrary nucleotide sequence to amplify random sections of genomic template DNA, are defined as arbitrary molecular markers. Typically, single PCR primers are used under relaxed stringency conditions, and no prior knowledge of DNA sequence is required. Following markers are considered under this category.

Random Amplified Polymorphic DNA Markers (RAPD)

In this technique the genomic DNA is amplified by PCR reaction implementing single nonspecific primers of generally 10 nucleotides long (10mers). These primers bind randomly on two different strands of genomic DNA, wherever they find complementary sequence. Any section of template DNA, where two primers present at appropriate distance in opposite direction, gets amplifies. A particular primer binds to its complementary sequence which present at different and multiple locations on genomic DNA of different

Figure 10. Analysis of microsatellite marker (RM-224) amplified products through capillary electrophoresis (five different plant samples i. e. 06, 07, 08, 09 and 10)

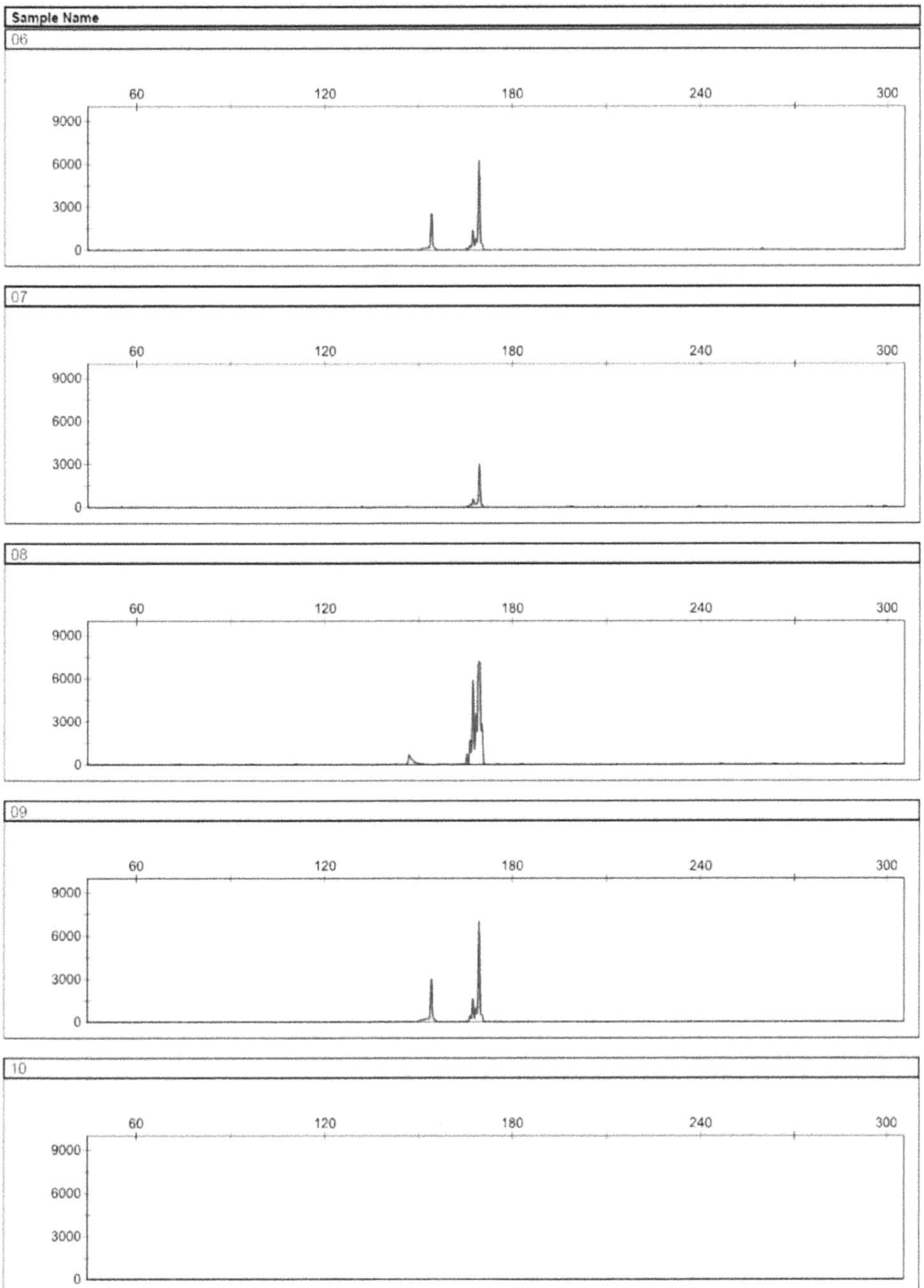

individuals (Collard et al., 2005). In this way by using same primer detection of polymorphism is possible.

First, the genomic DNA is extracted from various samples. This is amplified employing an arbitrary primer. Different arbitrary primers are used in different experimental set-ups. The amplification products are separated by agarose gel electrophoresis and stained with ethidium bromide for visualization of obtained banding pattern. For example, genomic DNA of five different plants (1, 2, 3, 4 and 5) is amplified using a RAPD marker i. e. OPAC-07. The obtained banding pattern is shown in Figure 11, which indicates that plant 5 belongs to a species different from rest four. Most RAPD fragments are inherited as dominant markers i. e. they are either present or absent. A fragment is seen in the homozygous (AA) as well as in the heterozygous (Aa) situation and only the absence of the fragment clearly reveals the genotype (aa). Main application areas include the identification of cultivars and clones, genetic mapping, marker assisted selection, population genetics and molecular taxonomy at the species level. RAPD works as efficient tool for identification of markers that are linked to agronomically important traits. These markers can detect genetic polymorphism and when linked to major genes it can be potentially used to identify morphological traits. The greatest advantage of RAPD is its technical simplicity without any requirement of prior sequence information. However, the use of RAPD is limited for population genetics and mapping studies because of dominant nature and less reproducibility of results between experimental replicates. In RAPD the banding pattern is highly influenced by PCR conditions which should be strictly maintained for all experimental replicates. The use of RAPD is further complicated by variation in band intensity. The brightness of a given band depends on several factors, including the degree of repetitiveness of the targeted DNA region, the extent of primer-template mismatch, and the presence or absence of competing target region in the genome. To calculate the expected number of bands per primer following equation can be used:

$$b = (2000 \times 4^{-2n}) \times C$$

where, b is the expected number of bands per primer, n is the primer length in nucleotides and C is the genome size in base pairs per haploid genome. For example, in a plant species such as maize (genome size of 6×10^6 Kb), 10.9 bands with a 100% homology between primer and template are expected per 10-nucleotide primer.

The other variants of RAPD are DNA Amplification Fingerprinting (DAF) and Arbitrarily Primed PCR (AP- PCR). DAF makes use of very short primers (often only 5 to 8 nucleotides long) at relatively high concentrations, with either low or high stringency annealing steps and two instead of three temperature cycles in the PCR. The resulting fragments are resolved in polyacrylamide gels and visualized by silver staining. While the AP-PCR make use of oligonucleotides of 20 or more bases, originally designed for other purposes, as primers. Two cycles with low stringency are followed by 30-40 cycles with high stringency. Radiolabelled nucleotides are included in the last 20 -30 cycles only. For convenience, all arbitrarily primed PCR techniques are termed as RAPD.

Figure 11. RAPD profile of five plants using arbitrary primer OPAC-07

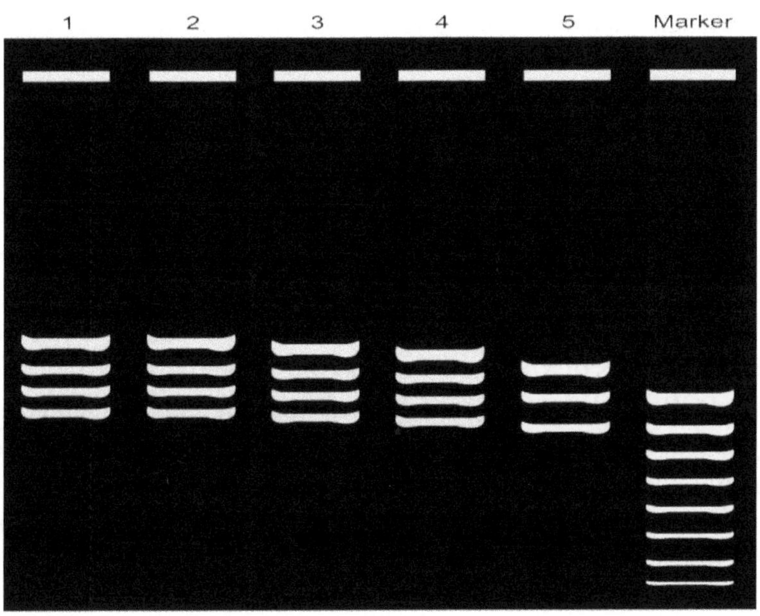

Sequence Characterized Amplified Regions (SCARs)

A modified RAPD derived molecular markers was developed by Paran and Michelmore (1993), which circumvent most of the drawbacks of RAPD. These are derived by selection, cloning and sequencing of desired RAPD fragment. This sequence information is used for designing specific primers which are longer (22-24 nucleotides), complementary to the ends of cloned

RAPD fragment. Implementing these primers with original template DNA, single loci is amplified which is known as sequence characterized amplified region (SCAR). The presence or absence of the band is the indication of sequence variation. These markers are better than RAPD markers in terms of reproducibility. In addition, digestion of these markers with a tetra cutting restriction enzymes converts them into a co-dominant markers and polymerization is analyzed either through denaturing gel electrophoresis or SSCP. Because of their locus specificity and co-dominant nature, SCARs exhibit several advantages over RAPDs. These include, mapping studies, map based cloning, and physical mapping. SCARs also implemented for comparative genomics or homology studies among related species.

Amplified Fragment Length Polymorphism (AFLP)

Amplified fragment length polymorphism (AFLP), discovered by Zabeau and coworkers in 1993, is a combination of RFLP and PCR. AFLP technology is applicable to all organisms without previous sequence information, and generally results in highly informative fingerprints. Principally the technique amplifies a desired restriction fragment obtained from genomic DNA. The procedure is summarized in Figure 12. There are two major steps. First, genomic DNA is digested with two different restriction enzymes to produce sticky ends. Among these two restriction enzymes, one is a rare cutter and another is a frequent cutter. In order to prevent the pairing between these sticky ends, adapters of a defined sequence are ligated to both ends. The next step is the PCR amplification of this restriction fragment, which is achieved by using specific primers according to the sequence of adapter and restriction site. An oligonucleotide with known sequence is added at the 3' ends of PCR primers, which therefore, can only initiate DNA synthesis from a subset of the restriction sites. Only those restriction fragments in which nucleotides flanking the restriction sites match the selective nucleotides will be amplified.

The AFLP procedure results in predominant amplification of restriction fragments on which a rare cutter sequence is present at one end and a frequent cutter sequence is present on the other end. A large number of restriction fragment bands are generated which facilitates the detection of polymorphism. Choosing the different base number and composition of nucleotides in the adapters, it is possible to control the number of DNA fragments that are amplified.

Figure 12. Principle of AFLP analysis (For details see text)

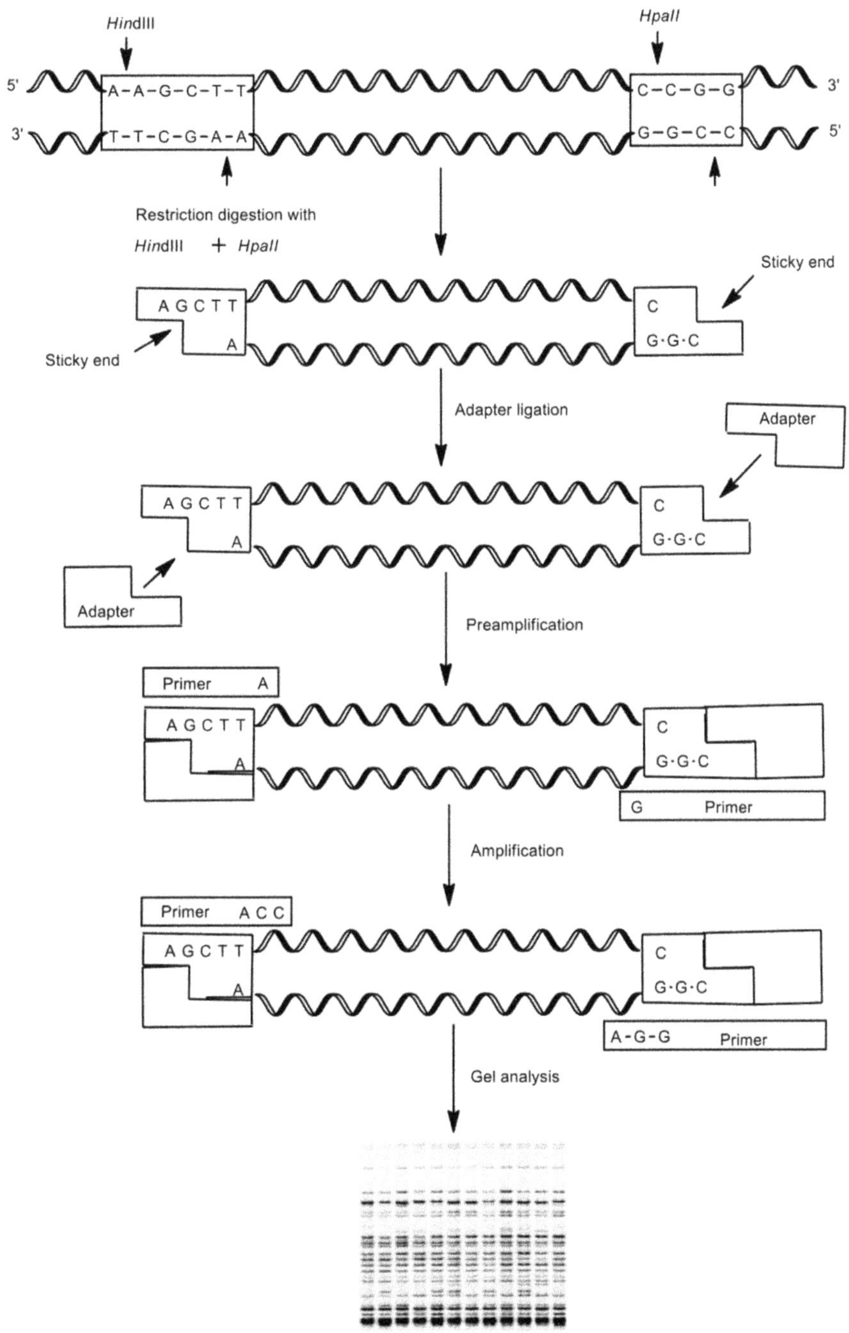

MOLECULAR MARKERS FROM ORGANELLE GENOME

In higher eukaryotic cell, chloroplast and mitochondria are two organelles which contain DNA as their own genetic material along with main nuclear genome. This DNA is known as organelle genome. This organelle DNA is circular in most of the cases along with few exceptions, double stranded and supercoiled. The mitochondrial and chloroplast genomes contain genes for the rRNA components of the ribosomes of these organelles, for many of the tRNAs used in organelle protein synthesis, and for few proteins that remain in the organelles and perform functions specific to the organelles. All other proteins required by these organelles are nuclear encoded, synthesized on cytoplasmic ribosomes, and transported into them. These organelle genomes vary significantly from that of nuclear genome, in the following manner: 1) show non-Mendelian inheritance, 2) show maternal inheritance, 3) present in multiple copies in several nucleoid regions, 4) differ in their GC content, therefore all type of genomes (nuclear, mitochondrial and chloroplast) are separable by CsCl density gradient centrifugation, 5) lower rate of mutation, 6) lack of introns, and 7) limited exposure to recombination.

Besides all these differences, like nuclear genome they also provide important molecular markers in the form of conserved and repetitive sequences. Therefore, organelle genomes can also be explored for species identification, taxonomic studies, genetic diversity analysis etc. This idea gives birth to a new technique known as "DNA barcoding". DNA barcoding is a diagnostic technique for species identification, using a standardized short DNA region.

The desirable features of an efficient DNA barcode are (1) easily retrievable with a single primer pair, (2) possibility of bidirectional sequencing with little requirement of manual editing of sequence traces, and (3) provide maximal discrimination among species (Kress, 2017). The idea of a DNA barcode is comparable with a supermarket scanner that identifies products using the black stripes of the Universal Product Code. The barcode of an unidentified specimen can be compared with the reference barcodes in database to find the matching species. DNA barcoding may serve as a potential tool for taxonomists to enhance their knowledge about taxonomy as well as for non-experts who need to make a quick identification. DNA barcoding has the potential for many useful applications in conservation, biodiversity inventories, forensic and trade surveillance.

The organelle genome evolves at a slower rate as compared to nuclear genes which makes it suitable for discrimination between two species. The intra species variability is less as compared to interspecies one. The DNA barcoding was first proposed at the University of Guelph, Ontario, Canada. It is well established in the case of animals and a universal barcode for animals was agreed in 2003. A 648-bp fragment of a mitochondrial gene which encodes an enzyme *cytochrome c oxidase I* (*COI*), serves as an effective DNA barcode for identification of animal species. Several studies have now proved that sequence variations in a ~650 bp region of *cytochrome C oxidase I* (*COI*) gene provides strong species level resolution for varied animal groups including birds, fishes, springtails, spiders and moths. However, no universally accepted barcode is available till now for plants. In plants, the evolution of mitochondrial genome is too slow to provide sufficient level of variations for species level discrimination. Therefore, the plastid genome is suggested as an alternative for DNA barcoding. Initially single regions in the plastid genome were suggested as DNA barcode (e.g. exon such as *rbcL, atpB, ndhF* and *matK* and non-coding regions such as the *trnL* intron and *trnL-F* intergenic spacer (Kress, 2017). Later the consortium for the barcoding of life (CBOL), via the Plant Working Group has recommend the 2-locus combination of *rbc L* + *matK* and a non-coding plastid region, the *trnH – psbA* spacer as the plant barcode. DNA barcoding in plants has been discussed in details in Chapter 8.

COMPARISON OF DIFFERENT GENETIC MARKERS

All the markers describe above have their own advantages and disadvantages. The choice of an appropriate marker depends on several factors including, the objective of study, basic characteristics of marker, level of genetic variability in the study material, available resources, cost and technical skills. A comparative assessment of these markers is given below which will assist the researchers to select the right marker.

Phenotypic markers have long been used to identify species, genera and families; to evaluate systematic relationships; and to discriminate cultivars, breeding lines, etc. In the case of qualitative traits which are governed by one or few genes, analysis of morphological characters is more pronounced than the extent of variation indicated by RAPD or other DNA based molecular marker. But these phenotypic markers are strongly influenced by the environment, therefore, special breeding programs and experimental designs are needed to distinguish genotypic from phenotypic variation. In addition, these markers are not applicable to all those organisms which lack identifiable morphological features.

Allozymes are associated with several advantages like low cost for chemicals and labor, the user-friendliness and co-dominant nature; however, they have number of limitations. With allozymes, a new allele will only be detected if it affects the electrophoretic mobility of the studied molecule. Only about 30% of nucleotide substitutions result in polymorphic fragment patterns, and allozyme analysis therefore underestimates the genetic variability. The interpretation of allozyme patterns is quite difficult with polyploid plants. In addition, for allozyme studies the collected sample has to be processed immediately because most of the enzymes are quite unstable. Although in the past allozymes have been used rather extensively for the discrimination of genotypes, they have now been superseded by DNA based markers.

CONCLUSION

Molecular markers have occupied center stage in plant breeding since late 1980s. the development of markers based on simple sequence repeats (SSRs) and single nucleotide polymorphisms (SNPs) and the availability of high-throughput (HTP) genotyping platforms have further accelerated the process of generating dense linkage maps and regular use of the molecular markers for marker-assisted breeding in many crop plants. Advancements in the sequencing technologies have led to the development of next generation sequencing (NGS) platforms that are substantially low cost with high throughput. However, despite the regular use of the molecular markers for genome-wide profiling and marker assisted selection (MAS), breeding of crops for yield, nutritive value, resistance against biotic and abiotic stresses has remained a challenge due to complex inheritance of these characters. Thus, it is important for the breeders to dissect these traits in the context of whole genome. Plant genomics has enormous potential to contribute towards crop improvement by providing extensive knowledge from the analysis of the genomes. The coming years are likely to see continued innovations in molecular marker technology to make it more precise, productive and cost effective in order to investigate the underlying biology of various traits of interest.

REFERENCES

Avise, J. C. (1994). *Molecular Markers: Natural History and Evolution*. Chapman & Hall. doi:10.1007/978-1-4615-2381-9

Boopathi, N. M. (2013). Success stories in MAS. In N. M. Boopathi (Ed.), *Genetic mapping and marker assisted selection: basic, practice and benefits* (pp. 187–192). Springer. doi:10.1007/978-81-322-0958-4_9

Caetano-Anolles, G., & Gresshoff, P. M. (1997). *DNA Markers: protocols, applications and overview*. Wiley-VCH.

Collard, B. C., Jahufer, M. Z., Brouwer, J. B., & Pang, E. C. K. (2005). An introduction to marker, quantitative trait loci (QTL) mapping and marker-assisted selection for crop improvement the basic concepts. *Euphytica, 142*(1-2), 169–196. doi:10.100710681-005-1681-5

Grover, A., & Sharma, P. (2016). Development and use of molecular markers: Past and present. *Critical Reviews in Biotechnology*, *36*(2), 290–302. doi:10.3109/07388551.2014.959891 PMID:25430893

Henry, R. J. (Ed.). (2012). *Molecular markers in plants*. John Wiley & Sons, Inc., doi:10.1002/9781118473023

Kress, W. J. (2017). Plant DNA barcodes: Application today and in the future. *Journal of Systematics and Evolution*, *55*(4), 291–307. doi:10.1111/jse.12254

Kumar, S., Banks, T. W., & Cloutier, S. (2012). SNP discovery through next-generation sequencing and its applications. *International Journal of Plant Genome*, *32*, 1–12. doi:10.1155/2012/831460 PMID:23227038

Litt, M., & Luty, J. A. (1989). A hypervariable microsatellite revealed by in vitro amplification of a dinucleotide repeat within the cardiac muscle actin gene. *American Journal of Human Genetics*, *44*, 397–401. PMID:2563634

Nadeem, M. A., Nawaz, M. A., Shahid, M. Q., Dogan, Y., Comertpay, G., Yildir, M., ... Baloch, F. S. (2018). DNA molecular markers in plant breeding: Current status and recent advancement in genomic selection and genome editing. *Biotechnology, Biotechnological Equipment*, *32*(2), 261–285. doi:10.1080/13102818.2017.1400401

Ng, W. L., & Tan, S. G. (2015). Inter-simple sequence repeat (ISSR) markers: Are we doing it right? *Academy of Science Malaysia (ASM). Science Journal*, *15*, 30–39.

Paran, I., & Michelmore, R. W. (1993). Development of reliable PCR-based markers linked to downy mildew resistance genes in lettuce. *Theoretical and Applied Genetics*, *85*(8), 985–993. doi:10.1007/BF00215038 PMID:24196149

Zargar, S. M., Raatz, B., Sonah, H., Nazir, M., Bhat, J. A., Dar, A., ... Rakwal, R. (2015). Recent advances in molecular marker techniques: Insight into QTL mapping, GWAS and genomic selection in plants. *Journal of Crop Science and Biotechnology*, *18*(5), 293–308. doi:10.100712892-015-0037-5

ADDITIONAL READING

Abduvakhmonov, J. V. (2016). *Microsatellite markers*. InTech. doi:10.5772/62560

Agarwal, M., Shrivastava, N., & Padha, H. (2008). Advances in molecular markers techniques and their applications in plant sciences. *Plant Cell Reports*, *2*(4), 615–631. doi:10.100700299-008-0507-z

Angaji, S. A. (2009). QTL mapping: A few key points. *International Journal of Applied Research in Natural Products*, *2*, 1–3.

Beissinger, T. M., Hirsch, C. N., Sekhon, R. S., Foerster, J. M., Johnson, J. M., Muttoni, G., Vaillancourt, B., Buell, C. R., Kaeppler, S. M., & de Leon, N. (2013). Marker density and read depth for genotypinh populations using genotyping-by-sequencing. *Genetics*, *193*(4), 1073–1081. doi:10.1534/genetics.112.147710 PMID:23410831

Belicuas, P. R., Guimaraes, C. T., Paiva, L. V., Dwarte, J. M., Maluf, W. R., & Paive, E. (2007). Androgenic haploids and SSR markers as tools for the development of tropical maize hybrids. *Euphytica*, *156*(1-2), 95–102. doi:10.100710681-007-9356-z

Chawla, H. S. (2002). *Introduction to Plant Biotechnology*. Oxford & IBH Publishing Co. Pvt. Limited.

Davey, J. W., Hohenlohe, P. A., Etter, P. D., Boone, J. Q., Catchen, J. M., & Blaxter, M. L. (2011). Genome-wide genetic marker discovery and genotyping using next-generation sequencing. *Nature Reviews. Genetics*, *12*(7), 499–510. doi:10.1038/nrg3012 PMID:21681211

Dhingani, R. M., Umrania, V. V., Tomar, R. S., Parakhia, M. V., & Golakiya, B. A. (2015). Introduction to QTL mapping in plants. *Annals of Plant Sciences*, *4*, 1072–1079.

Gilmartin, P. M., & Bowler, C. (2002). *Molecular Plant Biology*. Oxford University Press.

Grzebelus, D. (2006). Transposon insertion polymorphism as a new source of molecular markers. *Journal of Fruit and Ornamental Plant Research*, *14*, 21–29.

Hackett, C. A. (2002). Statistical methods for mapping in cereals. *Plant Molecular Biology*, *48*(5/6), 585–599. doi:10.1023/A:1014896712447 PMID:11999836

He, J., Zhao, X., Laroche, A., Lu, Z. X., Liu, H., & Li, Z. (2014). Genotyping by sequencing (GBS), an ultimate marker assisted selection (MAS) tool to accelerate plant breeding. *Frontiers in Plant Science*, *5*, 1–9. doi:10.3389/fpls.2014.00484 PMID:25324846

Jiang, G. L. (2013). Molecular markers and marker assisted breeding in plants. In S. B. Anderson (Ed.), *Plant breeding from laboratories to field* (pp. 45–83). InTech. doi:10.5772/52583

Jing, R., Bolshakov, V., & Flavell, A. J. (2007). The tagged microarray marker (TAM) method for high-throughput detection of single nucleotide and indel polymorphisms. *Nature Protocols*, *2*(1), 168–177. doi:10.1038/nprot.2006.408 PMID:17401351

Kalia, R. K., Raj, M. K., Kalia, S., Singh, R., & Dhawan, A. K. (2011). Microsatellite markers: An overview of the recent progress in plants. *Euphytica*, *177*(3), 309–334. doi:10.100710681-010-0286-9

Liu, Y., He, Z., Appels, R., & Xia, X. (2012). Functional markers in wheat: Current status and future prospects. *Theoretical and Applied Genetics*, *125*(1), 1–10. doi:10.100700122-012-1829-3 PMID:22366867

Lopez-Maestre, H., Brinza, L., Marchat, C., Kielbassa, J., Bastien, S., Boutigny, M., … Lacroix, V. (2016). SNP calling from RNA-seq data without a reference genome: identification, quantification, differential analysis and impact on the protein sequence. *Nucleic Acids Research, 44*, e148. doi:1093/nar/gkw655.

Mammadov, J., Aggarwal, R., Buyyarupa, R., & Kumpatla, S. (2012). SNP marker and their impact on plant breeding. *International Journal of Plant Genome*, *14*, 1–11. doi:10.1155/2012/728398

Morganet, M., Hanafey, M., & Powell, W. (2002). Microsatellite are preferentially associated with nonrepetitive DNA in plant genome. *Nature Genetics*, *30*(2), 194–200. doi:10.1038/ng822 PMID:11799393

Paran, I., & Michelmore, R. W. (1993). Development of reliable PCR-based markers linked to downy mildew resistance genes in lettuce. *Theoretical and Applied Genetics*, *85*(8), 985–993. doi:10.1007/BF00215038 PMID:24196149

Poczai, P., Varga, I., Laos, M., Cseh, A., Bell, N., Valkonen, J. P. T., & Hyvonen, J. (2013). Advances in plant gene-targeted and functional markers: A review. *Plant Methods*, *9*(1), 6–18. doi:10.1186/1746-4811-9-6 PMID:23406322

Salazar, J. A., Rasouli, M., Moghaddam, R. F., Zamani, Z., Imani, A., & Martinez-Gomez, P. (2014). Low cost strategies for development of molecular markers linked to agronomic traits in *Prunus*. *Agricultural Science*, *5*, 430–439.

Semagn, K., Bjornstad, A., & Ndjiondjop, M. N. (2014). An overview of molecular marker methods for plants. *African Journal of Biotechnology*, *2450*, 25–68.

Sharma, A., Namdeo, A. G., & Mahadik, K. R. (2008). Molecular markers: New prospects in plant genome analysis. *Physiological Reviews*, *2*, 23–34.

Silva, L. D., Wang, S., & Zeng, Z. B. (2012). Composite interval mapping and multiple interval mapping: Procedures and guidelines for Windows QTL cartographer. *Methods in Molecular Biology (Clifton, N.J.)*, *871*, 75–119. doi:10.1007/978-1-61779-785-9_6 PMID:22565834

Singh, B. D., & Singh, A. K. (2015). *Marker-assisted plant breeding: principles and practices*. Springer. doi:10.1007/978-81-322-2316-0

Singh, B. P., & Gupta, V. K. (2017). *Molecular markers in mycology: diagnostics and marker developments*. Springer. doi:10.1007/978-3-319-34106-4

Trick, M., Adamski, N. M., Mugford, S. G., Jiang, C. C., Febrer, M., & Uauy, C. (2012). Combining SNP discovery from next-generation sequencing data with bulked segregant analysis (BSA) to fine-map genes in polyploidy wheat. *BMC Plant Biology*, *12*(1), 14–24. doi:10.1186/1471-2229-12-14 PMID:22280551

Varshney, R. K., Hoisington, D. A., Nayak, S. N., & Graner, A. (2009). Molecular breeding: methodology and achievements. In D. J. Somers (Ed.), *Methods in molecular biology, plant genomics* (pp. 283–304). Humana Press.

Weising, K., Nybom, H., Wolff, K., & Kahl, G. (2005). *DNA fingerprinting in plants: principles, methods, and applications*. CRC press, Taylor & Francis Group. doi:10.1201/9781420040043

Wu, K. S., Jones, R., Danneberg, L., & Scolnik, P. A. (1994). Detection of microsatellite polymorphism without cloning. *Nucleic Acids Research*, *22*(15), 3257–3258. doi:10.1093/nar/22.15.3257 PMID:8065948

Yang, H., Li, C., Lam, H., Clements, J., Yan, G., & Zhao, S. (2015). Sequencing consolidates molecular markers with plant breeding practice. *Theoretical and Applied Genetics*, *128*(5), 779–795. doi:10.100700122-015-2499-8 PMID:25821196

Zane, L., Bargellioni, L., & Patarnello, T. (2002). Strategies for microsatellite isolation: A review. *Molecular Ecology*, *11*(1), 1–16. doi:10.1046/j.0962-1083.2001.01418.x PMID:11903900

APPENDIX

1. How morphological markers differ from molecular markers?
2. Explain why biochemical markers have become obsolete for using in crop improvement program?
3. What are the properties of an ideal molecular marker?
4. What are RFLPs and how did they arise?
5. What are some of the advantages of using RFLPs to diagnose human genetic disorders? What are the disadvantages?
6. Explain the statement that each RFLP is unique to a specific enzyme/probe combination.
7. How the techniques of RFLP differ from RAPD? Under what circumstances RAPD is preferred over RFLP?
8. How the techniques of RAPD differ from AFLP? Describe the conditions under which each one of them can be used efficiently.
9. What is a VNTR? How VNTRs can be identified?
10. Describe the difference between microsatellite and minisatellite? Which of the two is more useful in DNA fingerprinting and why?
11. Describe the general procedure used to produce a DNA fingerprint.
12. What are some of the practical applications of DNA fingerprints?
13. What the terms allozyme and isozyme signify? What are the advantages and disadvantages of using them as genetic markers?
14. Why RAPD works as dominant and SSR as co-dominant markers? Explain the principle behind this phenomenon.
15. What is a polymorphic marker? Why it is desirable for a molecular marker to be polymorphic?
16. If a statement says that "a marker will diagnose a gene with 99% accuracy", what does that mean in terms of genetic linkage?
17. What is a quantitative trait loci (QTL)? Describe how molecular markers can be utilize for the analysis of QTLs?
18. What are the basic criteria for selecting a DNA barcode? Why nuclear genes are not selected as DNA barcode?
19. Why chloroplast genome is preferred for the construction of DNA barcode in plants?
20. Describe how SNPs are classified.

Chapter 3
Marker-Assisted Breeding

ABSTRACT

Advancement in sequencing technologies has contributed towards identification and development of different types of molecular markers. Molecular plant breeding has contributed to a more comprehensive understanding of molecular markers and their role in identifying the genetic diversity within the crop plants. Marker-assisted breeding is basically the application of molecular markers, in combination with linkage maps and genomics, to alter and improve plant traits on the basis of genotypic assay. Several modern plant breeding strategies were developed which include marker-assisted selection (MAS), marker-assisted backcrossing (MABC), marker-assisted recurrent selection (MARS), and genome-wide selection (GWS) or genome selection (GS). The selection of right type of molecular markers is usually dependent on the breeding objectives. Similarly, selection strategies of molecular markers for qualitative and quantitative characters may differ. The procedure followed for marker assisted selection under various breeding objectives and conditions, for qualitative and quantitative traits are discussed in this chapter.

INTRODUCTION

Ancient farmers used to select best plants and save their seeds for cultivation in the following generation and thus considered to be the first 'plant breeders'. Archaeological evidence indicates that farmers used to employ selection pressure on the cultivated plants to meet their demands as early as 12,000

DOI: 10.4018/978-1-7998-4312-2.ch003

years ago. With the advancement of knowledge, plant breeding has evolved as a major discipline in plant biology.

By crossing two morphologically different parental genotypes, plant breeders could study the recombination and crossing-over events. Morphological markers such as flower color, flower shape, seed color, seed size, plant height, etc. were used to study the inheritance of genetic traits. However, morphological markers do not always inherited as simple Mendelian genes. Thus, over the years their usefulness in plant breeding programs has reduced considerably.

In higher plants, there exists enormous diversity (polymorphism) at the DNA level. In fact, in the natural population of plants, no two organisms are likely to be identical in their DNA sequence. . Molecular techniques have provided strategies to develop marker systems that detect such DNA variation, which can be used to assist traditional plant breeding. Once linkage between a marker locus and the gene for an agronomic trait of interest has been established, DNA-based tests can be used to enable more precise selection in plant breeding.

This powerful revolution has already demonstrated its impacts in the understanding of, and ability to manipulate, oligogenic and quantitative traits. The development and availability of abundant, naturally occurring, molecular genetic markers during last two decades has generated renewed interest in locating and measuring the effects of genes (polygenes or QTLs – quantitative trait (loci) controlling quantitative traits.

Molecular markers has now become an established powerful tools in plant breeding for indirect selection of difficult traits at the seedling stage, thus speeding up the process of conventional plant breeding. It also facilitates improvement of difficult traits that cannot be improved easily by the conventional plant breeding methods. A large number of genes and QTLs controlling agronomic traits and conferring tolerance to both abiotic and biotic stresses have been identified and tagged using molecular markers in several crop plants. In fact, the products of MAS have already been released as varieties in case of some cereal species.

The progress made in DNA marker technology has been remarkable and exciting in recent years. DNA markers have proved valuable tools in various analyses in plant breeding, for example, early generation selection, enrichment of complex F_1s, choice of donor parent in backcrossing, recovery of recurrent parent genotype in backcrossing, linkage block analysis and selection. Other main areas of applications of molecular markers in plant breeding include germplasm characterization/fingerprinting, determining seed purity, systematic sampling of germplasm, and phylogenetic analysis.

MARKER ASSISTED SELECTION (MAS)

Marker assisted selection is a concept which is being used by plant breeders to improve the properties of agronomically and medicinally important plants following the discovery of various molecular markers. The principle of this concept lies in the fact that markers show linkage with different agronomically important traits such as pest resistance, resistance to abiotic factors, qualitative and quantitative traits. Instead of looking for a trait, the breeder can select for an associated marker that can be detected vary easily in the selection scheme. The prerequisites for marker assisted selection in a plant breeding program are as follows (Boopathi 2013):

1. Marker(s) should show strong linkage (1cM or less) with the desired trait.
2. High level of polymorphism.
3. Even distribution across the whole genome (not clustered in certain region).
4. Co-dominance in expression (so that heterozygotes can be distinguished from homozygotes).
5. Clear distinct allelic features (so that the different alleles can be easily identified).
6. Single copy and no pleiotropic effect.
7. An efficient technique should be available to screen large population for a particular molecular marker.
8. The screening technique should be highly reproducible and amenable for automation.
9. Should not have any detrimental effect on phenotype.
10. It should be economical to use and be user friendly.

The molecular markers which had shown potential for MAS are RAPD, AFLP, RFLP, and microsatellite. These methods rely on two principles: (i) development of many (hundreds or even thousands) potentially polymorphic DNA fragments, (ii) methods for rapid visualization from single preparations of DNA (Henry, 2012).

This MAS has proved as a pavement for merging of biotechnology with conventional and traditional breeding. The interest of plant breeder in molecular markers revolves around following basic aims.

PREREQUISITS FOR EFFICIENT MARKER ASSAITED BREEDING PROGRAMME

For DNA-marker assisted breeding program elaborate equipment and facilities are required. The essential pre-requisites are as follows:

Reliable and Appropriate Markers

Availability of reliable and appropriate markers is critically important for the success of marker-assisted breeding program. Suitable markers should have attributes as indicated above. In addition, the markers should have close association with the target gene(s). This feature shall ensure success in selection of the target gene(s).

On the basis of specific purposes of their applications, each type of markers has advantages and disadvantages. Since SSR have most of the desirable features, they are often preferred over others. In the case of SNPs, more elaborate information on genetic variations due to single nucleotide DNA change among individuals is required. As more and more SNPs have become available in many plant species, they are also considered as important type of markers for MAS.

Quick DNA Extraction and High Throughput Marker Detection

Usually several hundreds of plants/individuals are screened for identification of desired marker pattern in most plant breeding programs. In addition, it is important for the breeders to get instant results to carryout selection process timely. Accordingly, it is essential to have a quick DNA extraction process and a high throughput marker detection system to handle large number of plant samples and for screening large-scale screening of multiple markers. Many laboratories have adopted 96- or 384-well plates for extraction of DNA from small samples. For marker detection high throughput PAGE and AGE systems are used. Automated marker detection systems are also available.

Genetic Maps

For MAS it is important to have linkage maps for detection of marker-trait associations and for selecting markers during breeding process. Therefore, it is important to develop a high-density linkage map. In case a particular region is found to be associated with the desired trait, it is possible to develop a fine mapping with the help of additional markers tightly linked with the gene controlling the trait. An ideal genetic map should have an adequate number of evenly-spaced polymorphic markers to help locate the desirable genes/ QTALs accurately (Dhingani, 2015, Angaji, 2016).

Marker-Trait Association

For MAS it is most important to know about the association between the trait(s) of interest and the marker. Only when the target trait is linked tightly with the gene(s) it can give positive results in the breeding program. Information about this aspect can be obtained through gene mapping through linkage and recombination analysis, association mapping, QTL analysis, analysis of the mutants, bulk segregant analysis etc. it is also important to whether the markers are linked in *Cis* or *Trans* (coupling or repulsion) position with the desired allele of the gene.

Data Processing and Management

Through quick and efficient data processing and management the breeder shall be able to generate useful reports for the breeding program. Usually in MAS program, large numbers of samples with multiple markers for each sample required to be screened together. This involves: labeling, storing, retrieving, processing, analyzing large data sets and also integrating data sets from other similar breeding programs. Useful bioinformatics and statistical tools are now available for this purpose.

Theoritical and Practical Aspects For MAS

In marker-assisted selection, detection of DNA marker and selection are integrated into classical breeding program. The basic procedure has been described taking into account a single cross as an example (Eathinton et al., 2007).

1. Select parents based on having DNA markers allele(s) (either in one or both the parents) and cross them.
2. Grow F_1 population and detection of marker alleles and thereby eliminate false hybrids.
3. Grow F_2 segregating population, screening and harvesting individual plants carrying the desired marker allele(s).
4. Grow $F_{2:3}$ plant rows, and screening individual plants having the marker(s). In case the F_2 plants are found to be homozygous for the marker, the F_3 individuals within a row of plants should be bulked and screened for confirmation of the presence of the marker. Select and harvest the plants having required marker alleles and other desirable characters.
5. Screen for markers in F_4 and F_5 generations and select in a manner described for $F_{2:3}$, giving more emphasis on superior individuals within homozygous rows/lines of markers.
6. Select and bulk the best seeds in $F_{5:6}$ and $F_{4:5}$ generations, based on phenotypic evaluation of target trait, performance of other traits and presence of the marker(s).
7. Evaluate the selected lines for yield, quality, resistance and other characters.

CRITERIA FOR SELECTION OF DNA MARKERS

The characteristics and basic requirements of DNA based markers described Chapter 2, are compared in Table 1. Locus specific microsatellites are often believed to be superior to at least RAPD, and often also to AFLP and ISSR for cultivar identification. The reason being: (1) at least in principle, alleles and genotypes can be assigned unambiguously, (2) primer sequences can easily be distributed among different laboratories, (3) high reproducibility-much higher as compared to RAPD, (4) more variability hence provide higher resolution, and (5) co-dominant nature. Microsatellites are most efficient for the discrimination of genotypes on a per-locus basis because of their multi allelic nature (Kalia et al., 2011).

AFLP is the most informative method for study of genetic variability and somaclonal variability during *in-vitro* cultivation because of the high multiplex ratio (i. e. the number of markers obtained in a single experiment). AFLPs scored highest for marker index (MI), Discriminatory power (D), and effective number of band patterns per assay (P), but quite low for expected heterozygosity averaged over all loci (Hc). Microsatellites had the highest Hc (Kebriyaee et al., 2012).

Genetic distances revealed by co-dominantly inherited microsatellite markers are mostly correlated with those obtained by dominant AFLP, RAPD and ISSR markers but correlations among the latter methods are generally stronger.

To evaluate the variation within population microsatellite markers are better compared to AFLP, RAPD and ISSR. In contrast, for analyzing among population differentiation dominant markers like AFLP, RAPD, and ISSR are more efficient. Marker assisted selection is becoming an increasingly important tool in plant breeding. Cost effectiveness is a major concern when screening large progenies, therefore, simple PCR based methods such as AFLP, RAPD and ISSR are generally preferred. Highly saturated genetic linkage maps have been produced for many species, often mainly based on dominant PCR-derived markers, which are required for gene identification and cloning experiments (Jiang 2013b).

The higher informativeness of microsatellite and AFLP markers makes them useful for many studies, but RAPD will remain attractive when financial investment is limited. AFLP can be more cost effective than microsatellite DNA analysis for population assignment studies. An overview about selection of marker according to purpose of study has been presented in Table 2. The organelle markers (DNA barcodes) are used for analysis of polymorphism at the intra specific level and phylogeographic studies because of their maternal inheritance.

HOW MANY QTLs SHOULD BE SELECTED FOR MAS

Basically, all the QTLs which are expected to contribute to the trait of interest should be included for the analysis. For most quantitatively inherited characters, say for example yield, number of QTLs or genes involved are numerous. Therefore, it is not possible to include all of them for any analysis, due to limitations of facilities and resources. Further, with the increase in the target loci involved the number of individuals in the population increases. On the other hand, the efficiency of MAS decreases with the increase in the number of QTLs. Heritability of QTLs will also decrease with the increase in the number of QTLs. Thus, for a highly complex character governed by many genes MAS will be less effective compared to a simple character controlled by a few genes. Apart from number of genes/QTLs, the efficiency of MAS

Table 1. Comparison of most widely used DNA markers in plants

Features	RFLPs	RAPDs	AFLPs	SSRs	SNPs	SSCPs	PCR-RFLPs
I. Technical requirement							
DNA required(pg)	10	0.02	0.5-1.0	0.05	0.05	0.05	0.5-1.0
DNA quality	High	High	Moderate	Moderate	High	High	High
Restriction enzyme	Yes	No	Yes	No	Yes	No	Yes
Radioactive detection	Yes	No	Yes/No	No	No	No	No
PCR-based	No	Yes	Yes	Yes	Yes	Yes	Yes
Specific primer	No	No	No	Yes	Yes	Yes	Yes
II. Technical characteristics							
Genomic abundance	High	High	High	Moderate to high	Very high	Very high	Very high
Genomic coverage	Low copy coding region	Whole genome	Whole genome	Whole genome	Whole genome	Whole genome	Whole genome
Nature of expression	Codomin-ant	Domina-nt	Domin- -ant	Codomin-ant	Codomin-ant	Codomin- ant	Codomin-ant
Number of loci	Small (<1000)	Small (<1000)	Modera-te (1000s)	High (1000s-10,000s)	Very high (>100,000)	Very high (>100,000)	Very high (>100,000)
Number of polymorphic loci analyzed	1.0-3.0	1.5-5.0	20-100	1.0-3.0	1.0		1.5-50
Type of polymorphism	Single base changes, indels	Single base changes, indels	Single base changes, indels	Chanes in length of repeats	Single base changes, indels	Single base changes, indels	Single base changes, indels
Type of probes/ primers	Low copy DNA or cDNA clones	10bp random nucleoti-des	Specifi-c sequen-ce	Specific sequence	Allelic-specific PCR primers	Allelic-specific PCR primers	Allelic-specific PCR primers
Cloning and/or sequencing	Yes	No	No	Yes	Yes	Yes	Yes
Reproducibility/ Reliability	High	Low	High	High	High	High	High
Accuracy	Very high	Very low	Medium	High	Very high	Medium	Very high
Effective multiplex ratio	Low	Moderate	High	High	High	Moderate to high	Moderate to high
Marker Index	Low	Moderat-e	Modera-te to high	High	Moderate	Moderate	Moderate
Genotyping throughput	Low	Low	High	High	High	High	High
Technical demand	Moderate	Low	Moderate	Low	High	High	High
Time required	High	Low	Modera-te	Low	Low	Low	Low
Ease to use	Not easy	Easy	Modera-te	Easy	Easy	Easy	Easy
Amenable to automation	Low	Modera-te	Modera-te	High	High	Moderate	Moderate
Development cost	Low	Low	Modera-te	High	High	High	High
Cost per analysis	High	Low	Modera-te	Low	Low	Moderate	Moderate

Table 2. Applicability of various DNA based markers for achieving different target

Type of markers	Appropriate target
Microsatellite	Cultivar identification, discrimination of genotypes
AFLP	Study of genetic variability and somaclonal variation, genetic distance measurement, marker assisted selection
Organelle DNA	Intraspecific polymorphism analysis, phylogenetic studies
RAPD and ISSR	Genetic distance measurement but required number of marker is more (up to 50), marker assisted selection
EST and SNP	Marker assisted selection

shall also depend on the design of the breeding program and the scheme of implementation. Taking the above facts into considerations, ideally three QTLs are recommended as feasible and appropriate choice. However, there exist reports on using more than three QTLs for improvement of crops effectively. With the advancement in automated detection, genotyping technologies and SNP markers, it is expected that it will be possible to handle more numbers of QTLs for selection in the near future.

In the case of multiple genes/QTLs, it is recommended to use 3-4 genes for selection. If the QTLs are linked to the markers and to 5-6 known genes, they can be selected directly. QTLs verified to possess medium to large effects at multi environmental locations should only be included. However, priority should be given to the major QTLs which accounts for greater proportion of phenotypic variations and can be easily detected under different conditions. It is also recommended that based on their relative importance to the breeding objectives, an index for selection of markers may be constructed. This will help to choose the markers according to the objectives of the breeding program (Xu, 2010).

HOW MANY MARKERS SHOULD BE USED IN MAS

It is generally believed that use of OTLs associated with more markers should result into greater success. Under limited resources and facilities, efficiency of any MAS breeding program is equally important. Therefore, under limited resources and facilities, for a single QTL, it is recommended to use two flanking markers that are tightly linked to the QTL of interest. Further, the markers should be <5cM in distance from the gene/QTL of interest so that no or very low frequency of recombination take place between them. High recombination frequency between the QTL and the marker shall reduce the efficiency of MAS, as it will alter the linkage association leading to selection error. This problem can be overcome by using two flanking markers instead of one marker, as two simultaneous crossovers on both side of the gene/QTL shall be required to affect the linkage relationship and selection error. It is well known that the frequency of double crossover is considerably rare, compared to single crossover.

In general, polymorphism nature of the marker should be carried out at the very stages of the growth of the plants, especially when MAS is applied in backcrossing and recurrent selection. When MAS is applied in backcrossing, individuals that carry the preferred marker alleles are usually used for the recurrent parent and/or for inter-mating between selected individuals or their progenies (Xu, 2010).

NUMBER OF GENERASTIONS OF MAS TO BE CONDUCTED

The generations for which MAS should be used shall vary with the number of marker used, and the closeness of association between the markers and the genes/QTLs of interest. Usually screening for marker is done for 2-4 generations in a segregating population. In situations where only a few markers are used and the markers are closely linked to the gene/QTL of interest, the number of generations required is few. If the status of the marker alleles of interest is found to be homozygous in two successive generations, further screening of the marker may not be required.

The presence of marker or QTL does not imply that the character of interest shall be expressed. Therefore, QTL data should be obtained from different environments and different populations, to understand the interactions between QTL x environment, QTL x genetic background, and QTL x QTL.

SITUATIONS WHERE MAS SHOULD BE ADOPTED

In practical plant breeding, adoption of molecular markers in the selection of traits of interest depends on several situations which include (Varshney, 2009):

1. When the gene of interest is recessive. Thus when it is present in heterozygous condition it is possible to select the allele and use fur further crosses to produce homozygous offspring having the desired character.
2. To select characters which are expressed late during the development of the plants. For example, flower and fruit characters.
3. When the expression of character is controlled by two or more linked genes. For example, expression multiple genes is required for development of resistance against a specific diseases or pests.
4. When special conditions are required to be created for the expression of the target gene(s). For example, inoculation with the disease causing spores for the expression of the target gene(s).

MARKER ASSISTED BACKCROSSING (MABC)

Marker assisted backcrossing (MABC) is the simplest and most widely used molecular plant breeding procedure. In MABC one or few genes of interest are transferred from one cultivar donor) to another agronomically superior cultivar or elite breeding line (recipient) to improve the specific trait in the recipient. In traditional backcrossing the progeny derived from backcrossing are selected in the basis of phenotypic expression of the character in all subsequent generations. However, in MABC the progeny derived from backcrossing are selected on the basis of presence of the marker associated or linked to the gene(s)/QTLs of interest. The procedure to be followed for MABC is as follows (Jiang, 2013a).

1. Select the donor (DP) plant, having the desired character (gene) and identify the DNA marker linked with the gene (allele). Select the recipient plant (RP), having the desirable agronomically superior characters.
2. Cross the DP and RP plants and raise the F_1 population. Select the plants having the marker allele(s) during early stages of growth, and discard the false hybrids.
3. Cross the selected F_1 plants with the RP.
4. Raise BCF_1 population. Screen and select for the presence of marker(s) at the early stage of growth. Cross the selected individuals, having the desired marker allele(s) in heterozygous state, to the RP.
5. Repeat the above step for 2-4 generations, on the basis of actual requirements and operational feasibility.

6. Raise the backcross population (say, BC_4F_1) and screen for the presence of the target marker(s) in individual plants. Discard plants having homozygous marker alleles from the RP.
7. Self the plants selected plants and harvest.
8. Raise the progenies of backcross selfing (say, BC_4F_2) plants. Screen and select plants having homozygous DP marker allele(s), evaluate and release.

The proportion of genome from the RP that will be present after n generations of backcrossing can be determined by the formula: $1 - (1/2)^{n+1}$ for a single locus. For k loci it can be calculated from: $[1- (1/2)^{n+1}]$. The proportion of the genome from the RP shall represent average values of the population, as some individuals shall have more RP genome than others. About 6-8 generations of backcrossing is required to recover full genotype of the RP. However, for the target gene-carrier chromosome this process may be slower than expected, due to linkage between the target gene and some undesirable characters. On the other hand the process may be accelerated by selecting flanking markers of QTLs. Basically, two types of selection can be made in MAC: foreground selection and background selection.

In foreground selection procedure, selection is made for the marker allele(s) of DP of the target locus in heterozygous condition till completion of the backcrossing. Selected plants are selfed and homozygous DP allele(s) of selected markers are selected from the progeny. All such plants are released after proper evaluation. The effective utilization of the foreground selection procedure depends on the number of genes/QTLS involved, association between marker gene and QTL, linkage distance between the market and QTL, and linkage between undesirable traits and the target gene/QTL (linkage drag).

In foreground selection procedure, the selection is made for the marker allele of RP for the desirable traits in the entire genome, except the target locus. In other words selection is made for against the undesirable genome of the DP, to hasten the restoration of the genome of RP and eliminated undesirable genes from DP. The number of background genes used in the background selection shall determine the progress in recovery of the RP genome. Larger the number of markers selected for the RP alleles, faster will be the recovery of RP genome. However, this will increase the population size and genotyping. The linkage drag can be efficiently managed by this method.

In practice, both foreground and background selections are conducted, either simultaneously or sequentially, under the same breeding program. Individuals having the desired marker alleles for the target character are selected first 9foreground selection), followed by selection of individuals having other marker alleles for the RP genome (background selection). Since selection of the target gene/QTL is the prime criteria for backcross breeding, the procedure described above is justified.

Through simulation it is possible to predict expected results of a typical MABC program by fixing the parameters. Expected results from 1000 replicates, where heterozygotes are selected at the target locus in each generations, and RP alleles were selected for 2 flanking markers each located 2 cM apart from the target locus, on 3 markers on non-target chromosomes is presented in Table 3. It is evident from Table 3 that with the combination of foreground and background selection can help to achieve recovery of the RP genome faster, which leads to considerable savings in time compared to conventional breeding procedure.

Table 3. Results expected of a MABC program in which foreground and background selections are adopted (Source: Jiang 2013, https://dx.doi.org/10.5772/52583)

Backcross generation	Number of individuals	% homozygosity of recurrent parent alleles at selected markers		% recurrent parent genome	
		Chromosome with target locus	All other chromosomes	Marker-assisted backcross	Conventional backcross
BC_1	70	38.4	60.6	79.0	75.0
BC_2	100	73.6	87.4	92.2	87.5
BC_3	150	93.0	98.8	98.0	93.7
BC_4	300	100.0	100.0	99.0	96.9

The population to be analyzed through MABC should have at least one genotype that contains all favorable alleles for a particular QTL. Progressively the number of QTLs may be increased, but should be restricted to 6 QTLs as it becomes difficult to handle the material beyond this. Moreover, larger proportion of unwanted genes would be transferred with more number of QTLS, due to linkage drag. In general, in the early back cross generations, most of the unwanted genes are located on non-target chromosomes, and shall get removed as the back cross generations progress. On the other hand, genes of the DP present on the target chromosome shall decrease much slowly, and many unwanted genes form DP may be present after several generations of backcrossing. For example, if the length of the genome is 3000 cM, and if we consider presence of 1% donor DNA fragments after 6 backcrosses, there will be 30 cM of chromosomal segment present. This segment obviously contains many unwanted genes, particularly if the DP is a wild relative.

The linkage drag can be substantially reduced through background selection procedure. For this, two flanking markers of the target gene should be used, and individuals that are heterozygous at the target locus and homozygous for the RP alleles at both the flanking markers should be selected. Use of closely linked flanking markers can reduce the linkage drag considerably. Although such condition shall demand larger population size and more genotyping, as the frequency of double crossover products shall be reduced. Thus to make the breeding program cost effective, it is important to pre-determine the effective minimum population size for each breeding program. Statistical tools are also available to assist in determining the minimum population size required for backcross breeding programs and to identify at least one individual which will be double recombinant, heterozygous at the target loci and homozygous for the RP alleles. When the flanking markers are closely linked to the target gene/QTL, it is unlikely to obtain double recombinants after one generation of backcrossing. Additional backcrossing shall be required to obtain double recombinant genotype. For example, in BC_1 single recombination on region I may be selected and in BC_2 recombination in region II may be selected. In this way, individuals having donor allele with two flanking markers can be obtained.

To accelerate the process of recovery of RP genome on non-target chromosomes in background selection method, homozygous recipient individual types are selected from the collection of markers located on non-carrier chromosomes. However, it is important to determine an appropriate number of markers to be used, for effectiveness and efficiency of the method. Incorporation of more markers does not imply that it will more beneficial.

Selection of large number of markers to cover the non-target chromosome is not recommended, unless fine-mapping of specific chromosome is desired. It is important to select appropriate number of markers and their specific position in the chromosome. Simulation experiments have suggested that for a chromosome of 100 cM size, utilization of 2-4 markers is sufficient. If the identified markers are optimally positioned along the chromosome, selection will be most effective. In practice, 2-3 markers per chromosome, distributed in all the chromosomes involved, should be ideal. In such a MABC breeding program, 3-4 backcross generations should be enough to achieve >99% of RP genome. Reduction in time due to MABC shall also contribute towards cost effectiveness of the breeding program. Moreover, background selection is more efficient in late backcross generations than early backcross generations (Jiang, 2013a).

APPLICATIONS OF MABC

Application of MABC is advantageous and essential in the following situations:

1. Characterization of the phenotype is difficult, expensive or impossible,
2. Low heritability of the target gene,
3. Special conditions are required for the expression of the target trait,
4. Recessive gene is involved in the expression of the trait,
5. The expression of the trait is manifested at later stages of development of the plant (seed, fruit, flower), and
6. Pyramiding of the genes is required for one or more traits.

Marker assisted back crossing is the most successful and widely used plant breeding tool till date. It has been used in wheat, rice, corn, soybean, millet, barley, tomato etc. for a variety of traits such as: disease and pest resistance, drought resistance and quality parameters. In corn, integration of *Bt* transgene into other corn cultivars having different genetic background was achieved through MABC. High lysine *opaque2* gene was also incorporated in corn through MABC. In rice, for the selection of aroma MABC has been used. In soybean, two QTL for seed protein content was transferred through MABC. In tomato, a strategy of MABC named advanced backcross-QTL (AB-QTL) was used to transfer disease resistance genes from wild relatives into elite genotypes. Later this strategy was also used in rice, wheat, maize, barley, soybean and cotton for transferring favorable genes from wild relatives to elite genotypes.

MAS FOR DEVELOPMENT OF RESISTANCE IN PLANTS

Through marker assisted selection, easy and rapid selection of desired gene which is responsible to provide specific kind of resistance is possible. In conventional breeding exercises resistance is developed by crossing of susceptible cultivar with resistant donor. These populations are then selected either under natural disease or pest hot spots or under artificially created conditions. Although these procedures have given excellent results, they are time consuming. Identification of a marker which is tightly linked to a resistance gene will help to select the plant carrying that particular gene without subjecting them to pathogen or insect attack in early generations. The breeder requires a small amount of DNA from individual plant to analyze the presence or absence of a particular marker band on the gel. The presence of marker band indicates the availability of resistance gene because of tight linkage among them. This procedure applies for the selection of both parent plants as well as progeny plants of different generations. Only materials in the advanced generations would be required to be tested in disease and insect nurseries. Thus, with MAS, it is now possible for the breeder to conduct many rounds of selection in a year without depending on the natural occurrence of the pest or pathogen as well (George et al., 2003, Gazal et al., 2016). DNA-based marker resources available for important crops are presented in Table 4.

Table 4. DNA-based marker resources available for important crops

Crops	SSRs	ESTs	Database
Rice	15687	1269116	http://www.gramene.org; http://www.sdwgi.com
Wheat	1603	1121459	http://www.gramene.org; http://www.sdwgi.com
Maize	2807	2019971	http://www.gramene.org; http://www.sdwgi.com; http://www.maizegdb.org
Barley	226	532161	http://www.gramene.org; http://www.sdwgi.com
Sorghum	260	242598	http://www.gramene.org; http://www.sdwgi.com
Soybean	3395	1454433	htttp://www.soybase.org; http://www.comparative-legume.org; http://www.sdwgi.com
Peas	-	10447	http://www.comparative-legume.org
Chickpea	698	34450	http://www.comperative-legume.org; http://www.icrisat.org; http://www.icarda.org
Common bean	-	107213	http://www.comperative-legume.org
Potato	1053	241130	http://www.bioinformatics.nl
Tomato	519	293182	http://www.sgn.cornell.edu; http://www.bioinformatics.nl
Brassica	2482	841970	http://www.icarda.org
Cucumber	200	18542	htttp://www.icugi.org; http://www.vegmarks.nivot.affrc.go.jp
Cotton	9358	376517	http://www.cottonmarker.org; http://www.sdwgi.com

MAS FOR PYRAMIDING OF MAJOR/MINOR GENES INTO A SINGLE CULTIVER

Pathogens and insects are known to overcome the developed resistance conferred by single gene in short period of time. In order to impart durability in developed resistance breeders always try to merge or accumulate multiple resistance genes from different wild varieties into a single cultivar. Such resistance, provided by multiple genes is long lasting utilization. Application of MAS for the search of resistance genes in wild cultivars is a promising approach. For example, resistance against early blight in potato is provided by the gene R and different allelic forms of this R gene such as Ra1, Ra2, Ra3 etc. confer different level of resistance in wild cultivars of potato. A microsatellite marker is reported to be linked with this R gene. Using this microsatellite marker polymorphism for gene R can be analyzed in local population and plants carrying any new allele of gene can be selected. Further, all these allelic forms of R gene can be accumulated in a single cultivar by making suitable crosses (Ye et al., 2008).

Different approaches are used for pyramiding of multiple genes/QTLs, such as multiple-parent crossing or complex crossing, back-crossing, and recurrent selection. The breeding strategy shall depend on the number of genes/QTLs required for improvement of the character, number of parents to be used, heritability of the gene(s), marker-gene interaction, expected duration to achieve the goal, and the cost. For example, if 3-4 desirable genes/QTLs are present in 3-4 different lines, the breeder can use three-way, four-way or double-crossing strategies for pyramiding. The convergent backcrossing or stepwise backcrossing may also be integrated for achieving pyramiding. However, if the number genes/QTLs are more than 4 for pyramiding, it would be ideal to adopt multiple crossing and/or recurrent selection methodology.

For marker-assisted backcrossing (MABC), three breeding strategies namely, stepwise, simultaneous/synchronized and convergent backcrossing can be used (Jiang, 2013a). For example, suppose a cultivar P_1 has all round superior traits but lacks a trait of interest, and 4 genes/QTLs has been identified in 4 different cultivars (P_1, P_2, P_3, P_4), the breeding procedures to be adopted has been presented in Figure 1a, Figure 1b, and Figure 1c.

In the process of stepwise backcrossing, an attempt has been made to transfer 4 target genes/QTLs to the recurrent parent K in order. In the first step of backcrossing, one gene/QTL is selected, followed by a different gene/QTL in the next step of backcrossing. The process continued till all the

Figure 1a. Procedure for stepwise backcrossing of MABC (Redrawn from: Jiang 2018, https://dx.doi.org/10.5772/52583)

Step 1 $K \times P_1^{AA}$

↓

$F_1 \times K$

↓

: :

BC_n^{Aa}

↓selfing

Step 2 $K^{AA} \times P_2^{BB}$

↓

$F_1^{Bb} \times K^{AA}$

↓

: :

BC_n^{Bb}

↓selfing

Step 3 $K^{AABB} \times P_3^{CC}$

↓

$F_1^{Cc} \times K^{AABB}$

↓

: :

BC_n^{Cc}

↓selfing

Step 4 $K^{AABBCC} \times P_4^{DD}$

↓

$F_1^{Dd} \times K^{AABBCC}$

↓

: :

BC_n^{Dd}

↓selfing

$K^{AABBCCDD}$

target genes/QTLs are introgressed into the recurrent parent (K). Since only one gene/QTL is selected at a time, it becomes easier to implement and the population size and genotyping remains small. However it takes longer time to complete the process.

In the simultaneous and synchronized backcrossing process, the recurrent should first be crossed to each of the donor parents to generate 4 single cross F_1s. Thereafter, two of the 4 single-cross F_1s are crossed to one another to

Figure 1b. Procedure for simultaneous or synchronize backcrossing of MABC (Redrawn from: Jiang 2018, https://dx.doi.org/10.5772/52583)

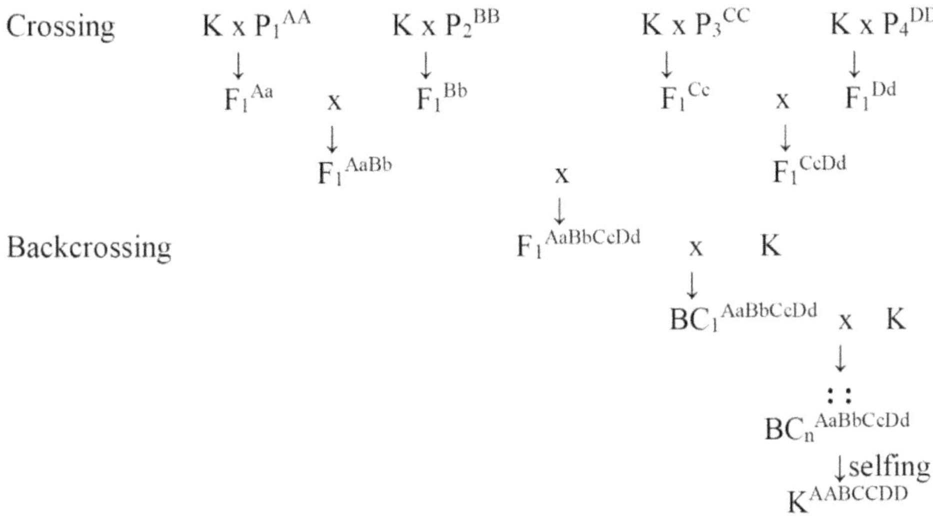

Figure 1c. Procedure for convergent backcrossing of MABC (Redrawn from: Jiang 2018, https://dx.doi.org/10.5772/52583)

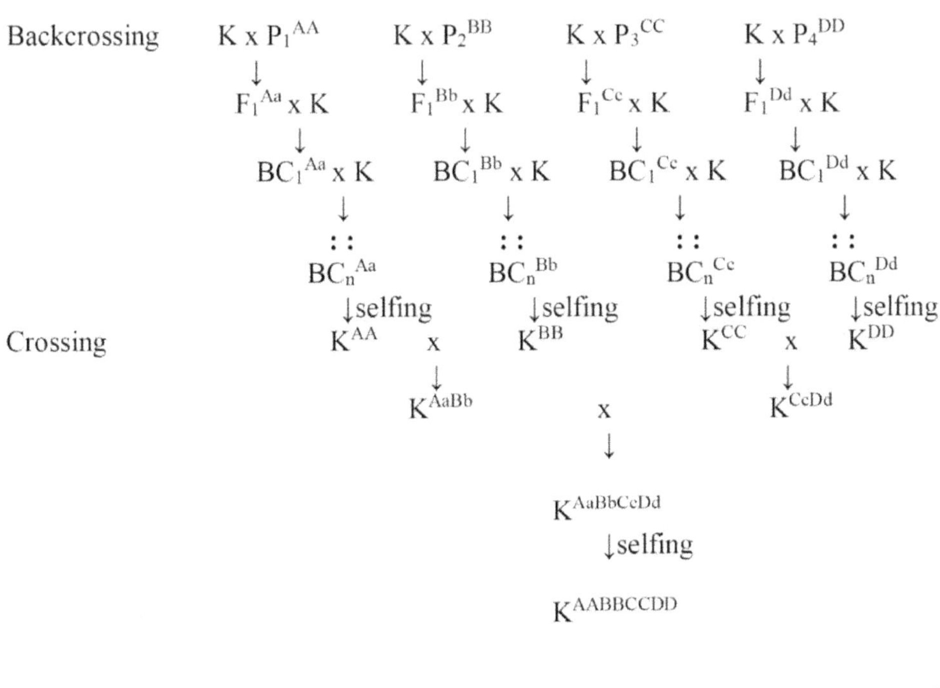

generate two double cross F_1s. They were then crossed again to generate an individual (hybrid) having all the 4 target genes/QTLs in heterozygous condition. This hybrid is then crossed back to the recurrent parent (K) till the recovery of recurrent parent's genome. Finally they are selfed. This method takes comparatively shorter time to complete the process. However, this procedure takes more genotyping and larger population size.

In convergent backcrossing combines the advantages of both synchronized and stepwise backcrossing methods. Initially the 4 target genes/QTLs are transferred from the donor to recurrent parent separately. Then backcrossing is made on the basis of the markers linked to the target genes/QTLs to generate 4 improved lines ($K^{AA,}$ K^{BB}, K^{CC}, and K^{DD}). Thereafter, 2 improved lines are crossed, and the 2 hybrids generated are intercrossed so that all the 4 genes/ QTLs are integrated (pyramiding) in one line ($K^{AABBCCDD}$). This method takes less time and fixation of the genes can also be carried out easily and thus preferred over other methods mentioned earlier (Jiang, 2013a).

Efficiency of gene pyramiding through marker assisted backcrossing (MABC) has been studied through computer simulations. Although simulation results supports the conformity of the gene pyramiding in several crop plants, their commercial exploitation is yet to be seen.

Marker assisted complex or convergent crossing (MACC) can also be used for pyramiding multiple genes/QTLs in plants. MACC is particularly applied in those cases where all the parents are improved cultivars, having good comprehensive performance and complimentary or different genes of interest. In this procedure, the hybrid derived from the convergent crossing is self-pollinated and selection for the marker-associated target traits is made for several generations. Through this the genetic stability lines with the desired markers can be obtained. The population size can be reduced by selecting different markers (based on their relative importance) in different generations. This will also help to avoid loss of important genes/QTLs. Thereafter, the lines with homozygous alleles should be identified and phenotypic evaluation should be carried out. MARS has been recommended for pyramiding QTLs for complex characters like grain yield, and abiotic/biotic resistance (Jiang, 2013a).

MAS FOR IMPROVING QUALITATIVE TRAITS (MAJOR GENES)

In plants, many important qualitative characters are controlled by major genes/ QTLs. Often these molecular markers are found to be linked with important qualitative traits such as linolenic acid content in soybean, starch content in wheat etc. MAS can be efficiently used for the identification of plants bearing the genes responsible for qualitative characters, and incorporating them in plants through suitable breeding program (Bilyeu et al. 2006, Pham et al. 2012) .

MAS FOR ABIOTIC RESISTANCE

Molecular markers having specific traits beneficial in improving drought responses such as osmotic adjustments, water use efficiency and efficient root system have been identified in various crops. These markers have been used for improvement in crop plants through MAS.

MAS FOR IMPROVEMENT OF QUANTATIVE TRAITS

Most agronomically important characters are polygenic and controlled by several QTLs. Therefore, improvement of such characters through MAS is difficult and complex. Each of these genes has relatively small effect on the overall phenotype and is highly influenced by the environmental factors. The efficiency of MAS can be reduced due to epistasis and QTL x E interaction. Often to augment such problems, repeated field trials under different environmental conditions are recommended. In plant breeding, application of QTL mapping information for quantitative characters has the following constraints (Collard et al. 2005).

1. Due to strong QTL x E interaction, the expression of the phenotype character may vary in different locations, and it may be difficult to ascertain the phenotype character.
2. Non availability of universal QTL markers which can be used across populations may distract some breeders from utilizing MAS.

3. Non availability of efficient statistical tools for QTL analysis, which leads to either overestimate or underestimate the number of QTLs involved.
4. In situations where the QTLs have limited effect on the character, large numbers of QTLs have to be identified. It may become difficult and complicated to achieve this goal.

Efficiency of MAS for quantitative characters can be increased by developing appropriate design of the field experiments. It is important to give attention to replications, experimental techniques adopted, sampling techniques, evaluation methods and analysis of data. Through composite interval mapping (CIM) it is possible to integrate data from different locations to estimate the QTL x E interaction. This helps in identifying stable QTLs in different environment. With the help of saturated linkage map it is possible to identify whether the targeted QTLs and linked QTLs in are in repulsion or coupling phases. For quantitative characters, usually only a few major QTLs are used for MAS. If many minor-effect QTLs are involved, it is recommended to consider gene pyramiding (Zargar et al. 2015).

GENOME SELECTION

Meuwissen (2007) defined the genomic selection (GS) or genome-wide selection (GSW) as the simultaneous selection of many markers, which cover the entire genome and all genes are expected to be in linkage disequilibrium with one of the markers. Thus the number of markers should be tens or hundreds of thousands and densely distributed. The genetic markers (GS genotyping data) distributed across the genome can be used to predict complex characters with accuracy. The genomic estimated breeding value (GEBV) can be calculated on the basis of genome-wide dense DNA markers, and used for selection of desirable individuals (Nakaya & Isobe, 2012). Testing and identification of a subset of markers is not required for GS. Thus, QTL mapping with populations derived from specific crosses is not required. However, GS models should first be developed. For this, phenotypes and genome-wide genotypes have to be investigated in a sub-set of population to predict significant relationship between phenotypes and genotypes through statistical approaches. Then, GEBVs can be used for the selection of desirable individuals. For accuracy, high density of the markers is necessary for the entire genome.

Success of the genome selection depends on the availability of high-throughput marker technologies, appropriate statistical methods and high-performance computing facilities. The feasibility of the approach has become more prominent due to discovery and development of large number of SNPs. The breeding value from genomic data can be estimated from the conditional mean of the breeding value of the genotype at each QTL. The conditional mean is calculated by using a prior distribution of QTL effects. However, in practice, estimation of breeding value is made by using the marker genotype instead of QTL genotypes. The use of QTL genotype shall be possible only with the increase in the number of SNPs (Goddard & Hayes, 2007).

Application of GS is more popular in animal breeding compared to plant breeding programs. Of late it has attracted the plant breeders and several studies have been conducted in plant systems. These studies have indicated that GS is superior to MARS and PS in terms of gain/unit cost and time. Accordingly, GS is emerging as the potential method for plant breeding. The major reason for not adopting GS, is the lack of sufficient knowledge on GS for its practical applications. With development of statistical methods for calculating GEBVs and availability of user friendly software packages, GS is becoming popular amongst plant breeders.

CHALLENGES AND PROSPECTS OF MARKER ASSISTED BREEDING

Marker-assisted breeding (MAB) has become a powerful and reliable tool as it has been successfully utilized in various fields of plant science, which include: genetic mapping, map-based discovery of genes, characterization of traits, improvement of crops, evaluation of germplasm etc. MAS has the following advantages over conventional breeding methods (Jiang 2013a).

1. MAB is not affected by the environment. Therefore, very useful for such traits which need special environmental conditions, e.g. disease/ pest resistances, abiotic stress resistance etc. MAB is also useful for low-heritable traits.
2. MAB can be carried at the early stages of development of the plants. This helps in reduction of time. For backcrossing and recurrent selection this feature is specifically important.
3. Since co-dominant markers are used in MAB, it helps in effective selection of recessive alleles of desired character residing in heterozygous

condition. Since selfing or test-crossing is not required, it helps in saving lot of time.

4. Individual genes/QTLs can be identified from a multi-genetically/QTL controlled character. Thus, it becomes easy to carry out MAB for gene pyramiding.
5. MAB is much efficient and effective in terms of predictability, reproducibility, and utilization of time and resources.

Although application of MAB continued to increase in various plant breeding programs, there has been constrains in using MAS and MABC for simply-inherited traits, such as monogenically or oligogenically inherited resistance genes against pests and diseases. Thus MAB is not always advantageous and faces the following constrain and challenges (Jiang 2013a).

1. Due to lack of polymorphism and non-reliable marker-trait association, all markers cannot be applicable across all populations. Use of multiple mapping populations may be used to understand the allelic diversity of the marker and the effect of genetic background.
2. All markers may not be breeder friendly. In such cases, non-breeder friendly markers should be converted to breeder-friendly markers. For example, RFLP to STS, RAPD to SCAR.
3. Detection of the efficiency of QTL depends on several parameters, such as, algorithms used, method of mapping used, number of polymorphic markers present, population size and type. Therefore, if the conditions of the parameters are not properly met, it may lead to imprecise estimates of QTL locations which in turn result in slower progress in the breeding program. Such issues may be resolved by using of high density markers with fine mapping and with large populations across multiple environments.
4. Recombination between the markers and the genes/QTLs of interest may lead to false selection. This can be resolved by using flanking markers on the target genes.
5. Require adequate facilities to carry out large breeding programs successfully, which may be expensive.
6. The breeder must have thorough knowledge about the breeding procedure, design the experiment and execute the same.

The MAB is not expected to replace the conventional breeding methodology, but can supplement substantially to crop improvement programs. Integration of conventional breeding programs with MAB should be the strategy for the improvement of any crop. It is expected that the drawbacks of MAB shall be overcome gradually, and it will be accepted as a useful breeding tool for crop improvement.

CONCLUSION

Marker-assisted breeding has been successful in introgressing and pyramiding major gene effect, but many challenges remain to be resolved before MAS can be used for breeding complex characters. It is expected that application of MAS will be for mono- and oligo traits that are difficult or expensive to screen through conventional methods. Advances in structural genomics has provided huge amount of sequence information which will help to breed for complex agronomic traits. Over the last decade, MAS technologies have become substantially cheaper and easier to apply at large scale.

Plants exhibit large changes in gene expression during different stages of development and when exposed to varying range of biotic and abiotic stresses. A new field of genetics has emerges which focuses on gene expression based on the extrapolation techniques of linkage and association analysis to the thousands of transcripts measured by microarrays. By dissecting the architecture of quantitative traits, it is possible to connect DNA sequence variations with phenotypic variations. It has been proposed to use dynamic mapping to understand gene expression at different developmental stages. As more and more information on dynamic properties of QTLs across different developmental stages becomes available, strategies for phenology-specific MAS and overall life cycle MAS can be developed. Advances in these areas are likely to have substantial impacts on our ability to deal with the effect of genotype by environment interaction.

Moreover, the genetic basis of complex traits and interactions between all related characters shall become much easier to understand. This will help to create accurate modeling of gene networks and robust simulation tools for designing target genomic ideotypes. Availability of such tools shall make plant breeding much easier in the early stages of the breeding programs. However, the tedious multi-locational replicated evaluation trials for screening elite breeding lines have to be carried out till alternative methods is developed.

REFERENCES

Angaji, S. A. (2009). QTL mapping: A few key points. *International Journal of Applied Research in Natural Products*, *2*, 1–3.

Bernardo, R., & Charcosset, A. (2006). Usefulness of gene information in marker-assisted recurrent selection: A simulation appraisal. *Crop Science*, *46*(2), 614–621. doi:10.2135/cropsci2005.05-0088

Bilyeu, K., Palavalli, L., Sleper, D. A., & Beuselinck, P. (2006). Molecular genetic resources for development of 1% linolenic acid soybeans. *Crop Science*, *46*(5), 1913–1918. doi:10.2135/cropsci2005.11-0426

Collard, B. C., Jahufer, M. Z., Brouwer, J. B., & Pang, E. C. (2005). An introduction to marker, quantitative trait loci (QTL) mapping and marker-assisted selection for crop improvement the basic concepts. *Euphytica*, *142*(1-2), 169–196. doi:10.100710681-005-1681-5

Dhingani, R. M., Umrania, V. V., Tomar, R. S., Parakhia, M. V., & Golakiya, B. A. (2015). Introduction to QTL mapping in plants. *Annals of Plant Science*, *4*, 1072–1079.

Eathinton, S. R., Crosbie, T. M., Edwards, M. D., Reiter, R. S., & Bull, J. K. (2007). Molecular markers in a commercial breeding programme. *Crop Science*, *47*, S154–S163. doi:10.2135/cropsci2007.04.0015IPBS

Gazal, A., Dar, Z. A., Wani, S. H., Lone, A., Shikari, A. B., Ali, G., & Abidi, I. A. (2016). Molecular breeding for enhancing resilience against biotic and abiotic stress in major cereals. *Society for the Advancement of Breeding Researches in Asia and Oceania (SABRAO). Journal of Breeding and Genetics*, *48*, 1–32.

George, M. L., Prasanna, B. M., Rathore, R. S., Settty, T. A., Kasim, F., Azrai, M., ... Hoisington, D. (2003). Identification of QTLs conferring resistance to downy mildews of maize in Asia. *Theoretical and Applied Genetics*, *107*(3), 544–551. doi:10.100700122-003-1280-6 PMID:12759731

Goddard, M. E., & Hayes, B. J. (2007). Genomic selection. *Journal of Animal Breeding and Genetics*, *124*(6), 323–330. doi:10.1111/j.1439-0388.2007.00702.x PMID:18076469

Henry, R. J. (Ed.). (2012). *Molecular markers in plants*. John Wiley & Sons, Inc., doi:10.1002/9781118473023

Jiang, G.L. (2013a). *Molecular markers and marker assisted breeding in plants*. doi:10.5772/52583

Jiang, G. L. (2013b). Molecular markers and marker assisted breeding in plants. In S. B. Anderson (Ed.), *Plant breeding from laboratories to field* (pp. 45–83). InTech. doi:10.5772/52583

Kalia, R. K., Raj, M. K., Kalia, S., Singh, R., & Dhawan, A. K. (2011). Microsatellite markers: An overview of the recent progress in plants. *Euphytica*, *177*(3), 309–334. doi:10.100710681-010-0286-9

Kebriyaee, D., Kordrostami, M., Rezadoost, M. H., & Samizadeh, H. (2012). QTL analysis of agronomic traits in rice using SSR and ALFP markers. *Notulae Scientia Biologicae*, *4*(2), 116–123. doi:10.15835/nsb427501

Meuwissen, T. (2007). Genomic selection: Marker assisted selection on a genome wide scale. *Journal of Animal Breeding and Genetics*, *124*(6), 321–322. doi:10.1111/j.1439-0388.2007.00708.x PMID:18076468

Nakaya, A., & Isobe, S. N. (2012). *Will genomic selection be a practical method for plant breeding? Annuals of Botany*. doi:10.1093/aob/mcs109

Pham, A. T., Shannon, J. G., & Bilyeu, K. D. (2012). Combination of mutant FAD2 and FAD3 genes to produce high oleic acid and low linolenic acid soybean oil. *Theoretical and Applied Genetics*, *125*(3), 503–515. doi:10.100700122-012-1849-z PMID:22476873

Varshney, R. K., Hoisington, D. A., Nayak, S. N., & Graner, A. (2009). Molecular plant breeding: methodology and achievements. In D. J. Somers (Ed.), *Methods in molecular biology, plant gen-omics* (pp. 242–254). Humana Press.

Xu, Y. (2010). *Molecular plant breeding*. CAB International. doi:10.1079/9781845933920.0000

Ye, G., & Smith, K. F. (2008). Marker-assisted gene pyramiding for inbred line development: Basic principles and practical guidelines. *International Journal of Plant Breeding*, *2*, 1–10.

Zargar, S. M., Raatz, B., Sonah, H., Nazir, M., Bhar, J. A., Dar, Z. A., ... Rakwal, R. (2015). Recent advances in molecular marker techniques: Insight into QTL mapping, GWAS and genomic selection in plants. *Journal of Crop Science and Biotechnology*, *18*(5), 293–308. doi:10.100712892-015-0037-5

ADDITIONAL READING

Abduvakhmonov, J. V. (2016). *Microsatellite markers*. InTech. doi:10.5772/62560

Bassi, F. M., Bentley, A. R., Charmet, G., & Crossa, J. (2016). Breeding schemes for the implementation of genomic selection in wheat (*Triticum* spp.). *Plant Science*, *242*, 23–36. doi:10.1016/j.plantsci.2015.08.021 PMID:26566822

Collard, B. C., & Mackill, D. J. (2008). Marker-assisted selection: An approach for precision plant breeding in the twenty-first century. *Philosophical Transactions Royal Society of Botany*, *363*(1491), 557–572. doi:10.1098/rstb.2007.2170 PMID:17715053

Comings, D. E., & MacMurry, J. P. (2000). Molecular heterosis: A review. *Molecular Genetics and Metabolism*, *71*(1-2), 19–31. doi:10.1006/mgme.2000.3015 PMID:11001792

Ebert, D., & Peakall, R. O. (2009). Chloroplast simple sequence repeats (cpSSRs): Technical resources and recommendations for expanding cpSSR discovery and applications to a wide array of plant species. *Molecular Ecology Resources*, *9*(3), 673–690. doi:10.1111/j.1755-0998.2008.02319.x PMID:21564725

Fu, Y. B., Yang, M. H., Zeng, F., & Billigetu, B. (2017). Searching for an accurate marker based prediction of an individual quantitative trait in molecular plant breeding. *Frontiers in Plant Science*, *8*, 1–12. doi:10.3389/fpls.2017.01182 PMID:28729875

Gupta, P. K., Langridge, P., & Mir, R. R. (2010). Marker-assisted wheat breeding: Present status and future possibilities. *Molecular Breeding*, *26*(2), 145–161. doi:10.100711032-009-9359-7

He, J., Zhao, X., Laroche, A., Lu, Z. X., Liu, H., & Li, Z. (2014). Genotyping by sequencing (GBS), an ultimate marker assisted selection (MAS) tool to accelerate plant breeding. *Frontiers in Plant Science*, *5*, 1–9. doi:10.3389/fpls.2014.00484 PMID:25324846

Jannink, J. L., & Walsh, B. (2002). Association mapping in plant population. In M. S. King (Ed.), *Quantitative genetics, genomics and plant breeding* (pp. 59–68). CAB International.

Lau, W. C., Rafii, M. Y., Ismail, M. R., Puten, A., Latif, M. A., & Ramli, A. (2015). Review of functional markers for improving cooking, eating, and the nutritional qualities of rice. *Frontiers in Plant Science*, *6*, 832–843. doi:10.3389/fpls.2015.00832 PMID:26528304

Liu, Y., He, Z., Appels, R., & Xia, X. (2012). Functional markers in wheat: Current status and future prospects. *Theoretical and Applied Genetics*, *125*(1), 1–10. doi:10.100700122-012-1829-3 PMID:22366867

Lubberstedt, T., Zein, I., Anderson, J. R., Wenzel, G., Krutzfeldt, B., Eder, J., ... Chun, S. (2005). Development and application of functional markers in maize. *Euphytica*, *146*(1-2), 101–108. doi:10.100710681-005-0892-0

Madhumati, B. (2014). Potential and application of molecular marker techniques for plant genome analysis. *International Journal Pure Applied Biosciences*, *2*, 169–188.

Mammadov, J., Agarwal, R., Buyyarapu, R., & Kumpatla, S. (2012). SNP markers and their impact on plant breeding. *International Journal of Plant Genomics*, *2012*, 1–11. Advance online publication. doi:10.1155/2012/728398 PMID:23316221

Moose, S. P., & Mumm, R. T. (2008). Molecular plant breeding as the foundation for 21st century crop improvement. *Plant Physiology*, *147*(3), 969–977. doi:10.1104/pp.108.118232 PMID:18612074

Nadeem, M. A., Nawaz, M. A., Shahid, M. Q., Dogan, Y., Comertpay, G., Yildiz, M., Hatipoğlu, R., Ahmad, F., Alsaleh, A., Labhane, N., Özkan, H., Chung, G., & Baloch, S. (2018). DNA molecular markers in plant breeding: Current status and recent advancement in genomic selection and genome editing. *Biotechnology, Biotechnological Equipment*, *32*(2), 261–285. doi:10.1080/13102818.2017.1400401

Ng, W. L., & Tan, S. G. (2015). Inter-simple sequence repeat (ISSR) markers: Are we doing it right? *ASM Science Journal*, *15*, 30–39.

Perez-de-Castro, A. M., Vilanova, S., Canizares, J., Pascual, L., Blanca, J. M., Diez, M. J., ... Pico, B. (2012). Application of genomic tools in plant breeding. *Current Genomics*, *13*, 179–195. doi:10.2174/138920212800543084 PMID:23115520

Poczai, P., Varga, I., Laos, M., Cseh, A., Bell, N., Valkonen, J. P. T., & Hyvonen, J. (2013). Advances in plant generated and functional markers: A review. *Plant Methods*, *9*(1), 6–18. doi:10.1186/1746-4811-9-6 PMID:23406322

Randhawa, H. S., Asif, M., Pozniak, C., Clarke, J. M., Graf, R. J., Fox, S., ... Singh, A. K. (2013). Application of molecular markers to wheat breeding in Canada. *Plant Breeding*, *132*, 458–471. doi:10.1111/pbr.12057

Semagn, K., Bjornstad, A., & Ndjiondjop, M. N. (2014). An overview of molecular marker methods for plants. *African Journal of Biotechnology*, *2450*, 25–68.

Singh, B. D., & Singh, A. K. (2015). *Marker-assisted plant breeding: principles and practices*. Springer. doi:10.1007/978-81-322-2316-0

Singh, B. P., & Gupta, V. K. (2017). *Molecular markers in mycology: diagnostics and marker developments*. Springer. doi:10.1007/978-3-319-34106-4

Wang, Y. H., Liu, S. J., Ji, S. L., Zhang, W. W., Wang, C. M., Jiang, L., & Wan, J. M. (2005). Fine mapping and marker-assisted selection (MAS) of a low glutinin content gene in rice. *Cell Research*, *15*(8), 622–630. doi:10.1038j.cr.7290332 PMID:16117852

Xu, Y., & Crouch, J. H. (2008). Marker-assisted selection in plant breeding: From publications to practice. *Crop Science*, *48*(2), 391–407. doi:10.2135/cropsci2007.04.0191

Xu, Y., Li, Z. K., & Thomson, M. J. (2012). Molecular breeding in plants: Moving into mainstream. *Molecular Breeding*, *29*(4), 831–832. doi:10.100711032-012-9717-8

Yamamoto, E., Matsunaga, H., Onogi, A., Kajiya-Kanegae, H., Minamikawa, M., Suzuki, A., Shirasawa, K., Hirakawa, H., Nunome, T., Yamaguchi, H., Miyatake, K., Ohyama, A., Iwata, H., & Fukuoka, H. (2016). A simulation based breeding design that uses whole genome prediction in tomato. *Scientific Reports*, *6*(1), 19454. doi:10.1038rep19454 PMID:26787426

Yang, H., Li, C., Lam, H., Yang, G., & Zhao, S. (2015). Sequencing consolidates molecular markers with plant breeding practice. *Theoretical and Applied Genetics*, *128*(5), 779–795. doi:10.100700122-015-2499-8 PMID:25821196

APPENDIX

1. Describe the prerequisites for conducting marker assisted selection.
2. Describe the basic procedure for conducting marker assisted selection for a single cross.
3. Describe the criteria to be considered for selection of DNA markers for marker assisted selection.
4. Explain how to determine the number of QTLs should be selected and used for marker assisted selection.
5. Explain how many generations should be conducted for marker assisted selection program.
6. Describe the situations where MAS should be adopted.
7. Describe the procedure to be adopted for marker assisted backcrossing (MAC).
8. Describe the situations where MABC is essential and advantageous for used.
9. Describe the procedure for developing resistance against pathogens in plants through MAS.
10. Describe the procedure for pyramiding of major and minor genes into a single pyramiding through MAS.
11. Describe the procedure for improving the quality traits in plants through MAS.
12. What is genome selection? How genome selection can be used in plant breeding.
13. Describe the challenges and prospects of marker assisted breeding for crop improvement.

Chapter 4
Molecular Markers for Plant Variety Identification and Protection

ABSTRACT

The identification of varieties of crop plants is important for their registration, breeding, seed production, and trade. The traditional approach to variety identification involves analysis and recording of their morphological characters, which is less informative, highly influenced by environmental factors and time consuming. Availability of molecular markers in large number in all the major crops has opened new avenue for their utilization in plant variety identification and protection. Molecular markers have the advantage of not being influenced by the environment and thus stable. Development of software to analyze and characterize the molecular markers has enhanced the process significantly. It also helps in protecting Plant Breeders Right. The establishment of genome and transcriptome sequencing projects for crops has generated a huge wealth of sequence data that could find much use in identification of plants varieties. In this chapter molecular basis of variety identification and their protection has been discussed.

DOI: 10.4018/978-1-7998-4312-2.ch004

INTRODUCTION

The prospect of utilizing molecular techniques in plant variety protection (PVP) has now been recognized by the breeders. The advantages of using molecular techniques in PVP are similar to its application in plant breeding programs. Among all the molecular markers, SNPs is the most sought after markers for PVP, as it is efficient and comparatively inexpensive. In several instances SSRs are also used, but there exist some limitations on its use, which are expected to be sorted out soon. Although traditional methods are still being used for PVP, molecular techniques can complement to achieve the goal more efficiently. Apart from description of the cultivars, supplementary information on isozyme analysis, RFLP, SSR, SNP or other genetic fingerprinting testing results can be provided during certification of the variety. Thus clear-cut and distinct data on molecular markers can be used for granting protection to plant variety.

If a cultivar is identified as unique based on molecular marker analysis, it is of no value unless the physical appearance is also unique. It may be possible to receive a PBR for an obsolete cultivar, without a physical verification. A DNA fingerprint library could prevent occurrence of such a possibility. However, significant efforts shall be involve to create such a database. Molecular techniques are particularly useful for crops having few variable morphological characters. Thus, under ideal conditions both botanical and molecular measurements should be applied to protect the cultivars.

CULTIVER IDENTIFICATION

Proper identification of the cultivars is very important for plant protection and production systems of Intellectual Property Rights (IPR) and Plant Biodiversity Register (PBR). Molecular marker techniques are relevant for PBR registration in the following manner (Ghosh et al., 2001): (1) genetic distance analysis between the candidate cultivar and the existing the pool of cultivars in order to define a set of comparison cultivars, (2) determination of the contribution of comparison cultivar for PBR registration, and (3) investigation of the use of DNA markers to resolve the identity of cultivars in those cases where infringement of PBR is claimed. Molecular techniques are particularly useful to resolve the issues related to infringement of cultivars, that is someone is selling someone else's cultivar (Becher et al., 2000). While in some species, evaluation of one gene or one trait may be sufficient to identify a cultivar, it may be required to analyze more than one gene or trait in some other species. Therefore, it has to be determined crop-by-crop (Korir et al., 2012).

SEED CERTIFICATION AND PURIFICATION

Maintenance of high quality of the seeds is the purpose of seed certification. This ensures high germination rate, seed health, and mechanical purity. An approved conditioner or grower of the seeds must process the seed certification process. The seeds must be sampled, tested and graded by accredited and recognized agencies. During certification process, verification of seed quality is done through traditional phenotypic inspection. Molecular techniques are not presently employed. However, there exists potentiality to use molecular techniques, as a control tool, to ensure the efficiency of the seed certification process. Application of molecular techniques will increase the level of confidence of the farmers regarding purity and security of the certified seeds (Gupta et al., 2001).

Seed purification process involves the following steps: selection of heads from the trial plots, growing single-head-derived breeder lines, discarding phenotypically poor lines, growing remaining breeder lines and further discarding on the basis of visual observations, and bulking the first breeders seeds. Molecular techniques can assist while discarding the lines. However, the necessary laboratory facilities should be available to the breeder to carry out the required process.

Molecular techniques can also be applied for purification of seeds, particularly for hybrid crops. In rice, when two-line hybrid system is applied, false hybrids may arise due to selfing of the female parents as a result of sterility instability of environment-genic male sterile lines, caused by fluctuations in the temperature. Thus the false hybrid may coexist with the real hybrid seeds and the purity of the hybrid will be at stake. Such phenomenon can happen in any other crop and therefore application of molecular markers is recommended to check the purity of hybrid seeds (Collard & Mackill, 2008).

BREEDING INFORMATICS

Molecular data having relevance to bioinformatics includes sequences of genome and cDNA, DNA markers and map, quantitative trait loci, candidate genes, physical maps constructed on the basis of chromosome breakpoints, gene expression, libraries of large inserts of DNA, and radiation hybrids. Flow of information from molecular markers to genetic maps to sequences has been used to establish the relationships. But, gaps exist between the sequence-based information and breeding related information such as phenotype, pedigree and germplasm. Even after availability of complete genomic sequence, it is possible to establish phenotyping on the basis of functional analysis in only 40% of the genes. Therefore it is essential to integrate breeding related information with genomics database to achieve meaningful results (Bernardo & Charcosset, 2006).

It is important to adopt a universally accepted system on information management to fulfill the requirements of modern plant breeding needs such as data acquisition, deposition, classification, integration, interpretation and utilization. All the information related to genotype, phenotype and environment should be considered together for integration, extraction, analysis and interpretation. A fully developed and stand-alone web-based information management system (IMS) has the following advantages (Zargar et al., 2015).

1. Provide a uniform information management system suitable for all types of breeding programs.
2. Provide efficient technology solutions for all types of breeding programs.
3. Accelerate breeding procedures by providing affordable IMS.
4. Provide single integrated source for all types of inquiries and ability to resolve problems more efficiently.
5. Stimulate collaborations by providing a platform for exchange of data having mutual interest.
6. Provide a platform for converting data into knowledge which are useful for different stake holders.

With the availability of such a system, along with computational biology and comparative genomics, breeders can create an intellectual property portfolio associated with their cultivars and hybrids.

SELECTION USING MOLECULAR MARKERS

Large numbers of molecular markers (several hundreds to thousands) are included in many molecular breeding programs, covering the entire plant genome. All these markers are used for genotyping and fingerprinting the accessions being used by the breeders. Such information proved the basis for selection of individuals to be included in the crossing programs having specific objective(s).

A database resource, PlantMarkers has been developed which can predict, analyze and display plant molecular markers. Techniques are available for identification of putative SNP, SST and COS markers from the available sequence databases. A web site at http://markers.btk.fi allows the users to search for species-specific markers, on the basis of specific criteria.

INFORMATION AVAILABLE IN THE SEQUENCE DATA

All typical sequence data are basically a string of nucleotide or amino acid residues. All such DNA or protein sequence data provide the following information: an accession number, source of the organism, name of the locus (gene), reference, key words applicable to the sequence, special features of the sequence e.g. coding region, introns, splice sites, mutation site, and the sequence. In the case of protein sequences, the sequences of amino acids are either obtained directly from the genomic sequences or from the cDNA sequences. In the case of protein sequences following additional information can be obtained: active site, structure and sequence of the motifs, domains, fingerprints, primary and 3-dimentional structure, structural properties, and classification according to the family.

Three major sources from which genome sequence data are accumulated are: whole genome sequence and assembly, genome survey sequencing, and ESTs. One of the challenges facing by the plant breeders is the conversion of complete genome sequence data to protein structure and its predicted functions.

Information on Expression

The variations in the expression patters of a particular DNA sequence across a population and its relationship between genes is an area of research for molecular plant breeders. While conduction such research it is important to have the following information: taxonomy, sex and developmental stages of the organism, growth conditions, organ and tissue from where the sample extracted, and protocol followed.

Microarrays derived from sequence data have been applied to measure the changes in gene expression that occur due to variations in ecological factors. Through analysis of microarrays it is possible to identify high-throughput transcriptional activity of the cell. This has led to development of several statistical-related disciplines within bioinformatics. With the expansion of microarray technology, cDNA array were developed for the analysis of gene expression in many crop plants. With continued improvement in microarray data production there will be significant improvement in data analysis and interpretation.

Plant Databases

Biological databases having large volumes of information about DNA, RNA and protein sequences and their functional and structural properties have been used to understand biological systematics. Since these databases also contain information on molecular markers, level of mRNA expression, concentration of metabolites, protein-protein interactions, and taxonomic relationships, it became most useful for systematic studies in plants.

The currently available molecular biology databases can be obtained at http://www.oxfordjournals.org/nar/database/a/. This has been classified into

Table 1. Major categories of the databases of molecular biology having relevance to plants

1. Nucleic acid (DNA) databases	**4. Protein sequence databases**
i. International Nucleotide Sequence Data Collaborations ii. Gene structure, introns and exons, splice sites base iii. Coding and non-coding DNA iv. Transcriptional regulator sites and transcription v. Nucleic acid structure	i. Protein properties ii. General sequence databases iii. Protein localization and targeting iv. Protein domain databases, protein classification i. Protein sequence motifs and active sites ii. Databases of individual protein families iii. Protein structure
2. RNA sequence databases	**5. Microarray data and other gene expression data**
3. Plant databases	**6. Genomics databases**
i. General plant databases ii. *Arabidopsis thaliana* iii. Rice iv. Other plants	i. Taxonomy and identification ii. General genomics databases iii. Genomic annotation terms, ontologies and nomenclature

6 categories (Table 1). A comprehensive list of database is also available at ExPASy Life Science Directory at http://expasy.ch/alinks.htmi.

Nuclear Sequence Databases

The International Nucleotide Database Collaboration is a joint venture of the NCBL of USA, European Bioinformatics Institute (EBI), and the DNA Data Bank of Japan. These repositories accept nucleic acid sequence data from the different laboratories and make them available freely. Each entry in the database is given a unique identification number, which is known as the Accession Number. To indicate any changes the particular sequence has undergone another code is assigned. This code is called Sequence Version and is composed of Accession Number followed by a period and Specific Version. The Nuclear Sequence Database was initiated 1980s and has grown exponentially over the years. Important DNA sequence databases are presented in Table 2.

Protein Sequence Databases

Protein sequence databases are classified as universal databases (covering information about proteins from all species), and specializes data collections (covering information about specific groups or families of proteins). The universal protein sequence databases can be divided into: simple archives of sequence data and annotated databases, where additional information is provided.

Established in 1984, Protein Information Resources (PIR) is the oldest protein sequence database, which was initiated by National Biomedical Research Foundation (NBRF) and been maintained by PIR. In 1986, Swiss-PROT an annotated universal protein sequence database was established, which was jointly maintained by Swiss Institute of Bioinformatics (SIB). Entries of the Swiss-PROT are thoroughly analyzed and annotated to ensure high level of quality of the database. Two classes of data can be distinguished through this database: core and annotated data. The core data comprises of sequence, citation and taxonomic information. The annotated data provide information about the functions of the protein, domains and sites, post-transcriptional modifications, secondary structure, quaternary structure, sequence conflicts, similarities with other proteins, diseases associated with deficiencies in the protein etc.

Table 2. DNA and protein sequence databases

Name of the database	Uniform Resource Locator (URL)	Description of the databases
DDBJ (DNA Database of Japan	http://www.ddbj.nig.ac.jo	One of the major databases of the International Nucleotide Sequence Database Collection
EMBL (European Molecular Biology Laboratory) Nucleotide Sequence Database	https://www.ebi.ac.uk/embl	Maintained at EBI (European Bioinformatics Institute) in collaboration with DDBJ and GenBank
GenBank	http://www.ncbi.nlm.nih.gov	Contains publicly available DNA sequences of different organisms
EXProt	http://www.cmbi.kun.nl/EXProt	Contains non-redundant protein sequence with experimentally verified functions
MIPS (Munich Information Centre for Protein Sequences)	http://mips.gsf.de	Contains protein sequences
NCBI Protein	http://www.ncbi.nlm.nih.gov/entrez/query.fcgi?db=Protein	Contains protein sequences derived from various sources
Patome	http://www.patome.org	Sequence data disclosed in patents and their other analyzed information
PIR-PSD (Protein Information Resources-Protein Sequence Database)	http://pir.georgetown.edu	Contains protein sequences derived from various sources
PRF (Protein Research Foundation)	http://www.prf.or.jp/en/index.shtm/	Contains protein sequences, literature and unnatural amino acids
RefSeq (Reference Sequence)	http://www.ncbl.nlm.nih.gov/RefSeq	Contains non-redundant sequence standards for genomic regions, transcripts and proteins
Swiss-PROT	https://www.expasy.org/sprot	Contains curated protein sequences, providing high level of annotation, minimal level of redundancy, and high level of integration with other databases
TCDB (Transporter Classification Database)	http://www.tcdb.org	Contains sequence, classification, structural, functional and evolutionary information about transport systems from many organisms
UniProt (Universal Protein Resources)	http://www.uniprot,org	Central repository of protein sequences and functions. World's most comprehensive catalogue of information on proteins

In 1996, Translation of EMBL nucleotide sequence database (TrEMBL), a supplement of Swiss-PROT was created, to make the availability of the new protein sequences as quickly as possible. This was necessitated as to generate high quality Swiss-PROT takes comparatively longer time.

For proper utilization of protein sequence database, the database should be non-redundant, comprehensive, well annotated, and up-to-date. But not a single protein database can satisfy these requirements. Therefore, users are forced to look for multiple databases. To overcome database of protein sequence and function called UniProt. UniProt is a central repository of protein sequence and function data, created by combining Swiss-PROT and TrEMBL and PIR databases. Important protein sequence databases are presented in Table 2.

General Plant Databases

The list of databases that covers multiple plant species is presented in Table 3. These databases contain information on physical and genetic mapping, sequencing, clustering, functional annotations, microarray analysis, signal transduction analysis etc. Some of these databases contain basic information such as C-value, promoter sequences, *cis*-element motifs, sonRNAs (small nucleolar RNAs), ncRNA (non-coding RNA), *cis*-acting regulatory elements/ repressors/ enhancers, mitochondrial protein and clusters of predicted plant proteins. Others provide specific tools for mining and managing of the data. These include: signal transduction analysis, plant EST clustering and functional annotation, analysis of functional properties of agriculturally important plants, classification of repetitive sequences, comparative genomics, retrieval of plant protease inhibitors and their genes.

Certain databases cover a specific group of plant species. For example, GrainGenes cover wheat, oats, barley, rye and triticale; TropGENE covers tropical crops; PLANTS cover vascular plants, mosses, liverworts, hornworts and lichens. Phytome provide information about genomic resources on angiosperms, built on publicly available sequences and information available on map from diverse plant species. Phytome contains functional and phylogenetic information which can be used for predicting protein sequences. It is basically designed to facilitate molecular plant breeding, functional genomics and evolutionary studies in plants. Some of the other databases support functions of comparative biology. For example, Gramene can be used to study comparative genome mapping of grasses.

Table 3. General plant databases

Name of the database	Uniform Resource Locator (URL)	Description of the databases
AgBase	http://www.agbase.msstate.edu	Curated resource for functional analysis of agricultural plants and animal gene products
BarleyBase	http://www.barleyase.org	Plant microarrays with integrated tools for statistical analysis
Cereal Small RNA database	http://sundarla.ucdavis.edu/ smrnas	Small RNAs expressed in rice and maize
CR-EST-Crop ESTs	http://pgrc.ipk-gatersleben.de/ cr-est	Sequence, classification, clustering and annotation of crop ESTs
CropNet	http://ukcrop.net	Genome mapping in crop plants of UK
DbEST	http://www.ncbi.nlm.nih.giv/ dbEST	Sequence data of 'single pass' cDNA and EST of several organisms
FLAGdb++	http://urgv.evry.inra.fr/project/	High-throughput functional analysis of fully sequenced genome of *Arabidopsis*
GeneFarm	http://urgi.versailles.inra.fr/	Annotated of Arabidopsis gene and protein families
GenoPlante-Info	http://www.genoplante.com	Genomics sequence, transcriptome, proteome, allelic variability, mapping and synteny and mutation data for rice, wheat, maize, rapeseed, pea, sunflower and *Arabidopsis*
GrainGenes	http://wheat.pw.usda.gov	Phenotypic and molecular information on wheat, oats, barley, rye and triticale
Gramene	http://www.gramene.org	Curated genome mapping database for grasses, with sequence-based maps, molecular markers, proteins, mutants, QTLs
ICIS (Internatio-nal Crop Infor-mation System)	http://www.icis.cgiar.org	Management and integration of global information on genetic resources and crop improvement for any crop
MIPSPlantsDB	http://mips.gsf.de/proj/plant/jst	Genomic database on rice, maize, Medicago, Lotus, Tomato, and *Arabidopsis*
PMIP	http://www.plantenergy.uwa.edu. au/applications/mpimp/ index.html	Mitochondrial protein import apparatus for wide range of organisms
PathoPlant	http://www.pathoplant.de	Plant pathogen interactions and components of signal transduction pathways related to plant pathogenesis
Pytome	http://www.hytome.org	Predicted protein sequences, protein family assignments, multiple sequence alignments, phylogenies, functional annotations from protein for large phylogenetically diverse plant taxa
PHYTOPROT	http://urgi.versailles.inra.fr/ phytoprot	Clusters of predicted plant proteins
PLACE	http://www.dna.affrc.go.jp/htdocs/PLACE	*Cis*-element motifs found in plant genes
Plant Genome C-value database	http://www.kew.org/genomesize/homepage.html	Genome size data of most plant species
Plant Genome Central	http://www.ncbi.nlm.nih.gov/genome/PLANTS/PlantList	Large scale sequences, genetic maps and EST sequences
Plant MPSS (Ma-ssive Parallel Signature Sequencing)	http://mpss.udel.edu	Largest set of tag-based gene expression data
Plant Ontology database	http://www.plantontology.org	Morphology and anatomy of plant and their developmental stages
PLANT-PIs	http://bighost.area.ba.cnr.it/ PLANT-Pis	Plant protease inhibitors and their genes
PlantGDB	http://www.plantgdb.org	Plant EST sequences that correspond to fragment of genes transcribed actively under specific conditions
PlantProm	http://mendel.cs.rhul.ac.uk/ mendel.php?topic=plantprom	Plant promoter sequences
PlantsP/PlantsT	http://plantsp.sdsc.edu	PlantsP focuses in proteins involved in phosphorylation process and PlantsT on membrane transport proteins
TAED (The Adap-tive Evolution Database)	http://www.bioinfo.no/tools/TAED	Phylogeny based tool for comparative genomics
TIGR plant repeat database	http://www.tigr.org/tdb/e2k1/ plant.repeats	Classification of repetitive sequences in plant genome
TropGENE DB	http://tropgeneedb.cirad.fr	Genetic and genomic information about tropical crops
UK CropNet Databases	http://ukcrop.net/db.html	Genome resource of Arabidopsis, Millet, Barley, Brassica

Individual Plant Databases

The list of databases that cover specific plants such as rice and other crops is presented in Table 4, and 5, respectively. Although the contents in these databases differ, they cover the following general subjects.

1. Genetic and cytogenetic maps of the chromosomes,
2. Genes, alleles and their products,
3. Genomic probes and nucleotide sequences,
4. Phenotypic traits, quantitative traits and QTLs,
5. Genotypes and pedigree of germplasm,
6. Information about biotic and abiotic stresses.

Arabidopsis and rice has been considered as model plant systems, and thus diverse information is available on these two plants. Two annotation related databases are available, one developed around contigs for high quality manual annotation (RAD) and the other allow integration of programs for prediction and analysis of protein coding gene structure (RiceGAAS).

The MaizeGDB (Maize Genome and Genomics Database) is a central repository for public maize information, which contains a series of computational tools which the breeders use to address their queries. Following information can be obtained from this database: data centers, ESTs, gene products, loci, maps, microarrays, metabolic pathways, phenotypes, probes, QTLs, sequences, SSRs etc.

The Dendrome database contains information about forest tree genome. It has several sub-branches and the primary genome database is called TreeGenes. TreeGene contain curated information about DNA sequences, genetic maps, markers, QTLs, ESTs and germplasm. The primary objective is to provide an efficient interface to compare between maps and for integration of expression and EST data.

PROSPECTS FOR BREEDING INFORMATICS

Molecular biology and informatics has vast scope for their application in plant breeding. Efficiency of the breeding programs shall depend on the accessibility of the information from the databases to the breeders, and their wise and effective utilizations. The databases should also be user-friendly.

Table 4. Databases on rice

Name of the database	Uniform Resource Locator (URL)	Description of the database
BGI-RISe (Beijing Genomics Institute Rice Information)	http://rise.genomics.org.cn	Comprehensive data on *O. sativa* on gene content, repetitive elements, gene duplication, and SNPs
IRIS (International Rice Information System)	http://www.iris.irri.org	Database for management and integration of global information on genetic resources
MOsDB	http://mips.gsf.de/proj/plant/jst/rice/index.jsp	Sequences of *O. sativa* genome, genes, genomics, mutants and expression profile
Oryza Tag Line	http://urgi.versailles.inra.fr/OryzaTagLine	Organizes data from T-DNA insertion lines of *O. sativa*
OryGenesDB	http://orygenesdb.cirad.fr	Rice genes, T-DNA and transposable elements flanking tags
Oryzabase	http://www.shigen.nig.ac.jp/rice/oryzabase	Rice classical genetics and recent genomics
RAD (Rice Annotation Database)	http://golgi.gs.dna.affrc.go.jp/SY-1102	High quality manual annotation of rice genome
RAP-DB (Rice Annotation Project Database)	http://rapdb.lab.nig.ac.jp	Provide access to annotation data of other rice genomics data. Serves as the hub for rice genomics
RetrOryza	http://www.retroryza.org	Completed resource on long terminal repeat-retrotransposition of rice
RiceGAAS (Rice Genome Automated Annotation System)	http://RiceGASS.dna.affrc.go.jp	Annotated genome sequences of rice
Rice Pipeline	http://cdna01.dna.affrc.go.jp/PIPE	Unique scientific resource of rice that pools publicly available data
Rice Proteome Database	http://gene64.dna.affrc.go.jo/RPD/main_en.html	Proteome database of rice
RMD (Rice Mutant Database)	http://rmd.ncpgr.cn	Rice T-DNA insertion lines generated by an enhancer trap system
RMD (Rice Mutant Database)	http://www.ricefgchina.org/mutant	Database on mutants of rice
WhoGA	https://rgp.dna.affrc.go.jp/whoga	Predicted genes and pseudogenes of rice with or without EST/full length cDNA support

Development of a universal database would require a universal language which can be applied to all plant species. Projects such as Gene Ontology and Plant Ontology are good beginning towards achieving this goal. The second universal language is also required so that breeders, curators, molecular biologists, bioinformaticians and tool developers can communicate easily among themselves. The prime role of breeders should be in developing universal databases or language.

Table 5 Databases of other plants (excluding rice)

Name of the database	Uniform Resource Locator (URL)	Description of the database
Barley Base	http://www.plexdb.org/plex.php?	Barley expression database for plant microarray data
Brassica BASC	http://bioinformatics.pbcbase.latrobe.edu.au	Multinational *Brassica* genome sequences
BeanGenes	http://beangenes.cws.ndsu.nodak.edu	*Phaseolus* and *Vigna* genome database
Cotton	http://cottondb.org	Genetic, genomic and taxonomic data of Cotton
CyanoBase	http://bacteria.kazusa.or.jp/cyano	Sequence and annotated data on *Cyanobacterial* genome
Dendrome	htttp://dendrome.ucdavis.edu	Forest tree genome database
Diatom EST Database	https://www.biologie.ens.fr/diatomics/EST	EST from two Diatom algae: *T. pseudonana and P. tricornutum*
ForestTreeDB	http://foresttree.org/ftd	EST sequence from tree species
ICIS	http://www.icis.cgir.org	Crop improvement and management system for crops and farming systems
Legume Information	http://www.comparative-legume s.org	EST sequence database and analysis system for legumes
MaizeGDB (Maize Ge-netics and Genomics Database)	http://www.maizegdb.org	Sequence, stock, phenotype, genotype, karyotype, chromo-somal maps of maize
MtDB	http://www.medicogo.org/MtDB	*Medicago truncatula* genome
Panzea	http://www.panzea.org	Genotype, phenotype and polym- orphic data of maize
PoMaMo (Potato Maps and More)	http://gabi.rzpd.de/PoMaMo.html	SNP and Indel data from diploid and tetraploid potato genotypes
RAPESEED	http://rapeseed.plantsignal.cn	EST, full length cDNA, unique serial analysis of gene expression (SAGE) tags, EMS mutations for *B. napus*
SGMD (Soybean Geno-mics and Microarray Database)	http://psi081.ba.ars.usda.gov/SGMD/default.htm	Genomic and microarray data of soybean
SGN (SOL Genomics Network)	http://sgn.cornell.edu	Genomic, genetic and taxonomic data on *Solanaceae* (potato, tomato, eggplant, pepper, petunia) and *Rubiaceae* (coffee).
Soybean Genome	http://www.soybeangenome.org	Genomics and its applications in soybean
SoyGD (Soybean Genome Database)	http://soybeangenome.siu.edu	Physical maps, BAC sequence and genetic maps of soybean
TED (Tomato Expression Database)	http://ted.bti.cornell.edu	Expression data on tomato
TIGR Maize Database	http://maize.tigr.org	Genomic sequences of maize
TomatEST DB	http://biosrv.ca.unina.it/tomatestdb.	Secondary database on EST and cDNA sequences of tomato

CONCLUSION

Identification of plant varieties has relevance in the areas of agriculture, research, protection of Plant Breeders Rights (PBR), and marketing. Application of molecular marker techniques for cultivar identification has been considered to be reliable, efficient and cost effective technology. They offer advantages over the traditional methods, with higher resolving power. Markers such as RFLP, RAPD, SSR, AFLP, SNP are used for DNA fingerprinting and comparative assessments have been made on the suitability of these markers for cultivar identification. Since it is possible to demonstrate distinctness, uniformity and stability (DUS) of the variety through molecular marker analysis, it has become increasingly important for fulfilling the requirements of PBR. Through molecular profiling it is possible to distinguish minor variants from the initial varieties and thus can be used to demonstrate DUS of the variety. The DNA marker data along with the morphological characters as now become the most powerful tool for plant variety identification and their protection.

REFERENCES

Becher, S. A., Steinmetz, K., Weising, K., Boury, S., Pelter, D., Renou, J. P., ... Wolff, K. (2000). Microsatellites for cultivar identification in pelargonium. *Theoretical and Applied Genetics*, *101*(4), 643–651. doi:10.1007001220051526

Bernardo, R., & Charcosset, A. (2006). Usefulness of gene information in marker-assisted recurrent selection: A simulation appraisal. *Crop Science*, *46*(2), 614–621. doi:10.2135/cropsci2005.05-0088

Collard, B. C., & Mackill, D. J. (2008). Marker-assisted selection: An approach for precision plant breeding in the twenty-first century. *Philosophical Transactions of the Royal Society of London. Series B, Biological Sciences*, *363*(1491), 557–572. doi:10.1098/rstb.2007.2170 PMID:17715053

Ghosh, S. K., Sarkar, C. K. G., & Datta, S. (2001). Molecular markers for cultivar identification and PBR. *Journal of Intellectual Property Rights*, *6*, 377–388.

Gupta, P. K., Roy, J. K., & Prasad, M. (2001). Single nucleotide polymorphism: A new paradigm for molecular marker technology and DNA polymorphism detection and emphasis on their use in plants. *Current Science*, *80*, 524–535.

Korir, N. K., Han, J., Shungguan, L., Wang, C., Kayesh, E., Zhang, Y., ... Fang, J. (2012). Plant variety and cultivar identification: Advances and prospects. *Critical Reviews in Biotechnology*, *8*, 1–15. PMID:22698516

Zargar, S. M., Raatz, B., Sonah, H., Nazir, M., Bhat, J. A., Dar, A., ... Rakwal, R. (2015). Recent advances in molecular marker techniques: Insight into QTL mapping, GWAS and genomic selection in plants. *Journal of Crop Science and Biotechnology*, *18*(5), 293–308. doi:10.100712892-015-0037-5

ADDITIONAL READING

Anderson, J. R., & Lubberstedt, T. (2003). Functional markers in plants. *Trends in Plant Science*, *8*(11), 554–560. doi:10.1016/j.tplants.2003.09.010 PMID:14607101

Archak, S., Gaikwad, A. B., Gautam, D., Rao, E. V. V. B., Swmy, K. R. M., & Karihaloo, J. L. (2003). DNA fingerprinting of Indian cashew (*Anacardium occidentale* L.) varieties using RAPD and ISSR techniques. *Euphytica*, *230*(3), 397–404. doi:10.1023/A:1023074617348

Arens, P., Bredemeijer, G., Smulders, M., & Vosman, B. (1995). Identification of tomato varieties using microsatellites. *Acta Horticulturae*, (412), 49–57. doi:10.17660/ActaHortic.1995.412.3

Aslam, S., Tahir, A., Aslam, M. F., Alam, M. W., Shedayi, A. A., & Sadia, S. (2017). Recent advances in molecular techniques for the identification of phytopathogenic fungi – a mini review. *Journal of Plant Interactions*, *12*(1), 493–504. doi:10.1080/17429145.2017.1397205

Cretazzo, E., Meneghetti, S., De Andre's, M. T., Gaforia, L., Frare, E., & Cifre, J. (2010). Clone differentiation and varietal identification by means of SSR, AFLP, SAMPL and M-AFLP in order to assess the clonal selection of grapevine: The case study of Manto Negro, Callet and Moll, autochthonous cultivars of Majorca. *Annals of Applied Biology*, *157*(2), 213–227. doi:10.1111/j.1744-7348.2010.00420.x

Curn, V., & Zaludova, J. (2007). Fingerprinting of oilseed rape cultivars. In S. Gupta (Ed.), *Rapeseed breeding: advances in botanical research* (Vol. 45, pp. 155–179). Elsevier Publications. doi:10.1016/S0065-2296(07)45006-6

Demirsoy, L., Demir, T., Demirsoy, H., Okumus, A., & Kaçar, Y. A. (2008). Identification of some sweet cherry cultivars grown in Amasya by RAPD markers. *Acta Horticulturae*, (795), 147–152. doi:10.17660/ActaHortic.2008.795.18

Dey, S. S., Singh, A. K., Chandel, D., & Behera, T. K. (2006). Genetic diversity of bitter gourd (*Momordica charantia* L) genotypes revealed by RAPD markers and agronomic traits. *Scientia Horticulturae*, *109*(1), 21–28. doi:10.1016/j.scienta.2006.03.006

Diaz, S., Pire, C., Ferrer, J., & Bonete, M. J. (2003). Identification of *Phoenix dactylifera* L. varieties based on amplified fragment length polymorphism (AFLP) markers. *Cellular & Molecular Biology Letters*, *8*, 891–899. PMID:14668912

Ding, X. D., Lu, L. X., Chen, X. J., & Guan, X. (2000). Identifying litchi cultivars and evaluating their genetic relationships by RAPD markers. *Redai Yaredai Zhiwu Xuebao*, *8*, 49–54.

Dje, Y., Tahi, C. G., & Zoro, B. (2010). Use of ISSR markers to assess genetic diversity of African edible seeded *Citrullus lanatus* landraces. *Scientia Horticulturae*, *124*(2), 159–164. doi:10.1016/j.scienta.2009.12.020

Eathinton, S. R., Crosbie, T. M., Edwards, M. D., Reiter, R. S., & Bull, J. K. (2007). Molecular markers in a commercial breeding programme. *Crop Science*, *47*, S154–S163. doi:10.2135/cropsci2007.04.0015IPBS

Foster, J. T., Allan, G. J., Chan, A. P., Rabinowicz, P. D., Ravel, J., Jackson, P. J., & Keim, P. (2010). Single nucleotide polymorphisms for assessing genetic diversity in castor bean (*Ricinus communis*). *BMC Plant Biology*, *10*(1), 13–23. doi:10.1186/1471-2229-10-13 PMID:20082707

Ghislain, M., Spooner, D. M., Rodríguez, F., Villamon, F., Nunez, J., Vasques, C., ... Bonjerbale, M. (2004). Selection of highly informative and user-friendly microsatellites (SSRs) for genotyping of cultivated potato. *Theoretical and Applied Genetics*, *108*(5), 881–890. doi:10.100700122-003-1494-7 PMID:14647900

Gupta, P. K., Langridge, P., & Mir, R. R. (2010). Marker-assisted wheat breeding: Present status and future possibilities. *Molecular Breeding*, *26*(2), 145–161. doi:10.100711032-009-9359-7

Jiang, D., Ye, Q. L., Wang, F. S., & Cao, L. (2010). The mining of citrus EST-SNP and its application in cultivar discrimination. *Agricultural Sciences in China*, *9*(2), 179–190. doi:10.1016/S1671-2927(09)60082-1

Kumpatla, S. P., & Buyyarapu, R. (2012). Genomics-assisted plant breeding in the 21st century: technological advances and progress. Retrieved from: www.intechopen.com

Lin, K. H., Lai, Y. C., Li, H. C., Lo, S. F., Chen, L. F. O., & Lo, H. F. (2009). Genetic variation and its relationship to root weight in the sweet potato as revealed by RAPD analysis. *Scientia Horticulturae*, *120*(1), 2–7. doi:10.1016/j.scienta.2008.09.008

Liu, Y., He, Z., Appels, R., Appeals, R., & Xia, X. (2012). Functional markers in wheat: Current status and future prospects. *Theoretical and Applied Genetics*, *125*(1), 1–10. doi:10.100700122-012-1829-3 PMID:22366867

Marinello, L., Sommella, M. G., Sorrentina, A., Forlani, M., & Porta, R. (2002). Identification of *Prunus armeniaca* cultivars by RAPD and SCAR markers. *Biotechnology Letters*, *24*(10), 749–755. doi:10.1023/A:1015516712754

Mcglaughlin, M. E., Riley, L., & Helenurm, K. (2009). Isolation of microsatellite loci from the endangered plant *Galium catalinense* subspecies *acrispum* (Rubiaceae). *Molecular Ecology Resources*, *9*(3), 984–986. doi:10.1111/j.1755-0998.2009.02545.x PMID:21564813

Meuwssen, T. H., Hayes, K., & Goddard, M. E. (2001). Prediction of total genetic value using genome-wide dense marker maps. *Genetics*, *157*, 1819–1829. PMID:11290733

Nadeem, M. A., Nawaz, M. A., Shahid, M. Q., Doğan, Y., Comertpay, G., Yıldız, M., Hatipoğlu, R., Ahmad, F., Alsaleh, A., Labhane, N., Özkan, H., Chung, G., & Baloch, F. S. (2018). DNA molecular markers in plant breeding: Current status and recent advancements in genomic selection and genome editing. *Journal of Biotechnology and Biotechological Equipment*, *32*(2), 261–285. doi:10.1080/13102818.2017.1400401

Nadia, H. (2011). Identification of plant species using traditional and molecular-based methods. In R. E. Davis (Ed.), *Wild Plants: Identification, uses, and conservation* (pp. 1–66). Nova Science Publications, Inc.

Naim, D. M., & Mahboo, S. (2020). Molecular identification of herbal species belonging to genus *Piper* within family Piperaceae from northern Peninsular Malaysia. *Journal of King Saud University*, *32*(2), 1417–1426. doi:10.1016/j.jksus.2019.11.036

Setoguchi, H., Mitsui, Y., Ikeda, H., Nomura, N., & Tamura, A. (2009). Development and characterization of microsatellite loci in the endangered *Tricyrtis ishiiana* (Convallariaceae), a local endemic plant in Japan. *Conservation Genetics*, *10*(3), 705–707. doi:10.100710592-008-9620-3

Tatikonda, L., Wani, S. P., Kannan, S., Beerelli, N., Sreedevi, T. K., Hoisington, D. A., Devi, P., & Varshney, R. K. (2009). AFLP-based molecular characterization of an elite germplasm collection of *Jatropha curcas* L.: A biofuel plant. *Plant Science*, *176*(4), 505–513. doi:10.1016/j.plantsci.2009.01.006 PMID:26493140

Varshney, R. K., Hoisington, D. A., Nayak, S. N., & Graner, A. (2009). Molecular plant breeding: Methodology and achievements. *Plant Genomics*, *513*, 283–304. doi:10.1007/978-1-59745-427-8_15 PMID:19347654

Ye, G., & Smith, K. F. (2008). Marker-assisted gene pyramiding for inbred line development: Basic principles and practical guidelines. *International Journal of Plant Breeding*, *2*, 1–10.

Zhang, D., Mo, X., Xiang, J., & Zhou, N. (2016). Molecular identification of original plants of *Fritillariae cirrhosae* bulbus, a traditional Chinese medicine (TCM) using plant DNA barcoding. *African Journal of Traditional, Complementary, and Alternative Medicines*, *13*(6), 74–82. doi:10.21010/ajtcam.v13i6.12 PMID:28480363

APPENDIX

1. Explain how molecular markers can be used for cultivar identification.
2. Explain how molecular markers can be used for seed certification and purification.
3. What are the advantages of web-based information management system (IMS) for cultivar identification and seed certification?
4. What types of information is available in sequence databases which are useful for cultivar identification and seed certification?
5. What are prospects of using sequence databases for cultivar identification and seed certification?

Chapter 5
Molecular Markers for Phylogenetic Studies and Germplasm Conservation

ABSTRACT

Application of molecular markers in phylogenetic studies has become increasingly important in recent times. Availability of fast DNA sequencing techniques and robust statistical analysis methods provided new momentum to this field. Different nuclear encoded genes (16S rRNA, 5S rRNA, 28S rRNA), mitochondrial encoded genes (cytochrome oxidase, mitochondrial 12S, cytochrome b, control region), and few chloroplast encoded genes (rbcL, matK, rpi16) have been used as molecular markers. This method allows researchers to obtain new evidence concerning their phylogeny and biodiversity. Measurement of genetic diversity is important for development of strategies for effective germplasm management. The DNA-based technologies can overcome all the limitations of traditional methods used for the estimation of genetic diversity. This chapter deals with historical developments of molecular phylogeny, use of molecular markers in phylogeny, and evolution of phylogenetic tree building methods.

DOI: 10.4018/978-1-7998-4312-2.ch005

INTRODUCTION

The term phylogeny in short means the history of descent of a group of taxa from their common ancestors. This may also include information about the order of branching and the time of divergence. The term is also used to understand the genealogy of genes from a common ancestral gene. In molecular phylogeny, homology of DNA and protein sequences is used to determine the relationship among organisms and genes. Any dissimilarity in the sequence of DNA or protein indicates divergence due to molecular evolution over a period of time. In the traditional phylogenetic approach morphological characters of the organism are used, while in molecular phylogenetic approach sequence of the nucleotides in DNA and RNA and amino acids in protein are used. Both the traditional and molecular phylogenetic approaches are important and they can complement each other. Molecular phylogenetic approach is particularly helpful in those cases where important morphological characters of any organism are not available.

Due to the presence of unimaginable amount of diversity within the living organisms in the nature, it is really difficult to ascertain the exact phylogenetic relationship among them. The diversity is not limited to the morphological characters, but also to the ultra-structure of the cellular components, and biochemical and molecular components. Organisms may have similar morphological features, but may have diverse biochemical and molecular components.

Basically three types of information are necessary to determine the evolutionary history and phylogenetic relationship of every organism: morphological features (phenotype), cellular ultra-structures, and biochemical and molecular features (protein and DNA/RNA sequences). Availability of large amount of DNA/RNA sequence data, and robust mathematical and statistical tools has made the analysis of the date much easier.

TRADITIONAL VS MOLECULAR PHYLOGENETIC STUDIES

During 384-322 B.C., extensive morphological and embryological studies were carried out on marine organisms by Aristotle, for their proper classification. Thereafter, Linnaeus developed binomial system of nomenclature during 18th century. That was the beginning of taxonomy which led to development of phylogenetic trees. In the 19th century Charles Darwin developed the process of branching and divergence. With the advancement in the application of molecular markers, it became apparent that molecular markers shall play a major role in the phylogenetic studies. Technological advancement has made sequencing of DNA/RNA and protein much easier and less expensive. Accordingly large sequencing data of DNA/RNA and protein from diverse organisms became available. This has opened-up great prospect for their utilization in phylogenetic studies. Earlier, sequencing of the DNA segment of large and small sub-units of rRNA obtained through reverse transcriptase was used for phylogenetic studies. However, this method was found to be error-prone, compared to the DNA sequences obtained from the nuclear genes which cod e for rDNA (Nadia 2011).

PHYLOGENETICS AND MOLECULAR CLOAK

Studies on amino acid sequence of hemoglobin in different animal species by Zuckerkandl and Pauling (1965) has provided remarkable information. They observed that hemoglobin molecule of human and mouse differ by only 16 amino acids, between human and horse by 18 amino acids, between horse and mouse by 22 amino acids, while between human and shark by 79 amino acids. These observations imply that there has been a constant rate of substitution of amino acids over time. To explain this phenomenon, Zuckerkandl and Pauling (1965) proposed the molecular clock hypothesis. According to the hypothesis, the difference in amino acids between different organisms correlates with the evolutionary time scale. The difference in amino acids between mammals are less compared to that between the mammals and fish (shark). Apparently, a biomolecule has been acting like a molecular clock. Greater will be the distance between two organisms, if the differ by more number of amino acids and vice versa. Thus the distance between them can be estimated in the evolutionary timescale. Using this hypothesis it has been estimated that humans and apes diverged about five million years ago. However, validity of the hypothesis has been questioned as changes in biomolecules can occur at different rates.

Conclusions drawn from inheritance of single marker gene or protein sequence shall reflect evolution of that particular gene. Other genes present in an organism may show different evolution rates or history. Therefore interpretations based on a single gene may not reflect the actual evolutionary history of an organism. Another problem which interferes with the interpretation is horizontal or lateral gene transfer. Horizontal gene transfer refers to transfer of gene(s) between unrelated organisms. Such transfer is common in bacteria, and also been reported in eukaryotes. Determination of phylogeny of organisms becomes difficult or complicated due to horizontal gene transfer. Therefore it is important to distinguish between vertically inherited (genes inherited from parents to offspring) and horizontally inherited genes in any organism. The largest set of genes inherited together is considered to be vertically inherited. For this, analysis of large number of genes is required as opposed to studying inheritance of single marker gene. Therefore, convincing results on evolutionary status of an organism can only be obtained by considering evolution of multiple genes of the organism.

ADVANTAGES OF USING MOLECULAR MARKERS IN PHYLOGENETIC STUDIES

Mutation of the genes and the rate at which they mutate play an important role while studying phylogenetic studies of organisms. Some genes mutate more frequently than others. Different genes accumulate mutations at different rates. Accumulation of the mutant gene shall only be possible, if the change in the gene can be tolerated by the organism. For example, if some amino acids are replaced in the histone molecule, it becomes non-functional. On the contrary, the ITS (internal transcribed spacer) of rRNA can carry out its functions even if some nucleotides are changed. Thus, rate of accumulation of mutant ITS (internal transcribed spacers) shall be much higher than mutant histone(s) (Grechko 2002).

Compared to fossil records, molecular data are numerous and much easy to access. Lack of sample biasness in molecular data helps to correct the gaps in fossil records. On the other hand, insufficient data on morphological parameters may not be suitable to construct a phylogenetic tree. Molecular data, which are large in number and occur in various forms, can fill the gaps in morphological data for phylogenetic studies.

POTENTIAL GENES FOR PHYLOGENETIC STUDY

For phylogenetic studies, all genes and biomolecules are not suitable. It is important to screen molecular sequences for their ability to resolve relationships within a group of organisms. The methods of screening include:

assess the ability of a gene to confirm certain well-established phylogenetic relationships, and reconstruct fossil-based pair wise difference curve. Through this it is possible to estimate the rate of potentially informative character changes during the geological interval when a clade underwent phylogenetic divergence. Once identified, such genes serve as the molecular fossils, and can be utilize to study the evolutionary history of the gene.

PROPERTIES OF IDEAL MARKER GENES FOR PHYLOGENETIC STUDIES

Ideal marker genes should have the following properties (Grechko 2002):

1. The ideal marker gene should have single copy rather than multiple copies. The mitochondrial and chloroplast satisfies this condition.
2. Knowledge about the alignment of the sequence of the marker genes in the chromosomes is important before the phylogenetic studies. Therefore, such alignment studies should be simple and easy. Due to insertions and deletions, length of the same gene may vary among members of an organism. All ambiguous alignment should be avoided.
3. The rate of mutation of the marker gene should be optimum. Any gene showing faster rate of mutation may be ambiguous, as the second mutation at a particular site may be a reverse mutation.
4. Primers should be such that they amplify the marker gene selectively. They should not amplify non-specific genes, as this may lead to contamination.
5. High variations in the base sequence among the taxa are not preferable, as they may not reflect the actual ancestry.

MOLECULAR MARKERS USED IN PHYLOGENETIC STUDIES

Highly conserved coding regions (18S, 26S rDNA) are useful primarily at the family level and above, whereas rapidly evolving regions such as ITS are often best suited for compairing species and closely related genera (Figure 1). Some of the very important molecular markers used in phylogenetic studies are as follows (Soltis et al. 1992, 2012, Grechko, 2002).

Single Copy Nuclear Genes

Single copy nuclear genes or single-copy nuclear gene families have been recognized as one of the ideal molecular markers for determining phylogenetic relationship of plants species. Amplification and sequencing of these genes has remained easy due to uniqueness and high sequence conservation across species. Being nuclear genes, single-copy genes follow bi-parental inheritance pattern, and thus better suited for hybridization, and determination of speciation of closely related species, compared to organelle genes. Use of unlinked nuclear single-copy genes has the potential to reflect true species relationships compared to organelle genes. Although, single-copy genes have been used widely for angiosperms, its application in gymnosperms has been initiated recently. Since single-copy genes expresses more broadly compared to non-single copy genes, they can be easily detected by transcriptome sequencing, which helps to identify suitable molecular markers easily.

Nuclear Ribosomal Genes

For phylogenetic studies, the ribosomal RNA is considered to be the best target, as it is universal, and consist of both highly conserved and variable domains. The constituents of ribosomes are: rRNA and protein. All ribosome has two sub-units: small and large. The small sub-units contain a single RNA molecule (16S rRNA in prokaryotes and 18S rRNA in eukaryotes), whereas large sub-unit of prokaryotes contains two rRNA (5S and 23S rRNAs) and three sub-units in eukaryotes (5S, 5.8S and 25S/28S rRNAs). The core regions of the small and large sub-units contain 10 and 18 variable sites, respectively. Moreover, rRNA genes are evolving at a much slower rate compared to protein coding genes.

16S rRNA: The 16S rRNA has been observed to be highly conserved, but its rate of evolution varies in different organisms. This phenomenon has been exploited by the researchers to distinguish between different bacterial groups. The length of the 16S rRNA gene is about 1550 bp and have both conserved and variable regions. The variable regions are interspersed between the conserved regions. It is possible to amplify the variable regions by using primers of conserved regions. Comparison of 16S sequences has led to understand the relationships between different bacterial species. Analysis of 16S and 23S rRNA has become the most

Figure 1. Taxonomic level of utility of various nuclear (nDNA), mitochondrial (mtDNA), and chloroplast (cpDNA) genes and DNA regions used in phylogenetic reconstruction in plants. The shaded box represents the taxonomic zone that is not presently well covered by nuclear gene sequences. ? refers to genes that have rarely used; ------ designates the approximate upper or lower limits of applicability
(Redrawn from: Soltis et al.1998, Molecular Systematics II, Springer, New York)

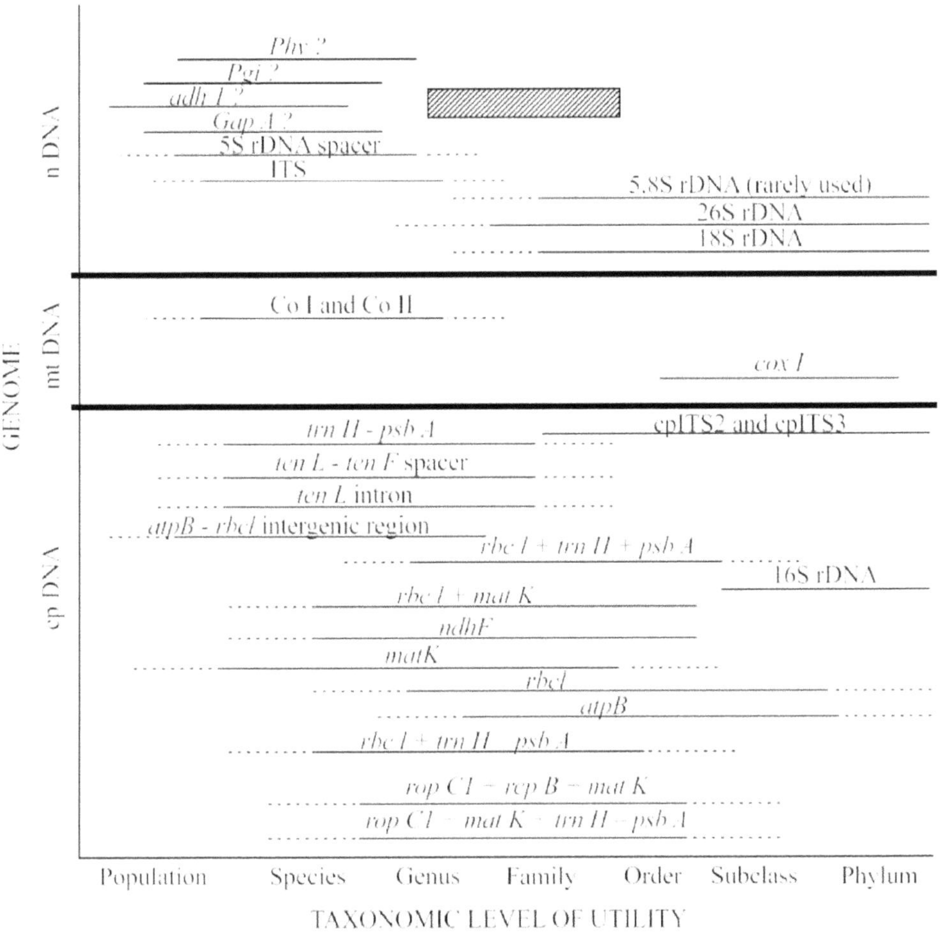

important tool for phylogenetic studies in bacteria, particularly to identify non-culturable bacteria.

5S rRNA: The 5S rRNA is present universally in all ribosomes, barring in the mitochondria of some fungi, most protists and higher animals. The 5S rRNA has about 120 nucleotides and is highly conserved. Although it is a rapidly evolving molecule, reliability of the analysis based on using 5S rRNA has been question due to presence of only 129 bp. It has been suggested that 5S rRNA sequence data do not have sufficient power to resolve the phylogenetic relationships at the taxonomical level.

28S rRNA: The sequence of 28S rRNA gene is now available for many organisms. It has about 811 bp and present in some eukaryotes. All phylogenetic studies made so far have used the D2 expansion region of the 28S rRNA. Phylogenetic studies reported so far using 28S rRNA has remained unresolved. With the availability of more information it may be useful in the future.

Mitochondrial Genes

Mitochondrial genes can be effectively used to resolve species-level phylogenies. Mitochondrial genes are separated by large non-coding DNA regions, and have variable gene order. Mitochondrial genes can undergo rearrangement frequently and many rearranged forms can exist within the same cell. Developments in molecular biological techniques have made use of mitochondrial DNA popularity in phylogenetic studies.

Cytochrome oxidase I and II (COI and II): The COI and COII genes codes for cytochrome c oxidase complex having 7 polypeptide subunits. They are the components of electron transport chain and found in mitochondria and bacteria. The COI and COII genes consist of approximately 894 bp and 485 bp, respectively. The rate of evolution of COI gene is comparatively slower than other protein coding mitochondrial genes, and therefore widely used in phylogenetic studies. Because of their usefulness, large sequence data has been generated for both COI and COII genes. It has been observed that combination of COI and 12S rRNA sequence data usually fulfills the requirements to distinguish the taxa of interest at different taxonomic levels. Recently COI sequence data has been recommended for to be used as 'DNA barcode' for animal kingdom.

Mitochondrial 12S: Sequence of mitochondrial 12S rRNA gene is also extensively used in phylogenetic studies. The length of the 12S gene is about 450 bp and can be easily amplified through universal primers. In several instances, analysis of 12S sequence data did not support the conventional view of phylogeny. Therefore, it may be interesting to study 12S gene sequence data to relate with the present concepts on phylogenetic relationship in many organisms.

Cytochrome b: the Cytochrome b gene is about 1,143 bp long and is reported to be one of the most important marker genes to study the phylogenetic relationships among closely related taxa. The strength of its utility normally found to be lineage-dependent, and tend to decline with evolutionary depth.

Non-Coding Area of mtDNA: The major non-coding region of mtDNA is known as control region, which is about 1.0 kb long. This region is involved in regulation and initiation of mtDNA replication and transcription. The mutation rate in mtDNA is about 10^{-9} per site per year in the nuclear genes. Generally, occurrence of point mutation is the significant characteristics of this region, which is mainly due to transitions. Presence of variable number of tandem repeats (VNTR) in the control region is observed to be common phenomenon. Further, this region also shows high rate of heterogeneity, tandem repeats, and insertion and deletion of nucleotides. These characteristics make the control region a good candidate for phylogenetic studies.

Chloroplast Genes

Chloroplast genome (DNA) is comparatively smaller than the mitochondrial and nuclear genomes. Most of the genes in cpDNA are assumed to be conserved. Thus the rearrangements and nucleotide substitution in cpDNA is considered to be less. Therefore, cpDNA is considered to be ideal to study phylogenetic relationships in plants. However, selection of appropriate gene(s) having sufficient length and appropriate mutation rate is considered to be crucial for the success of such studies. At present genes of chloroplast genome found to be suitable for phylogenetic studies include: *rbc*L, *mat*K, *rpl*16, *ndh*F, *atp*B and some others. Brief accounts of some of these genes are as follows.

rbcL gene: A single copy of the *rbc*L gene is part of the chloroplast genome of plants, which is about 1428 bp long. The gene is responsible for production of the ribulose 1, 5-bisphosphate carboxylase/ oxygenase (rubisco), which is the first enzyme of C3 cycle in plant. The secondary structure of the enzyme is known, which is present in many copies with less deletions and insertion. Thus it becomes easy to study and align. The large subunit of rubisco is encoded by *rbc*L, while *rbc*S gene, present in the nucleus, encodes for the smaller sub-unit. The *rbc*L gene is the most frequently sequenced gene and most widely used for phylogenetic studies in plants, particularly angiosperms. Since *rbc*L gene is conserved, show moderate substitution rate and can be rapidly aligned across diverse taxa, it is preferred for taxonomical studies in plants. However, it has been shown that 16S rRNA gene is the most conserved followed by 23S rRNA gene, and therefore they are preferred over *rbc*L gene for studying higher levels of phylogeny in plants.

matK Gene: The *mat*K gene is encodes a maturase enzyme, involved in splicing type ll introns from RNA transcripts, is about 1500 bp long. The gene has been reported to have relatively high rates of substitutions, due to transverse mutations. It has been shown to be useful for resolving intraspecific and intergeneric relationships among flowering plants.

ndhF gene: The *ndh*F gene is present in the small single copy region of the chloroplast genome, which is about 1100 bp long. It codes for subunit F of NADP dehydrogenase, and used to reconstruct relationships between 282 taxa from 78 families of monocots. The *ndh*F provides more than twice informative characters compared to *rbc*L, and nearly as many when *rbc*L, *atp*B and 18S rRNA are combined. This is because of the fact that the *ndh*F is substantially longer and evolves twice as fast.

rpl16 gene: The *rpl*16 is a non-coding intron region of the chloroplast genome with a size of about 1059 bp. It has been shown that *rpl*16 intron has lower transition/ transversion ratio but higher nucleotide divergence and genetic distance in grass family. Like other non-coding regions, *rpl*16 also has complicated evolution pattern and more frequent insertion/ deletion events than coding regions. Some of the other marker genes used for phylogenetic studies of plants are shown in Table 1.

CONSTRUCTION OF PHYLOGENETIC TREE

Basically through phylogenetic studies it possible to reconstruct the events of the past from available evidences. The results obtained through molecular phylogenetic analysis can be presented as a phylogenetic tree. The phylogenetic tree provides an overall idea about a given species in terms of its relationships with other related species (Mondini et al. 2009). It is rather difficult to construct a perfect phylogenetic tree, which will reflect the exact evolutionary history of a group of plants. A phylogenetic tree can be both rooted and un-rooted. The exponential relationship between the possible numbers of phylogenetic trees for 'n' taxa is given by:

$N = (2n-3)!2n^{-2}(n-s)!$, for rooted trees, and

Table 1. Some of the important molecular markers used in phylogenic studies in plants

Gene	Description
*Atp*B	Beta sub-unit of ATP synthase
*Dna*A	Initiation of DNA
EF-1α	Elongation factor-1α in protein synthesis
FtsZ	Role in cell division
*Gap*A	Codes for glyceraldehyde phosphate dehydrogenase
*Glt*A	Encodes citrate synthase
*Gro*EL	Encodes bacterial heat shock protein
ITS	Non-functional RNA situated between structural ribosomal RNAs precursor transcript
lux	Encodes proteins involved in luminescence
Nuclear H3	Codes for protein associated with DNA
PEPCK	Codes for phosphoenolpyruvate
*pyr*H	Codes for uridine monophosphate kinase
*Rec*A	Role in recombination
*rpo*A	Encodes α-subunit of RNA polymerase
*Rpo*B	Coding region located in plastid genome
*rpo*C1	Coding region located in plastid genome
*trn*H-*psb*A	Non-coding intergenic spacer region located in plastids
U2 *sn*RNA	Components of spilceosome
Wsp	Encodes a major cell surface coat protein

$N = (2n-5)!/2n^{-3}(n-3)!$, for un-rooted trees.

Accordingly, even for studying only 10 taxa, millions of possible phylogenetic tree topologies shall be available. Therefore, it is important to identify and select the best method, which will provide the maximum information. The phylogenetic trees can be drawn either as 'phylogram' (a phylogenetic tree in which the branch lengths represent the amount of evolutionary divergence) or a 'cladogram' (a phylogenetic tree in which the branch lengths are not proportional to the number of evolutionary changes and thus have no phylogenetic meaning). Following steps should be followed for construction of phylogenetic trees: 1) selection of the gene family or the organisms, 2) selection of the molecular markers, 3) amplification of the markers, 4) study of multiple sequence alignments, 5) selection of the evolutionary model, 6) phylogenetic analysis, and g) construction and evaluation of phylogenetic tree.

Selection of Molecular Markers

Either or both nucleotide sequence data or the protein sequence data can be used for selection of molecular markers. For closely related organisms nuclear sequence data is preferred. Genes that evolves slowly are preferred for studying widely divergent groups, while non-coding mtDNA is preferred for studying individuals of a population. Due to codon degeneracy, protein sequences are more conserved. Usually third position of a codon may show variation.

Multiple Sequence Alignment

After selection of the molecular markers to be used, it is important to determine the DNA sequence of the marker genes experimentally. For this the total DNA of the organism should be isolated and then the chosen markers should be amplified using marker specific primers through PCR. Many well-known universal primers have been descried in the literature. Alternatively, specific primes can be designed and utilized for specific markers. The amplified PCR products should then be sequenced. After obtaining the sequence data of the markers, they should be aligned with the sequence of the same markers of closely related species. Multiple alignments is the most critical step as it establishes corresponding positions of individuals in the evolutionary process. Various alignment programs such as, T-coffee, ClustalW, Multialin etc. are available for multiple alignment studies. Information on secondary structures may also help in alignment studies. To obtain information on the secondary structure, researchers use programs such as Praline. Some other programs such as, NorMD, Rascal, Gblock etc. are used to improve the alignment results, which can rectify the errors and remove poorly aligned positions.

Selection of Evolutionary Model

Selection of a proper substitution model that can provide ideas about the evolutionary process, by taking into account multiple substitution events, is the next step to follow. It may so happen that the observed number of substitutions for a particular locus of interest does not represent the actual evolutionary process. For example, when a substitution is detected as from T to G, the nucleotide may have actually undergone several transitional steps in-between, as T→ C→A→G. Similarly, back mutations may restore the original nucleotide as T→A→T. On the other hand, identical nucleotide sequences observed in an alignment study may be due to parallel mutations, on both the organisms studied. Such events obscure the true estimation of the evolutionary distance between the sequences of different individuals. This effect is known as 'homoplasy' (observed sequence similarity that is a result of convergence or parallel evolution, but not direct evolution), and such effects should be corrected so that true evolutionary distances can be ascertained. For correction of 'homoplasy', certain statistical models such as, Substitution models and Evolutionary models are used. Two of the common Substitution models are as follows (Lio & Goldman 1998, Emerson 2001).

Jukes-Cantor Model: According to this model the purines and pyrimidines are substituted with equal probability. Through this model reasonably close sequences can only be analyzed.

Kimura Model: According to this model mutations through transition should be more frequent than mutations through transversion. Accordingly, the differential mutation rates of transitions and transversions are taken in account, which is more realistic. In the case of protein sequencing discrepancies in the calculating the evolutionary distances from an alignment analysis can be corrected through JTT or PAM amino acid substitution matrix. Alternatively, protein equivalents of Kimura model and Jukes-Cantor model can be used for the correction of the evolutionary distances.

Tree Building Process: Several methods are available for development of evolutionary trees. It is important to conduct extensive experiments to generate meaningful data, which is time consuming. The task becomes more exhaustive with the increase in the number of taxa. Brief description of these methods is as follows.

Methods Based on Character: All mutational events accumulated on the sequence are taken in to account in these methods. This helps to retail all information without any loss. It also provide information on 'homoplasy' and ancestral states of the sequences. These methods produce more authentic phylogenetic trees compared to the distance based methods. The 'maximum parsimony' and 'maximum likelihood' methods fall under this category.

Methods Based on Distance: By utilizing these methods, it is possible to calculate the evolutionary distance between sequences, based on the observed distance, after corrections made through different models. They are subdivided in to two algorithms: optimality based and clustering based.

EVALUATION METHOD FOR PHYLOGENETIC TREE

The validity of the newly constructed phylogenic trees should be tested before this can be accepted for further studies. Various statistical tests are available for testing the reliability of the constructed trees (Lio & Goldman 1998). For testing the reliability, Bootstrapping and Jackknifing tests are conducted, while to confirm whether the newly constructed tree is better than the existing ones, the Kishino-Hasegawa test, Bayesian analysis and Shimodaira-Hasegawa tests are adopted. In Bootstrapping test, pieces of varying size of the target sequence to be tested are selected randomly, and a new phylogenetic analysis is performed to generate a tree. To check the robustness of the tree, it is recommended to bootstrap the phylogenetic tree for 500 – 1000 times. Thereafter, the bootstrap results are compared with the original tree. When the branch point score is found to be 90% and above, the predicted tree is considered to be accurate. In the Jackknifing test, half of the data set is used to construct the phylogenetic trees by following the same methodology as that of the original. The Kishino-Hasegawa analysis is used specifically for evaluation of maximum parsimony trees. In this analysis, a t-value is calculated with the following formula, and the then the t-distribution is evaluated by checking whether the values fall within the significant range at <0.05,

$$t = \text{Pa} - \text{Pt} / \text{SD} / \sqrt{n}$$

Where, n = number of informative sites, Pa = average site-to-site difference between the two trees, Pt = total difference of branch lengths of the two trees, SD = standard deviation.

The Bayesian analysis uses Markov Chain MontoCarlo (MCMC) procedure, which is considered to be very fast. The Shimodaira-Hasegawa test is used mostly for Maximum likelihood trees, where the Chi-square test is applied to test the goodness of fit.

DNA BARCODE FOR PHYLOGENETIC STUDIES

Application of DNA barcode in phylogenetic studies has now been a well-established technique. Although analysis of DNA barcodes for animal system has become more relevant than plant systems, it is recognized as a powerful techniques for plant systematics also, particularly in those situations where conventional methods does not provide conclusive results (for details see Chapter 8).

MARKER-BASED MANAGEMENT OF PLANT GENETIC RESOURCES

Usually evaluation of the plant germplasm is carried out by studying phenotypic (morphological) traits. The evaluation process can be complimented through application of marker assisted germplasm evaluation (MAGE) technique. Through MAGE it is possible to define the genetic architecture of germplasm resources by identifying the alleles associated with traits of economic importance. It is possible to characterize germplasm on the basis of genes, genotypes and genomes through MAGE, which is more informative than classical morphological data. MAGE also can be used to resolve the issues such as, genetic diversity, duplication, identity, contamination and integrity of regeneration through. In the case of vegetatively propagated plants, determination of ploidy level is not possible through phenotypic observations. However, molecular markers can be used to resolve the ploidy levels on such plants e.g. potato, sweet potato, taro, sugarcane etc. Information generated through analysis of molecular markers can also be used for identification of useful genes, contained in any collection of germplasm. Thus, MAGE can be used for various applications of germplasm management such as, acquisition, distribution, maintenance, identifying germplasm redundancy, genetic shifts, screening for novel genes, construction of heterotic groups, and specialized applications.

Molecular markers linked with known genes (alleles) associated with good agronomic traits can be used to identify, select and manage these genes (alleles) or traits. Molecular markers that represent multiple loci, showing multiple bands, such as AFLP or RAPD, are normally difficult to link to specific loci or alleles. Therefore, in such cases it is important to convert them to locus specific markers such as SSRs, STSs or SNPs. Markers not identified to specific region of the chromosomes can also be used for background examinations. The only criterion that has to be fulfilled by such markers is high genome-wide polymorphism (Tatikonda et al. 2009).

Multivariate analysis of the DNA genotypes is the basis for MAGE process. The number of markers to be used to conduct a MAGE successfully shall depend on the type of marker being used. Roughly, the number of markers that are required to detect linkage disequilibrium between any two markers in a genome should be the guiding principle to find out the number of markers required for a genome-wide MAGE. This estimate shall vary on the basis of the crop. With large scale development of SNP markers, involving whole genomes, it is expected that all candidate genes derive from germplasm collections can be realistically analyzed in the future (Tautz et al. 2010).

MOLECULAR MARKERS AND GENETIC DIVERSITY

Comparison between organisms is now possible from genomic level, through FISH (fluorescent in situ hybridization), to the level of single nucleotides, through DNA sequencing and SNP. The purposes for which molecular markers are used for the study of genetic diversity are as follows.

1. To find the frequency of deviation of individual loci in different genotypes,
2. To characterize the molecular variation of the loci within and between population,
3. To classify the germplasm accessions on the basis of genetic distance and construct phylogenetic trees,
4. To determine the heterozygotic groups for hybrid crops,
5. To analyze the correlation between the genetic distance and performance of the hybrid, heterosis and specific combining ability,
6. To compare the genetic diversity among different groups of population of a crop's germplasm.

Knowledge about the range of diversity and the genetic structure is important for better management of the germplasm of any crop. Although there exist differences in opinion about what level of diversity should be maintained, it has been generally agreed that the first priority should be given to taxon-specific markers and then to estimate the degree of differentiation between the units.

Diversity studies are usually made by using molecular markers that are neutral, i.e. not present within the expressed regions of DNA. On the other hand, relation between the molecular markers and quantitative traits has not been studied thoroughly. This is one area where emphasis should be given for more effective utilization of molecular markers for assessment of biodiversity and their conservation. Application of molecular markers has enhanced our knowledge on spatial and temporal patterns of genetic variations and also about the evolutionary mechanisms that generate variation.

It has been confirmed that the genetic diversity in modern varieties have declined compared to wild relatives and land races. This is a matter of concern and therefore, more emphasis should be given for conservation of land races and wild relatives for their future use (Zhang et al. 2012).

Factors Responsible for Genetic Diversity

The degree of polymorphism varies between species and between loci. Several factors have been identified which contribute towards genetic diversity in plants. These include, mutation rate, population size, outcrossing, recombination, positive trait selection, line selection, diversification selection, balancing selection, background selection, population structure, sequence error and PCR problems. Kimura in 1969 defined the natural theory of evolution which states that the level of polymorphism (θ) should be the product of the effective population size (N_e) and the rate of mutation (μ) with $\theta = 4 N_e\mu$. However, in plants there exists little or no empirical proof. Background selection is considered to be one of the major factors which contribute towards nucleotide diversity. Strong selection pressure contributes towards decreasing nucleotide diversity in plants species. In balancing selection, maintenance of multiple alleles is favored, which also contributes towards increasing diversity. It has been suggested that intra-genomic recombination and outcrossing rates also defines the diversity in plants.

Measurement of Genetic Diversity

Measurement of genetic diversity is important for development of strategies for effective germplasm management. The traditional methods for the estimation of diversity are often found to have several limitations. The DNA-based technologies can circumvent the problem, as such techniques can directly identify the diversity at the gene level and its expressed products, instead of using anonymous sequence differences, among accessions (Hogland 2009, Avolio et al, 2012) . This will not only provide an indication of genetic diversity and relationships among accessions, but also increase in the content of information for the accessions. The allelic diversity can be expressed in the following manners:

1. Percentage of polymorphic loci (number of polymorphic loci/ total number of loci analyzed x 100,

2. Mean number of alleles per locus (total number of alleles detected/ the number of loci assayed x 100,
3. Total gene diversity (H) or average expected heterozygosity which can be calculated by the following formula:

$$H = 1 - \sum_{lj=1}^{m} \sum P_{ijlm}^{2}$$

where, P_{ij} = frequency of the jth alleles at the ith of m loci, and

4. Polymorphic information content (PIC), which refers to the relative value of each marker with respect to the amount of polymorphism exhibited and can be estimated by the following formula:

$$PICi = 1 - \sum_{j=1}^{m} P_{ij}^{2}$$

where, Pij = frequency of the jth alleles at the ith of m loci.

The number of loci involved and sample size (which include: number of progeny studied per plant and number of plants studied per population, number of populations studied per taxon) shall affect the variances of all the above mentioned estimates.

While applying molecular marker data, it is important to select the proper similarity s or dissimilarity coefficient ($d = 1- s$). The selection of these matrixes is dependent on the, objectives of the study, properties of the marker system, genology of the germplasm, operational taxonomic unit, and preconditions for multivariate analysis.

CLASSIFICATION OF GERMPLASM

The strategy for numerical classification of the germplasm should be such that it produces maximum variability among different groups, while minimum variability is produced within each group. It has been observed that two-stage sequential clustering strategy, which uses both continuous and categorical variables, tends to form more homogeneous groups of individuals compared to other clustering strategies. The three-way data having genotype x environment attributes can be analyzed by adopting the sequential clustering strategies.

In cluster analysis, information on genetic markers, germplasm collections, and taxa are arranged in hierarchy, known as dendrogram or phenogram, by an agglomerative algorithm. The hierarchical information obtained through this analysis is highly dependent on the algorithm used and similarity matrix. The arithmetic mean (UPGMA or WPGMA) is the most frequently used clustering methods. Commercial packages such as NTSYS (https://www.. exetersoftware.com/cat/ntsyspc/ntsyspc.html), and PowerMaker (https:// statgen.ncsu.edu/powermaker/.) can be used for this purpose.

Germplasm accessions can be classified in to different heterotic groups, with high level of similarity within each group, through DNA-based markers. Divergence at molecular marker loci can also be used for heterotic group formations.

PHYLOGENETICS

Molecular taxonomical studies have substantially improved our understanding about the primary, secondary and tertiary gene pools in many crop plants. Molecular evolutionary studies shall assist to identify crop ancestors, existence of any genetic bottlenecks, and to introduce useful variations. It will also help to discriminate between recently developed hybrids from intermediate taxa originated from convergent-parallel evolution and recombinational speciation.

During recent past, several studies have been made to re-evaluate the taxonomic relationships for many crops, through molecular markers and genomic sequences that cover specific traits such as: flower and seed production, response to photoperiod, period of seed maturity etc. Such studies have helped to focus on areas of the genome where information on diversity has special importance for that trait.

The whole genome analysis can also be used for phylogenetic diversity studies and construct phylogenetic trees thereof. Based on the gene order, gene content, evolutionary distances between orthologous, and concatenated alignments of sequences of orthologous proteins, it is possible to construct the phylogenic trees. Encouraging results have been obtained from all such studies, as additional information can be generated to understand phylogenetic relationships.

MOLECULAR MARKERS FOR IDENTIFICATION OF GERMPLASM REDUNDANCIES

In almost all germplasm collections, there exist duplications, which need to be excluded, so that more and more unique accessions can be accommodated. Duplications generally accumulate due to the different names given to the same cultivars. Pedigree-related cultivars, isogenic lines and sibling lines may represent another type of duplications, as they are genotypically duplicated at most of the loci. All such duplications can easily be identified through molecular techniques.

Frequencies of alleles at all genetic loci can be compared in all germplasm collections, and thereby it is possible to identify the distinctive alleles, allele combination, and their frequency patterns for a given population. Specific regions of the chromosome which contain highest changes in allele frequency between collections can also be identified. Such analysis provides information about specific genomic regions where selection pressure resulted in to specific allele combinations that distinguishes a group of accessions with more diversity from those with less diversity. Modern varieties are developed by using small number of superior accessions and therefore tend to loose diverse alleles that may be important for future breeding strategies. Important genes/alleles can be recovered from those germplasm accessions which are considered to be wild or ancestors of the crop plants. Table 2 shows the list of lost or underrepresented alleles with frequencies less than 2% in rice cultivars of USA, which clearly indicate that rice cultivars in USA have been developed from a small number of superior accessions.

MOLECULAR MARKERS TO STUDY GENE FLOW AND GENETIC DRIFTS

During medium and long term storage genetic profile of the germplasm may change. The changes may be due to mutations, chromosomal aberrations and shift in gene frequencies. Germplasm of self-pollinated crops may also contain some amount of heterozygosity which acts as buffer for genetic diversity and genetic drifts. Proper identification and selection of heterogeneous accessions may help to develop regeneration strategies without loosing allelic diversity. In general, traditional cultivars possess higher levels of heterozygosity.

Table 2. List of markers (SSR and RFLP) lost or underrepresented in the rice cultivars of USA but most frequent in World germplasm collections

Chromosome number	Marker	Size (bp)	World frequency (%)	USA frequency (%)
1	CDO118	17000	49.5	1.6
1	RM259	156	20.4	0
2	RM207	131	21.7	0
3	RM7	181	31.5	1.6
5	RM233B	138	41.9	1.7
7	RM11	143	29.6	1.6
9	CDO1058	4100	55.9	1.6
9	RM205	123	46.4	1.6
9	RM219	216	21	0.9
9	RM257	149	21.5	0
12	RG901X	4600	43.6	1.6

Often deviations in allelic frequencies in the offspring compared to their parental accessions have been observed during regeneration of germplasm, due to genetic drifts. This may result in to fluctuations in the frequency of alleles from generation to generation. Gene flow often leads to either diversification or assimilation of crops and a combination of both. To monitor the gene flow among cultivars over long period of time (vertical flow) or in short period (horizontal flow), molecular markers can effectively be used.

In vitro cultures of plant cells and tissue can also lead to genetic shifts. Genetic stability of the germplasm maintained *in vitro* through tissue culture usually been monitored by cytological analysis of the chromosomes, as it was believed that that primary causes of somaclonal variations is due to change in the number and morphology of the chromosomes. However, it has now been shown that mobilization of the transposable elements can also create such variations. Therefore, it has been suggested that genetic stability of tissue culture materials should be analyzed through transposon-based molecular markers.

ALLELE MINING

Allele mining can be used to identify or capture diversity that might not exist in a particular germplsm pool of the existing breeding lines. In other words, novel alleles hidden in genetic diversity can be utilized through allele mining. Two approaches are used for allele mining: re-sequencing, TILLING (Target Induced Local Lesions In Genomes) and EcoTILLING. The foundation of re-sequencing method is based on whole genome sequencing using gene-based markers. The basic challenge is to identify and establish the alleles which are functionally different from the wild type alleles, and that the new alleles are functionally beneficial for the trait targeted. The methods usually used to ascertain this include: marker-based backcrossing (MABC), transient expression assay, transformation, and association analysis. Experiments are now being carried out to establish relationships between haplotype SNPs and changes in the phenotypes, which might be useful to the breeders. Such studies have led to development of bioinformatics tools that can compare variations in sequences with variations in protein, in order to predict which haplotype SNP variants have the maximum likelihood of providing benefit to the targeted trait. The SNPs in promoters and non-coding regions may also contribute towards predicting beneficial phenotypic traits. TILLING combines an efficient chemical mutagenesis technique with a sensitive screening technique that can identify point mutations in a target gene. The method identifies the formation of DNA heteroduplexs and bubbles at the mismatch position of two DNA strands. The sites are cleave by single stranded nucleases and analyzed. EcoTILLING is a modified version of TILLING, which is used to study the allelic variations in natural populations, instead of induced variations (Comai & Henikoff 2006, Kurowska et al. 2011).

Allele mining has also potential for using in diverse core sub-sets of germplasm accessions including developed lines and wild relatives. If a gene of interest is identified and sequenced, the same gene can be re-sequenced in all other individuals of a sub-set. Any change in the DNA sequence found in the germplasm accessions can be considered as a new allele, and should be evaluated for altered phenotype, and if found to be beneficial, should be included in the breeding programs. Such alleles may not ever be found through conventional screening, either because its effect may be masked in unsuitable genetic background, or its effect may be too small to be detected without employing specific conditions.

PURIFICATION OF COLLLECTED GERMPLASM

During conservation of germplasm, the off-type plants (phenotypically different from the typical type or plants developed by breeder) may arise and contaminate the main accessions. When the proportion of off-type plants become sufficiently high, it becomes difficult to differentiate them from the typical plants. Uniformity of the crop will be reduced due to presence of off-type plants and that may affect the quality and productivity of the crop. If phenotypically distinguishable, off-type plants can easily be rouged out, in small populations. On the other hand, there could be off-type plants, which are genetically different but not possible to distinguish phenotypically. By adopting molecular technologies it is possible to easily distinguish both phenotypic and genotypic off-types plants from the typical types. It has been suggested that high-resolution molecular markers such as SSRs and SNPs should be used to distinguish between two plants having similar genetic backgrounds. Thus molecular markers can effectively contribute towards the purification process of the germplasm collections.

FUTURE PROSPECTS

At present large number of markers are available for phylogenetic studies. However, one should not be confined within these genes, as there is a need to identify many more molecular markers to study phylogenetic relationships accurately, in the huge plants kingdom. The number of molecular marker genes can be increased through nuclear genome sequencing and expressed sequence tag (EST) projects. It is also important to develop and improve the algorithms applicable to various analytical software. Genes controlling the physiological aspects of the plants such as: salt tolerant genes, heat shock genes, receptor genes, homeotic genes, cell division genes etc. have great potential to be used as marker genes, as they show great homology over a large number of plants. With time more and more sequence data from diverse organisms shall be available. Accordingly, it will be necessary to develop software to analyze and link different databases. Processing the large data into useful classification concepts shall be the biggest challenge.

CONCLUSION

Phylogeny depicts the history of decent of a group of taxa such as species from their common ancestors, which include the order of branching and the time of divergence. In molecular phylogeny, the relationships between organisms or genes are studied through comparison of homologues of DNA or protein sequences. Similarities and dissimilarities among the sequences shall indicate genetic closeness or divergence, which occur due to molecular evolution during the course of time. By comparing homologous molecules from different organisms, it is possible to establish their degree of similarity, thereby establishing or revealing a hierarchy of relationship, and help to create a phylogenetic tree. Combination of morphological traits and molecular markers can be used to determine the phylogenetic relationships of organisms. The efficiency of identification of duplications and maintenance of germplasm has been greatly enhanced with the application of molecular markers.

REFERENCES

Avolio, M., Beaulieu, J. M., Lo, E. Y. Y., & Smith, M. D. (2002). Measuring genetic diversity in ecological studies. *Plant Ecology*, *213*(7), 1105–1115. doi:10.100711258-012-0069-6

Comai, L., & Henikoff, S. (2006). TILLING: Practical single-nucleotide mutation discovery. *The Plant Journal*, *45*(4), 684–694. doi:10.1111/j.1365-313X.2006.02670.x PMID:16441355

Emerson, B. C., Ibrahim, K. M., & Hewitt, G. M. (2001). Selection of evolutionary models for phylogenetic hypothesis testing using parametric methods. *Journal of Evolutionary Biology*, *14*(4), 620–631. doi:10.1046/j.1420-9101.2001.00306.x

Grechko, V. V. (2002). Molecular DNA markers in phylogeny and systematics. *Russian Journal of Genetics*, *38*(8), 851–868. doi:10.1023/A:1016890509443 PMID:12244688

Hogland, J. (2009). How to measure genetic variation. In J. Hogland (Ed.), Evolutionary conservative genetics. Oxford Scholarship Online. doi:/9780199214211.003.0002 doi:10.1093/acprof:oso

Kurowska, M., Daszkowska-Golec, A., Gruszka, D., Marzec, M., Szurman, M., Szarejko, I., & Maluszynski, M. (2011). TILLING- a shortcut in functional genomics. *Journal of Applied Genetics*, *54*(4), 371–390. doi:10.100713353-011-0061-1 PMID:21912935

Mondini, L., Noorani, A., & Pagnotta, M. A. (2009). Assessing plant genetic diversity by molecular tools. *Diversity (Basel)*, *1*(1), 19–35. doi:10.3390/d1010019

Nadia, H. (2011). Identification of plant species using traditional and molecular-based methods. In R. E. Davis (Ed.), *Wild Plants: Identification, uses, and conservation* (pp. 1–66). Nova Science Publications, Inc.

Soltis, D. E., Soltis, P. S., & Doyle, J. J. (2012). *Molecular systematics II: DNA sequencing*. Springer.

Soltis, P. S., Soltis, D. E., & Doyle, J. J. (1992). *Molecular systematics in plants*. Chapman and Hall. doi:10.1007/978-1-4615-3276-7

Tatikonda, L., Wani, S. P., Kannan, S., Beerelli, N., Sreedevi, T. K., Hoisington, D. A., Devi, P., & Varshney, R. K. (2009). AFLP-based molecular characterization of an elite germplasm collection of *Jatropha curcas* L.: A biofuel plant. *Plant Science*, *176*(4), 505–513. doi:10.1016/j.plantsci.2009.01.006 PMID:26493140

Tautz, D., Ellegren, H., & Weigel, D. (2010). Next generation molecular ecology. *Molecular Ecology*, *19*, 1–3. doi:10.1111/j.1365-294X.2009.04489.x PMID:20331765

Tiwari, J. K., Singh, B. P., Gopal, J., Poonam, P., & Patil, V. U. (2013). Molecular characterization of the Indian Andigena potato core collection using microsatellite markers. *African Journal of Biotechnology*, *12*, 1025–1033.

Zuckerkandl, E., & Pauling, L. (1965). Evolutionary divergence and convergence in proteins. In V. Bryson & H. J. Vogel (Eds.), *Evolving genes and proteins* (pp. 97–165). Academic Press. doi:10.1016/B978-1-4832-2734-4.50017-6

ADDITIONAL READING

Abdurakhmonov, I. Y., & Abdukararimov, A. (2008). Application of association mapping to understand the genetic diversity of plant germplasm resources. *International Journal of Plant Genomics*, *17*, 4927–4938. PMID:18551188

Arif, I. A., Bakir, M. A., Khan, H. A., Ahmed, A. H., Al Homaidan, A. A., Bahkali, A. H., ... Shobrak, M. (2010). A brief review of molecular techniques to assess plant diversity. *International Journal of Molecular Sciences*, *11*(5), 2079–2096. doi:10.3390/ijms11052079 PMID:20559503

Aslam, S., Tahir, A., Aslam, M. F., Alam, M. W., Shedayi, A. A., & Sadia, S. (2017). Recent advances in molecular techniques for the identification of phytopathogenic fungi – a mini review. *Journal of Plant Interactions*, *12*(1), 493–504. doi:10.1080/17429145.2017.1397205

Dong, W., Liu, J., Yu, J., Wang, L., & Zhou, S. (2012). Highly variable chloroplast markers for evaluating plant phylogeny at low taxonomic levels and for DNA barcoding. *Public Library of Science (PLoS). ONE*, *7*(4), e35071. doi:10.1371/journal.pone.0035071 PMID:22511980

Elshire, R. J., Glaubitz, J. C., Sun, Q., Pollard, J. A., Kawamoto, K., Buckler, E. S., & Mitchell, S. E. (2011). A robust, simple genotyping-by-sequencing (GBS) approach for high diversity species. *Public Library of Science (PLoS). ONE*, *6*(5), e19379. doi:10.1371/journal.pone.0019379 PMID:21573248

Fan, F., Cui, B., Zhang, T., Borges, R. S., Matioli, F. F., Fontes, M. R. M., & Marino, C. L. (2014). LTR-retrotransposon activation, IRAP marker development and its potential in genetic diversity assessment of masson pine (*Pinus masoniana*). *Tree Genetics & Genomes*, *10*(1), 213–222. doi:10.100711295-013-0677-x

Hillis, D. M., & Moritz, C. (1990). *Molecular systematics*. Sinauer Associates.

Kalendar, R., Antonius, K., Smykal, P., & Schulman, A. H. (2010). IPBS: A universal method for DNA fingerprinting and retrotransposition isolaton. *Theoretical and Applied Genetics*, *121*(8), 1419–1430. doi:10.100700122-010-1398-2 PMID:20623102

Kalendar, R., Flavell, A. J., Ellis, T. H., Sjakste, T., Moisy, C., & Schulman, A. H. (2011). Analysis of plant diversity with retrotransposen-based molecular markers. *Heredity*, *106*(4), 520–530. doi:10.1038/hdy.2010.93 PMID:20683483

Karp, A., Seberg, O., & Buiatti, M. (1996). Molecular techniques in the assessment of botanical diversity. *Annals of Botany*, *78*(2), 143–149. doi:10.1006/anbo.1996.0106

Li, Z., De-La-Torre, A. R., Sterck, L., Cancvas, M., Avila, C., Merino, C., ... de Peer, Y. V. (2017). Single-copy genes as molecular markers for phylogenomic studies in seed plants. *Genome Biology and Evolution*, *9*(5), 1130–1147. doi:10.1093/gbe/evx070 PMID:28460034

Lin, K. H., Lai, Y. C., Li, H. C., Lo, S. F., Chen, L. F. O., & Lo, H. F. (2009). Genetic variation and its relationship to root weight in the sweet potato as revealed by RAPD analysis. *Scientia Horticulturae*, *120*(1), 2–7. doi:10.1016/j.scienta.2008.09.008

Liu, W., Shahid, M. Q., Bal, L., Lu, Z., Chen, Y., Jiang, L., ... Lu, Y. (2015). Evaluation of genetic diversity and development of a core collection of wild rice (*Oryza rufipogon* Griff.) populations in China. *Public Library of Science (PLoS). ONE*, *10*(12), e0145990. doi:10.1371/journal.pone.0145990 PMID:26720755

Liu, Y., Xue, J. Y., Wang, B., Li, L., & Qiu, Y. L. (2011). The mitochondrial genomes of the early land plants *Treubia lacunose* and *Anomodon rugelli*: Dynamic and conservative evolution. *Public Library of Science (PLoS). ONE*, *6*, e258336.

Lu, Y., Ran, J. H., Guo, D. M., Yang, Z. Y., & Wang, X. Q. (2014). Phylogeny and divergence times of gymnosperms inferred from single-copy nuclear gene. *Public Library of Science (PLoS). ONE*, *9*, e107679. doi:10.1371/journal.pone.0107679 PMID:25222863

McCullum, C. M., Comai, L., Greene, E. A., & Henikofff, S. (2000). Targeting induced local lesions in genomes (TILLING) for plant functional genomics. *Plant Physiology*, *123*(2), 439–442. doi:10.1104/pp.123.2.439 PMID:10859174

Nadeem, M. A., Nawaz, M. A., Shahid, M. Q., Doğan, Y., Comertpay, G., Yıldız, M., Hatipoğlu, R., Ahmad, F., Alsaleh, A., Labhane, N., Özkan, H., Chung, G., & Baloch, F. S. (2018). DNA molecular markers in plant breeding: Current status and recent advancements in genomic selection and genome editing. *Journal of Biotechnology & Biotechological Equipment*, *32*(2), 261–285. doi:10.1080/13102818.2017.1400401

Naeem, M., Ghouri, F., Shahid, M. Q., Iqbal, M., Baloch, F. S., Chen, L., ... Rana, M. (2015). Genetic diversity in mutated and non-mutated rice varieties. *Genetics and Molecular Research*, *14*, 17109–17123. doi:10.4238/2015. December.16.11 PMID:26681058

Naim, D. M., & Mahboo, S. (2020). Molecular identification of herbal species belonging to genus *Piper* within family Piperaceae from northern Peninsular Malaysia. *Journal of King Saud University*, *32*(2), 1417–1426. doi:10.1016/j. jksus.2019.11.036

Navascues, M., & Emerson, B. C. (2005). Chloroplast microsatellite: Measures of genetic diversity and the effect of homoplasy. *Molecular Ecology*, *14*(5), 1333–1341. doi:10.1111/j.1365-294X.2005.02504.x PMID:15813774

Nawaz, M. A., Yang, S. H., Rehmen, H. M., Baloch, F. S., Lee, J. D., Park, J. H., & Chung, G. (2017). Genetic diversity and population structure of Korean wild soybean (*Glycine soja* Siebb. and Zucc.) inferred from microsatellite markers. *Biochemical Systematics and Ecology*, *71*, 87–96. doi:10.1016/j. bse.2017.02.002

Poczai, P., Varga, I., Laos, M., Cseh, A., Bell, N., Valkonen, J. P. T., & Hyvonen, J. (2013). Advances in plant generated and functional markers: A review. *Plant Methods*, *9*(1), 6–18. doi:10.1186/1746-4811-9-6 PMID:23406322

Setoguchi, H., Mitsui, Y., Ikeda, H., Nomura, N., & Tamura, A. (2009). Development and characterization of microsatellite loci in the endangered *Tricyrtis ishiiana* (Convallariaceae), a local endemic plant in Japan. *Conservation Genetics*, *10*(3), 705–707. doi:10.100710592-008-9620-3

Song, Q., Hyten, D. L., Jia, G., Quigley, C. V., Fickus, E. W., Nelson, R. L., & Cregan, P. B. (2015). Fingerprinting soybean germplasm and its utility in genome research. *G3: Genes, Genomes. Genetics*, *5*, 1999–2006. PMID:26224783

Uzun, A., Yesiloglu, T., Aka-Kacar, Y., Tuzcu, O., & Gulsen, O. (2009). Genetic diversity and relationships within citrus and related genera based on sequence related amplified polymorphism markers (SRAPs). *HortScience*, *121*, 306–312.

Wang, Y., Ghouri, F., Shahid, M. Q., Naeem, M., & Baloch, F. S. (2017). The genetic diversity and population structure of wild soybean evaluated by chloroplast and nuclear gene sequences. *Biochemical Systematics and Ecology*, *71*, 170–178. doi:10.1016/j.bse.2017.02.008

Wang, Y., Shahid, M. Q., Ghouri, F., Lin, X., Yuan, C., Qi, G., ... Dong, Y. (2015). Evaluation of the geographical pattern of genetic diversity of *Glycine soja* and *Glycine max* based on four single copy nuclear gene loci for conservation of soybean germplasm. *Biochemical Systematics and Ecology*, *62*, 229–235. doi:10.1016/j.bse.2015.09.006

Xu, Y. (2014). *Molecular plant breeding*. CAB International.

Zeng, W., Zhou, B., Lei, P., Zeng, Y., Liu, Y., Liu, C., & Xiang, W. (2015). A molecular method to identify species of fine roots and to predict the proportion of a species in mixed samples in subtropical forests. *Plant Science*. Advance online publication. doi:10.3389/ fpls.2015.00313 PMID:25999977

Zhang, N., Zeng, L., Shan, H., & Ma, H. (2012). Highly conserved low-copy nuclear genes as effective markers for phylogenetic analysis in angiosperms. *The New Phytologist*, *195*(4), 923–937. doi:10.1111/j.1469-8137.2012.04212.x PMID:22783877

APPENDIX

1. Explain the molecular clock hypothesis.
2. Explain the terms "horizontal gene inheritance" and "vertical gene inheritance".
3. What are the advantages of using molecular markers in phylogenetic studies?
4. Describe the properties of ideal marker genes for phylogenetic studies.
5. Which ribosomal nuclear genes are used for phylogenetic studies? State their characteristics.
6. Which mitochondrial genes are used for phylogenetic studies? State their characteristics.
7. Which chloroplast genes are used for phylogenetic studies? State their characteristics.
8. Describe the steps involved in construction of phylogenetic trees.
9. Explain how the validity of the newly constructed phylogenetic tree can be tested?
10. Explain how molecular markers can be used for efficient management of genetic resources?
11. Explain how genetic diversity can be measured through molecular markers?
12. Explain how molecular markers can be used to identify germplasm redundancy?
13. Explain how allele mining can be used to identify novel alleles hidden within the existing breeding lines?

Chapter 6
Plant DNA Barcoding

ABSTRACT

DNA barcoding has evolved as an effective species identification tool in diverse areas such as phylogeny, ecology, population genetics, and biodiversity. In this approach, a short DNA sequence from a standardized locus is employed for species identification. The technique is simple, time and cost effective, and accurate. Selection of correct DNA marker is the main criterion for success in DNA barcoding. Compared to animals, DNA barcoding is more difficult in plants, as there are multiple consensuses about selection of barcoding markers for plants DNA barcoding. Some common plant barcoding markers are chloroplast genes such as matK, rbcL, ropC1, ropB, and trnL; chloroplast intergenic specers trnH-psbA, atpF-atpH, and pdbK-psbI; and the nuclear ribosomal internal transcribed spacer (ITS). These markers can be used alone or in combinations with other markers or spacers. In this chapter, the basic requirements, selection of markers, databases, advantages, and limitations of DNA barcoding have been discussed.

INTRODUCTION

Identification of the biodiversity of our planet has been a major challenge for the biologists. The system of classification of biodiversity was initiated by Carl Linnaeus in about 250 years ago and so far taxonomists have been able to describe 1.7 million species. However, it represents a tiny fraction of the actual biodiversity present. Although traditional morphology based identification and classification is of sole importance, it becomes challenging

DOI: 10.4018/978-1-7998-4312-2.ch006

task due to non-availability of expert taxonomists. Identification becomes much difficult, when the sample to be identified is not in intact shape or when a sample contains mixture of minute organism with very few identifiable traits. Hebert et al. (2003) reported that a small fragment of DNA can be successfully employed for identification of a species. Although the term barcoding was used in this sense for the first time by Arnot et al (1993), the revolutionary finding of Hebert et al. (2003) of using the 648 base pair-long mitochondrial gene, *CO1* in animals as marker formed the basis of DNA Barcoding. DNA Barcoding can thus be defined as an approach where a short DNA sequence or sequences from a standardized locus (or loci) are employed as species identification tool.

This system circumvents the lapses associated with morphology based identification systems by providing a better taxonomic resolution. Moreover, it has gained considerable validation as a suitable tool to identify species, delimit species and define species boundaries. With the immense progress in DNA barcoding research area, the pros and cons have been emerged. DNA barcode based Next Generation Sequencing (NGS) is becoming new platform to enhance the level of identification and classification owing to its cost-effective and parallel sequencing at a single run.

The overwhelming popularity of DNA Barcoding is not only because of its discriminating power but also due to the availability of three modern innovations techniques: (1) molecularization: use of different molecular markers which are variable and thus useful as a discriminator; (2) computerization: the revolution in data storage, analysis, sharing and annotation, and (3) standardization: application of the idea to vast groups of organisms which are not deeply related.

BARCODING PROJECT HAS MULTIPLE COMPONENTS

The Barcode of Life Database (http://www.barcodeoflife.org/) recognises four basic components involved in a barcoding project:

1. **Specimens**: Specimen collection, storage still processing and proper description etc. are initial processes involved in a barcoding project. Stored identified specimens in museums, zoos, herbaria and seed bank etc. are the guidance source for description and identification of specimens.
2. **Analysis in Laboratory**: Different barcoding protocols are necessary to obtain DNA barcode sequences from different specimens. The first

task is to find an appropriate method of DNA isolation which is based on the type of the specimen to be processed. The second task is to select appropriate primers and PCR condition. This is followed by amplification of the region of interest and DNA sequencing. The process can be completed in few hours within a reasonable cost. One has to submit the sequence data obtained to a barcoding database for subsequent analysis.

3. **The Database**: The most crucial component of the Barcoding project is a common platform of a public reference library which is a repository of identified barcodes. This database can be used as a guide to identify unknown specimens based on known specimen information. Currently two major barcode databases serve this purpose.

The International Nucleotide Sequence Database (INSD): This is a collaborative approach maintained for a very long time by three leading sequence database groups of world which are GenBank in the U.S., the European Molecular Biology Lab (EMBL) in Europe and the DNA Data Bank of Japan (DDBJ). They have collaboration with Barcoding initiative CBOL (The Consortium for the Barcode of Life) and accept the CBOL's data standards for barcode records.

Barcode of Life Database (BOLD): This is a huge barcoding database which guides barcoding researchers with information of existing projects, collection process, sample processing and primer information and to analyse barcoding data. This successful venture is maintained by University of Guelph in Ontario.

4. **Analysis of Barcoding DATA**: The new barcodes isolated are identified by finding the closest matches in the reference record present in the barcoding databases. CBOL is working to improve the ways of barcoding data analysis, display and use and arranged a Data Analysis Working Group.

ADVANTAGES OF DNA BARCODING

The advantages of DNA barcoding are as follows:

1. **Simple:** As evident from the components described above, DNA barcoding is a simple process where information and guidance about all

the processes involved like DNA isolation, primers, PCR condition, data analysis etc. are in standardised form and available in public databases.

2. **A Multidisciplinary and Technically Advanced Approach:** The prerequisite of DNA barcoding approach is to assemble the reference DNA barcode data which in turn, depends on the conventional methods of DNA isolation, amplification followed by sequencing. The reference library is extremely useful in rapid identification of low taxonomic level taxa using specific short-DNA sequences i.e. mini-barcode 100 bp, 300 bp.

An increasing number of techniques (e.g. silicon-based microarrays, nylon-membrane-based macroarrays, bio-barcode amplification assay (BCA), fluorescent DNA barcode-based immunoassay and the Luminex system of DNA-tagged polystyrene beads (patented) sorted by flow cytometry) have recently been developed as a next step of DNA barcoding.

High-throughput sequencing enables the most efficient means of rapid barcode-based species identification (in case of mixed samples such as stomach contents, food, and blood or water columns). SASI-Seq (Sample assurance Spike-Ins) is one of the simple and inexpensive approaches to check the samples quality. Sample mix up or identification of cross-contaminants could be easily achieved through high throughput sequencing methods. The full viral genome sequencing has also been carried out using multi-faceted approach that includes nucleic acid preparation, Illumina Seq and a novel iterative sequence classification algorithm.

Moreover, the use of cutting edge sequencing techniques in metagenomics could be promising in DNA barcoding as a new approach called metabarcoding. It can be used for exploration of both culturable as well as non-culturable microbes. Conservation of entire morphological reference for species is one of the conditions for submission of data to BOLD database for which new techniques have been developed for extraction of non-destructive DNA from the specimens.

3. **Studying Species Biology and Ecology**: DNA barcoding has emerged as an effective tool in ecology and species biology. This approach is now widely used for understanding species interactions and detecting the presence of elusive and endangered species. For example, long range wolf colonization of France and Switzerland has been found out using extracted DNA barcodes from faeces and hair samples.

Moreover, the tool is widely used to elucidate the transmission pathways to study the interactions among the beetles (Lecythidaceae) and their endosymbiotic yeasts. It allowed the detection of North American bullfrog (*Rana catesbeiana*), one of the world's worst invasive species in southwest France using water samples from ponds. The ecology and biology of fungus, invasive plant species and microbes has also been studied using this powerful tool. Thus, DNA barcoding has immense potential to solve important ecological problems.

4. **Accurate**: With some rare exceptions, DNA barcoding is generally accepted as an accurate tool for species identification. It is a regulatory tool used by the U.S. FDA.
5. **Rapid:** Once the specimen to be worked with is available, getting a barcode sequence is a matter of few hours. Barcoding service providers like CCDB provides species identification in two-three days.
6. **Cost Effective**: Getting a barcode within a reasonable cost is also a factor for wide acceptance of it. Sequencing services are rapid and reasonable and other factors involved can be managed in a reasonable amount of money.

DNA BARCODING AND ITS SYMBOLOGY

DNA barcoding is a rapid and accurate species identification tool that uses short, standardized, conserved gene regions as internal species tags called DNA barcodes. It performs species level identification by sequencing a specific barcode region derived from unknown specimens and comparing with the reference library of sequences of known specimens. Since every species most likely have a unique DNA signature code, this approach is based on the postulate that inter-species variation exceeds intra-species variation.

An ideal DNA barcoding system should serve following criteria:

1. Variation of barcode region between species but nearly identical within species.
2. Enough phylogenetic information for easy identification and classification of unknown species.
3. Robustness in highly conserved regions and highly consistent DNA amplifications and sequencing.
4. Short target gene that can amplify the degraded DNA.

The main goal of DNA barcoding is to identify unknown, cryptic, microscopic and other species with inaccessible or complex morphology as well as discovery of new species. It has proved to be effective amongst various groups of animals such as birds, fishes, spiders, lepidopterans, etc. Apart from this, it has also been applied in plants, bacteria, fungi, mollusks, macroalgae and protists taxonomic classification.

It is a broadly accepted species identification tool for budding researchers and specialists. It also complements taxonomy, molecular phylogenetics and population genetics. The integrated information on species distribution, structure and their genetic diversity will thus enhance the speed and efficacy of local population studies.

Presently used DNA sequences are not amenable for documentation or data storage. Despite the development of variant barcode technology in manufacturing and retailing industries, no such type of barcode has been assessed for its suitability in representing the DNA barcode sequences. In contrast to one dimensional barcodes (1D), two dimensional barcodes (2D) have better potential in DNA barcoding applications. A study reveals that the use of PDF417 digital barcodes has high longevity for digital imaging of DNA sequences. Moreover, various 2D barcodes (e.g. Data Matrix, CodaBlock-F, the Aztec Code, PDF417, PDF417 Truncated, QR2005 code, and QR code) have been evaluated as symbol-logy for barcoding by using five broadly accepted plant and animal barcodes sequences (COI, *rbc*L, *mat*K, ITS2 and *psb*A-*trn*H). Among all, Quick Response (QR) code was reported as the most accurate symbology for DNA barcodes due to its good compression efficiency and ease of scanning and decoding.

MARKER SELECTION FOR DNA BARCODING

Factors to be considered during selection of marker for DNA barcoding are described in the following section.

- **Universality of PCR Amplification:** In a typical barcoding approach, universal primers are used for amplification of the DNA region of interest. Thus the ease of amplification with universal primer is a key factor guiding marker selection. Markers should have conserved flanking region for this purpose (Hollingsworth et al. 2009). Universality also helps in sequencing.

- **Power of Species Differentiation:** Although we want the flanking region to be conserved for universal primer binding, the internal region should be sufficiently variable for proper identification of a species. Low level of intraspecific variation is desirable to reduce confusion (KEW Web).
- **Length:** A DNA marker should be short so that it can be amplified even in sub optimal conditions. Short markers can be sequenced in one reaction as after a particular length poor quality chromatograms are obtained. Also short length marker can be isolated easily from degraded sample (Kew Web).
- **Variability:** Marker should be less variable in length and structure. This reduces sequence quality and creates additional problems (Hollingsworth et al. 2009).
- **Complementation Among Markers:** This is important for marker selection in plants and not in animals because for animals we have the universally accepted CO1 marker. Complementation is necessary in plants to increase the probability of correct identification (PCI) (Erickson et al. 2008). Complementation in plant barcoding includes selection of unlinked marker to address the correlation problem in chloroplast markers. Complementation addresses the problems like low recovery or less sequence variation in one marker by using another marker that does not have these problems.
- **Bioinformatics Analysis:** Markers should be easily alignable, because sequence alignment is a key factor for comparisons and further analyses. They should have more substitutions than insertions and deletions (Erickson et al. 2008).
- **Presence of a Reading Frame:** This is only a desirable quality and does not matter much even if the criteria are not fulfilled. In this approach, the presence of nonsense substitutions can be used for evaluation of the quality of sequencing reactions and sequence editing (chase et al. 2007). This is also helpful for alignment purpose. A list of popular plant DNA Barcoding markers is presented below (Table 1).

DIFFERENT MARKERS USED FOR PLANT DNA BARCODING

The mitochondrial *CO1* gene proved to be useful in DNA barcoding of various vertebrates and invertebrates, but this marker is not very useful for barcoding in plants, because in plants it evolves very slowly and thus cannot provide necessary resolution (Kress and Erickson 2007). Plant barcoding is a complicated procedure with multiple consensuses about selection of barcoding markers. Various strategies have been proposed, such as use of a single chloroplast region, nuclear ribosomal ITS region, which is fast

Table 1. Commonly used DNA Barcoding markers in plants

Gene	Primer	Direction	Sequence 5'-3'
matK	2.1	F	CCTATCCATCTGGAAATCTTAG
	2.1a	F	ATCCATCTGGAAATCTTAGTTC
	5	R	GTTCTAGCACAAGAAAGTCG
	3.2	R	CTTCCTCTGTAAAGAATTC
	matK 390 F	F	CGATCTATTCATTCAATATTTC
	1326R	R	TCTAGCACACGAAAGTCGAAGT
rpoC	1	F	GTGGATACACTTCTTGATAATGG
	2	F	GGCAAAGAGGGAAGATTTCG
	3	R	TGAGAAAACATAAGTAAACGGGC
	4	R	CCATAAGCATATCTTGAGTTGG
rpoB	1	F	AAGTGCATTGTTGGAACTGG
	2	F	ATGCAACGTCAAGCAGTTCC
	3	R	CCGTATGTGAAAAGAAGTATA
	4	R	GATCCCAGCATCACAATTCC
ITS	AB101	F	ACGAATTCATGGTCCGGTGAAGTGTTCG
	AB102	R	TAGAATTCCCCGGTTCGCTCGCCGTTAC
ITS2	ITS-S2F	F	ATGCGATACTTGGTGTGAAT
	ITS4	R	TCCTCCGCTTATTGATATGC
trn H–*psb*	psbAF	R	GTTATGCATGAACGTAATGCTC
	trnH2	F	CGCGCATGGTGGATTCACAATCC
	psbA	R	CGAAGCTCCATCTACAAATGG
	trnH (GUG)	F	ACTGCCTTGATCCACTTGGC
	psbA501f	R	TTTCTCAGACGGTATGCC

Note: Additional information about different primers for barcoding can be retrieved from: http://www.boldsystems.org/index.php/ Public_Primer_PrimerSearch.

evolving, and finally a combination of different regions (Kress et al. 2007, Chase et al. 2007).

Chloroplast Genes

Chloroplast genes comprise most of the plant DNA barcoding markers. They have various advantages such as:

1. They are present abundantly and thus isolation and amplification is easier.
2. Chloroplast primarily contains single copy genes and there are less chance of getting a mixed product
3. The nucleotide substitution rate is conserved in chloroplast, and
4. With more and more chloroplast genome being sequenced, sufficient background information is available for chloroplast genome.

Some important chloroplast barcoding markers are described in the following section.

matK: This gene is now extensively used for Barcoding and probably the best to be used. The *mat*K gene, which was known as *orf*K before is used very commonly as a reliable marker in plant molecular systematics and evolution (Liang and Hilu 1996). This gene is located within two introns of the chloroplast gene *trn*K and approximately 1500 base pairs in length. One interesting fact about this gene is that it is present in parasitic *Epifagus*, a taxon that lost nearly 65% of the chloroplast genes. The retaining of the matK gene proves its functional significance in plants. Even the two exons of the *trn*K gene flanking the *mat*K region were lost, but *mat*K is present with a large deletion in coding region. Comparatively *matK* is one of the fast evolving plastid gene and successfully used for species discrimination among angiosperm species (Lahaye et al. 2008). About universality of *matK* primers, there are mixed reports ranging from successful amplification to intermediate kind of response. Thus a lack of consensus about universality leads to reservations about this locus by some researchers. The use of best currently available 'universal' primer pair for *matK* (3F/1R) on diverse samples sets results in PCR and sequencing success of 70% in angiosperms. Use of a secondary internal primer pair 390F/1326R is reported to increase amplification and sequencing success by 10% more than the existing one.

In gymnosperms, the success rate (83%) is reasonable but in cryptogams it is very limited (10%) and not satisfactory even after use of multiple primer sets. Amplification of matK has been reported to be problematic in many experiments.

rbcL: *rbcL* is the best characterized gene among all chloroplast genes. This gene is approximately 1430 base pairs in length. It is a single copy gene and generally free from length mutations except at the far 3' end. The rate of evolution is conserved. The *rbc*L gene codes for the large subunit of ribulose 1, 5 bisphosphate carboxylase/oxygenase (RUBISCO or RuBPCase). The sequence data of the *rbc*L gene are widely used in phylogenetic study throughout the seed plants. However, the resolution ability of *rbc*L below the family level is often very poor and thus not very suitable for species level phylogeny.

This coding gene has been proposed as a potential plant barcode by many research groups, usually in conjunction with one or more other markers. Large amount of existing sequence in public databases is one of the additional advantages of this region. There are more than 10,000 *rbcL* sequences reported in GenBank. Universality of amplification is the most promising aspect of *rbc*L. Additional improvements in primer designing has made the recovery of high-quality bidirectional sequences possible across land plants (Fazekas et al. 2008).

rpoC1: This is a slow evolving marker and thus bears less potential. However, it was able to discriminate among species in many groups of plants. The universality of this marker with potential of amplification with limited range of PCR conditions and primer sets is a great advantage. Probably it is the only barcoding locus which can be routinely amplified and sequenced using a single primer pair and standard reaction conditions in all samples in all taxonomic groups Hollingsworth et al. 2009).

rpoB: This is regarded as a poor barcoding marker. It is amplifiable with a limited range of PCR conditions and primer sets. Although it is a slow evolving gene, it can discriminate among species in many groups of organisms particularly when used in combination (e.g., *rpoC1+rpoB+mat*K *)* with other locus (Chase et al. 2007).

trnL: (have a P6 loop with great potential) The plastid intron in trnL region has also been suggested as an potential DNA barcoding sequence. However, slow rate of evolution, a genuine problem in chloroplast region leads to poor resolution. It is surprisingly highly conserved even after being a non-coding region. The presence of a small stem-loop structure called P6 loop within this intron bears great potential for species discrimination. P6 has a variable minibarcode of 10-143 base pairs which has conserved priming sites in both sides and thus bears great potential to be used for barcoding of degraded DNA samples (Hollongsworth et al. 2009).

The collaborative approach of RBG, Kew has suggested six plastid coding regions (*acc*D, *mat*K, *ndh*J, *rpo*B2, *rpo*C1, and *ycf5*) as potential plant barcodes (http://www.rbgkew.org.uk/barcoding/index.html), but comparisons of their effectiveness is not available. The three regions, *accD*, *ndhJ*, and *ycf5* were deleted from the list of proposed barcodes because of their absence from certain plant groups. The *ycf5* might have been deleted from the bryophyte plastid genomes and *accD* is missing in the grass family while *ndhJ* is found to be absent from *Pinus* and might be truncated or non-functional in some orchids. Many other regions are proposed but there is no single consensus about using one particular marker like CO1 in animals.

Chloroplast Intergenic Spacers

- ***trn*H-*psb*A:** The non-coding *trn*H-*psb*A intergenic spacer is proposed to be the most viable candidate for a single-locus barcode for land plants (Kress and Erickson 2007). This is one of the most variable non-coding regions in chloroplast genome of the angiosperms and contains the highest percentages of variable sites. Thus a high level of species discrimination is possible through this marker. The potential problem is the presence of high rates of insertions/deletions which creates problem in sequence alignment. Great length variation is seen even within closely related taxa and no shared sequence remains in taxonomically distant taxa.

Very short (less than 300 bp) *psb*A-*trn*H spacer is reported in some taxa, whereas in some genus like in orchids, the spacer is much longer because it contains copies of *rpl22* and *rps19* making it greater than 1,000 bp. The pseudogene *rps19* is also found between *trn*H and *psb*A in maize, rice and wheat. Various representatives of Commelinales, Dioscoreales, Liliales of Commelinales, Dioscoreales, Liliales and Zingiberales also have a copy of *rps19* in this position, but it is absent in representatives of Acorales or Alismatales. Probably the *rps*19-*trn*H cluster was duplicated early in the evolution of monocots and in some monocots a copy of *rps19* is positioned between *trn*H and *psb*A. The length of this spacer is more than 1000 base pairs and this might create a problem while isolating this marker from fragmented or degraded DNA. Although this non-coding region is more rapidly evolving than genes, this is not true always. In some groups of mosses, *mat*K and *rpo*C1 was found to contain more variable positions than this spacer. Two other spacers with potential application are *atp*F-*atp*H and *psb*K-*psbI*. The later one has good discriminatory power.

Combinations of Various Chloroplast Markers and Spacers

As no plant DNA barcoding marker is universally accepted because of various lacunae, a combination of various markers was suggested by different working groups. Although the cost will increase, but this approach has the potential of providing better results because markers will complement each other. The general principle of having a multi-locus barcode has been accepted by CBOL. Some popular combinations used in this approach are as follows.

- ***rbc*L + *mat*K:** The combination of chloroplast *rbc*L and *mat*K marker was suggested as the standard barcode fo r land plants by the CBOL working group (Janzen et al. 2009). This combination is preferred for its universality (*rbc*L), sequence quality, discrimination (*mat*K), and comparatively lower cost involved. Species discrimination was successful in 72 per cent of cases in this CBOL approach. Because of its great potential, this combination was provisionally adopted by the Consortium for the Barcode of Life.
- ***rbc*L + *trn*H-*psb*A:** The species discrimination success was increased to nearly 88 per cent using the combination of highly variable non-coding *trn*H-*psb*A spacer with coding loci like rbcL. This is significant because no single marker has the ability to discriminate among species in a pair in more than 79% of genera.
- ***rpo*C1 + *rpo*B + *mat*K:** This three locus combination was suggested by Chase in 2005. Although slow evolving, the markers *rpo*C1 and *rpo*B are easily amplifiable and provide limited resolution. The *mat*K region complements for it being more variable.
- ***rpo*C1 + *mat*K + *trn*H-*psb*A**: Replacing *rpo*B with *trn*H-*psb*A might provide additional species level resolution. Size variation in *trn*H-

*psb*A which creates problem in alignment and its long length might be a hurdle for isolation from degraded samples.

Nuclear Markers

Theoretically barcoding should be based on nuclear DNA markers because they are inherited from both parents. This might provide much more information than barcoding based on chloroplast DNA markers which are inherited from only one parent. The nuclear ribosomal DNA (nrITS) is used routinely for DNA barcoding. nrITS is also popular phylogenetic marker and has a long record of its use for flowering plants (Baldwin 1992). It is the desired choice when direct sequencing is possible.

Issues with ITS

There are some issues which raise question about wide use of ITS marker. Presence of multiple, divergent paralogs makes nrITS unacceptable as a standard barcoding region across all land plants as it might be potentially misleading. The recovery rate (amplification and sequencing) might also not be satisfactory always. Another minor concern is that nrITS of fungal contaminants might get amplified.

Alternate Sources of Plant DNA Barcoding Markers

Several alternative source of plant DNA markers used for DNA barcoding are discussed in the following section.

- **Complete Plastid Genome**: This alternative was suggested by various research groups (Nock et al. 2011) Getting a complete plastid genome sequence is not that difficult with next generation sequencing facility. The cost and assembly process might be the matter of concern but the potential of this approach cannot be ignored.
- **Low or Single Copy Nuclear Genes:** Nuclear markers for DNA barcoding might be a potential alternative solution. Nuclear markers such as waxy, leafy, alcohol dehydrogenase and phytochrome genes etc. are being tried for phylogenetic studies in various groups with different degrees of success (Small et al. 2004). The potential problems involved in this approach might be mutations in primer site, gene duplication,

insertion of transposable elements, recombination and heterozygosity etc.

DNA BARCODING IN PLANT IS MORE CHALLENGING THAN IN ANIMALS

Finding a good marker in plants is a challenging task as there is no *CO1* like-alternative. Till now, there is no consensus has been made about a particular plant DNA barcoding marker. As mentioned above, various chloroplast genes, introns, spacers, nuclear markers and combination of various markers have been proposed. All of these have limitations. Thus search for a *CO1*-like marker is still desirable.

The low rate of nucleotide substitution in plant mitochondrial genomes restricts the use of *CO1* as a universal plant barcode (Fazekas et al. 2008). In addition various factors in plant like polyploidy, hybridization, uniparental inheritance of chloroplast and mitochondria make it more complicated (Hollingsworth et al. 2011) .

INTERNATIONAL BODIES WORKING FOR DNA BARCODING

Consortia: The largest consortia for DNA Barcoding are as follows:

- ***IBOL*** (the International Barcode of Life Project): IBOL is the largest biodiversity genomics initiative ever undertaken. It includes hundreds of biodiversity experts in the fields of genomics; technologists and ethicists from 25 nations working together to construct a rich DNA barcode reference library. In their first phase of operations (2010-2015), IBOL has a plan to barcode five million specimens representing five lakh species. In the process of construction of the barcode library, other initiatives like biodiversity conservation, monitoring of ecosystem, forensics and agricultural pest control and invasive species etc. are also in focus.
- ***CBOL*** (the Consortium for the Barcode of Life): CBOL is an international initiative with a goal of developing DNA barcoding

as a global standard for the identification of biological species. It was established in 2004 with support from the Alfred P. Sloan Foundation. CBOL promotes DNA barcoding through various working groups, networks, workshops, conferences, and training etc. CBOL is a huge initiative with 200 member organizations from 50 countries and operates from Washington, DC.
    - ***ECBOL*** (the European Consortium for the Barcode of Life): ECBOL was established with the initiative of the European Distributed Institute of Taxonomy (EDIT).

DNA Barcode Databases

The two central DNA barcode databases are as follows:

- **BOLD** (the Barcode of Life Data Systems) at the University of Guelph is a public workbench for barcoding projects. In this platform, one can assemble, and analyse their data records in BOLD. BOLD is a very informative database where one can get information about existing barcode, suitable primers and published literature etc. (url: http://www.boldsystems.org/).
- **INSDC** (International Nucleotide Sequence Database Collaboration): It is a huge database which is a 18 year old collaboration of GenBank, EMBL and DDBJ. They are the permanent public repository for barcode data records in addition to many other resources. (url: http://insdc.org).

Other Organisation Working in the Field of DNA Barcoding

- **CCDB** (Canadian Centre for DNA Barcoding): It was established in 2006 within the Biodiversity Institute of Ontario (BIO) at the University of Guelph. CCDB is the world's first and largest high-throughput DNA barcoding facility. As per web site approximately 1,442,414 barcodes have been produced till now by CCDB (url: http://ccdb.ca/).
- **SwissBOL** (Swiss Barcode of Life): It is a national barcoding approach of Switzerland (url: http://www.swissbol.ch/).
- **RBG Kew:** Kew is a member of CBOL. They have collaboration with ten other organizations and support from the Alfred P. Sloan and Gordon and Betty Moore Foundations, A huge collaborative work was

initiated for investigating DNA regions for their potential as barcodes for all land plant species (url://www.kew.org/).

APPLICATIONS OF DNA BARCODING

DNA barcodes can be useful in many fields such as ecology, biomedicine, epidemiology, evolutionary biology, biogeography, conservation biology, bio-industry and bio-monitoring. Some of the applications of DNA barcoding are as follows:

1. **Species Identification:** The main goal of Barcoding is identification of recognized species and discovery of novel genotypes that may form the basis of subsequent species discovery. It facilitates the discovery of many unknown species and identification of pathogenic species with ecological, medical and agronomical significance and thus serves as a promising tool in epidemiology studies.
2. **Biodiversity Assessment and Conservation:** Conservation of biodiversity is under severe threats due to habitat degradation and change in the environment caused by human activities. DNA Barcoding is a useful tool for assessment of diversity for past and present organisms. Accordingly, DNA barcoding has great potentiality to monitor the extent of biodiversity before and after conservation action, more accurately and rapidly. It is also possible to generate useful data for estimation of phylogenetic diversity for setting conservation priorities.
3. **Diet Analysis:** To understand the food chain and the functioning of the ecosystem as a whole.
4. **Complement Other Field of Studies:** The vast data generated in barcoding projects complements other area of studies like molecular phylogenetics and population genetics etc.
5. **Identification of Poisonous Plant:** Barcoding is successfully employed for identification of poisonous plants, which is very important for food safety.
6. **Authentication of Natural Health Products**: This application is important for legal, economic, health and conservation issue.
7. **Forensic Science:** Barcoding is successfully employed in forensic science because only a small amount of DNA might help in solving cases.
8. **Detect Immature Specimen:** DNA barcode is an important tool to associate adults with immature specimens (fish larvae, fungal sexual

stage) and in cases, when morphological traits fail to discriminate species (e.g. red algal species and fungal species).

9. **Identification of the Cryptic Species:** The existence of many ''cryptic'' fungal species, which are morphologically alike, emphasizes that molecular information is essential.
10. **Species Interaction and Patterns of Associations:** DNA barcoding can be used to study species interactions. Community ecologists have used plant DNA barcoding to understand the factors (such as diversity pools and functional traits), which control the assembly of species into ecological communities. In 2005, Webb and Donoghue developed a tool called 'Phylomatic' to estimate phylogenetic trees for plant communities. Application of the tool has greatly boosted in 2009, after publication of the first community phylogeny based on DNA barcode sequence data for the trees in the forest dynamics plot in Panama. To determine whether a species in a community is related closely rather than by chance (phylogenetic clustering), related distantly rather than by chance (phylogenetic over-dispersion), or distributed randomly across the plant tree of life can be ascertained by establishing a DNA barcode library of the species assemblages and constructing a phylogenetic tree based on the sequence data. It has been used to study the host specificity of Australian beetles and their interaction with host plants.
11. **Species-Species Contamination:** It has been useful in identifying species-specific contaminants in fish consumed as food that may cause health problems.
12. **Authentication of Seafood Species:** Authentication of seafood species, which is commercially important having high market value, is one of the areas where DNA barcoding has been attempted. This approach aims to circumvent several issues related to food safety, species misbranding and accurate identification and conservation of wildlife and sustainable fishery fields.
13. **Food Traceability:** It is widely used to trace food to identify biological specimens of both raw and processed food.
14. **Authentication of Natural Health Products (NHPs):** NHPs derived from animals or plant origin has become possible by reliable DNA barcoding method. This reduces the chances of misrepresentation and substitution of unprotected species in the commercial market.
15. **Ecological Health:** `Bio-monitoring´ or `Bio-assessment´ is the method of evaluation for overall ecological health of surrounding environment (e.g. streams, reservoirs, wetlands, estuaries) on the basis of change

in the composition and structure of the benthic (bottom dwelling) community. DNA barcoding using COI has also gained much interest in the field of environmental biomonitoring, by providing a level of data standardization that lacked in the previous environmental assessments, using aquatic life forms such as in-stream alga (diatoms or soft-bodied algae) or benthic invertebrates.

16. **Authentication and Quality Control:** Authentication and quality control of aesthetically potent materials (plants or animals) can be done through DNA barcoding tools. An integrated MMDBD (Medicinal Materials DNA Barcode Database) have been developed for the collection and storage of DNA barcodes of these materials and act as a resource for researchers of various fields such as conservation, forensic, systematic study and herbal industry.
17. **Identification of Illegal Timber Products:** It is now possible to identify species of trees used in the various timber products through DNA barcoding. Accordingly, DNA barcode is used to monitor illegal timber trade in regions of biodiversity hotspots. Once the DNA barcode libraries for endangered taxa are developed, it will be possible to monitor all illegal trade of timber related species.
18. **Combining with Ethno-Botanical Knowledge:** To facilitate to understand the traditional ecological knowledge of a particular region, efforts have been made to combine local ethno-botanical knowledge, linguistics, and cultural history, with DNA barcode documents of the regional flora. Although, collection and documentation of such knowledge have been going on for centuries, inclusion of DNA barcode data facilitates accurate identification and collection of local flora. This will also help to understand how the indigenous people used to name the plants, classify and utilize them.

LIMITATIONS OF DNA BARCODING

Some of the limitations of DNA barcoding are as follows.

Exploration of Microdiversity

Barcoding the microdiversity (i.e. archea, bacteria, unicellular fungi and protists) is one of the major challenges for the BOL objectives. The diversity of microscopic and sub-microscopic communities is masked by the dominant populations which lead to unexploration of microbes. It can be explored using COI-based identification system but for macroscopic life, multi-locus barcoding system may be needed.

The use of SSU has been reported in identification of human parasite *Blastocyctis hominis*. Similarly other barcode regions (18S rRNA, 28S rRNA, ITS etc.) are being developed to study the diversity of protozoans, archea etc. Advanced methods such as metagenomics and community genomics have been emerging as new fields to study microdiversity, population studies and evolutionary relationships (Tringe et al. 2005). But the expensive metagenomics approach makes DNA barcoding as a useful tool for studying microdiversity.

Non Availability of Universal Barcode for Some Taxa

COI serves as a universal DNA barcode in animals but its utility in protists, fungi and plants is quite challenging. Amphibians have high variability in mtDNA and its COI priming sites and therefore large-scale effort is needed to barcode the amphibians using the same primary barcode region of COI. COI fails to distinguish closely related animal species requiring nuclear regions e.g. *Cytb* and *Rhod* to identify all teleost fish species.

In algae, there are shortcomings with the primers for COI and therefore, universal plastid amplicon (UPA) was reported as an alternative tool, although having a drawback of low interspecific divergence. COI has failed in barcoding of some protists (some group of amoeba) and elicits the use of an alternative barcode. It is also not suitable for DNA barcoding of plants and fungi due to low rate of molecular evolution.

To address all these issues, alternative stretches of DNA have been proposed as barcodes for the barcoding of these organisms. Combination of two plastid regions *matK* and *rbcL* is recommended as universal barcodes for terrestrial plants by CBOL Plant Working Group in 2009. But it is not still widely accepted for all plant species. Best results have been obtained using the internal transcribed spacer (ITS) region of the ribosomal DNA as a barcode for fungi. However, ITS offers less remarkable result in the DNA barcoding of *Penicillium* moulds. Additional markers (single or in combination) are now being explored to resolve the fungal diversity. The development of universal barcode is need of the hour towards the exploration of biodiversity.

Barcoding of Formalin Preserved Specimens

Preserved tissues and specimens are vast repositories of genetic information which form the basis for construction and development of barcode libraries. Formalin-preserved specimens provide a great resource for this approach (Zhang 2010). However, extraction of DNA and amplification of the target gene from these specimens is quite challenging.

Tissue lysis is the primary obstacle for formalin preserved tissues which affect the integrity of the extracted DNA. Formalin preserves morphological structure more effectively by fixing proteins through peptide linkage conjugation but degrades DNA. It also affects the downstream applications such as PCR amplification and sequencing reactions due to the crosslink formations by formaldehyde.

Unfavourable preservation condition is a major constraint on DNA quality. Use of alternative preservatives such as ethanol is preferred than formalin as it denatures proteins (at high concentration ≥95%) that may degrade DNA. For DNA barcoding, ethanol at a concentration of 95% is recommended for storage. Fresh sample and alcohol preservation is more preferable for DNA barcoding studies than formalin preservation to maintain maximum integrity.

Presence of Barcoding Gaps

Constructing a database for DNA barcoding is essential where each taxon is optimally represented. Lack of reference data may lead to the mis-interpretation of barcode-based identification. Limited geographical sampling may lead to `Barcoding gaps' which should be considered during the database construction phase. This may lead to failure of more recently diverged species (sister-species) identification (Wiemers and Fiedler 2007). The unknown specimens may be problematic if belongs to under-described part of biodiversity. Collaboration with expert taxonomists for identification of undocumented voucher specimens is essential for proper construction of barcode database.

Delimitation of Species

Species delimitation refers to the process which determines the species boundaries and discovery of new species. The accuracy of species identification using DNA barcoding has been criticised by several taxonomists that depends on the overlap between intraspecific and interspecific variation. This approach is based on the hypothesis that intraspecific divergences will be less than interspecific divergences which is absent in several taxa. This implies a query whether DNA barcoding can attain its goal to unravel the biodiversity. It is possible only if the target region is conserved among the members of a species and differs from others (Prendini 2005).

Utility of Mitochondrial Genome

The utility of a single mitochondrial gene in DNA barcoding to delineate species boundaries has also triggered several criticisms. Classifications solely based on the one character (i.e. mitochondrial gene) are inadequate for species identification. Nuclear loci are only required to resolve phylogenetic relationships in several cases where one host species can bear different symbionts which may lead to intraspecific (i.e. interpopulation) variation in mtDNA sequences.

Paraphyly, polyphyly, heteroplasmy (e.g. Mussels) or presence of nuclear pseudogenes in mitochondrial genome i.e. nuclear mitochondrial DNA (NUMTs) are some of the artefacts associated with mitochondrial genes. This may cause the overestimation of sample divergence and thus obscure the real evolutionary history of organisms. It has been reported that heteroplasmy level differs among tissues and DNA extracted from large tissues displays less polymorphism (Magnacca and Brown 2010).

Methodological Advances

There are two other drawbacks of DNA barcoding: (1) damage of specimen during DNA extraction method and (2) nature of sample. A simple, rapid and reliable DNA extraction protocol is essential to obtain efficient yields of DNA for further process.

Various non-destructive DNA extraction protocols have been proposed which enable the determination of barcode with minimal damage of museum specimens or small insects (Acarina, Araneae, Coleoptera, Diptera, Hemiptera, Hymenoptera, etc.). However, these methods have been useful in some insects; DNA extraction is challenging in plants (e.g. Malvaceae, Asteraceae) having mucilaginous compounds, polysaccharides, phenolic compounds and other secondary metabolites. These compounds co-precipitate with DNA and thus hinder the extraction of high quality DNA. In addition, they complicate the downstream applications such as polymerase chain reaction (PCR) or other enzymatic reactions.

To overcome these issues, several modified protocols have been reported that enable effective DNA extraction from problematic plant samples (Souza et al. 2012). Recent technical advancements such as highly efficient DNA polymerase and a DNA-repairing enzyme allow the extraction and amplification of DNA from historical museum specimens as well as fossilized samples.

Distance or Character Based Barcoding Analyses

Enumeration of molecular distance is proposed to be a standard method for barcode analysis (Herbert et al. 2003). But it has lately been criticized and should not be considered as it is a phenetic measure non-indicative of common origin.

On the contrary, character-based analysis has proven to be an effective tool for species identification and discrimination (Bergmann et al. 2009). It addresses all the objectives with no loss of information as multiple nucleotide differences are reduced to a single distance-based measure. This approach has been implemented through development of Characteristic Attributes Organization System (CAOS) software. Some studies have been reported on character- based barcoding analysis but conflict still lies between distance and diagnostics. However, the character- based barcoding is currently proposed to meet the requirement.

FUTURE PROSPECTS

It is expected that plant DNA barcoding will expand in two key areas: development of comprehensive plant DNA barcode library for global use and identification of new markers for accurate identification of plant species. Efforts have already been made towards developing DNA barcode libraries in forestry species from different regions around the world. Although, popularization and development of the global plant DNA barcode library is one of the biggest challenges, it is important to know that several laboratories are engaged in this direction throughout the globe.

In 2012, Taberlet and coworkers developed a modification in DNA barcoding called ‘metabarcoding’ or ‘eDNA’, in which genetic markers are used for the identification of organisms from samples derived from soils, coral reefs and sea water. Usually very short and unique genetic markers are required for successful identification of organisms in such environments, due to degraded DNA in such samples. They are called "mini-barcodes", which are a sub-region of a standard marker. Metabarcoding is rapidly evolving due to significant improvement of the methodology adopted for recovering, amplification and sequencing of short DNA fragments.

It has been suggested by Cossac and coworkers 2016, that genome skimming (shallow sequencing approaches aiming to uncover conserved ortholog sequences for phylogenomic studies) by combining the nuclear and plastid regions as an "extended DNA barcode", may be the ultimate solution for species identification. On the other hand in 2015, Li and co-workers have advocating for the use of super-barcodes with the design and selection of "specific barcodes" loci for individual group of species. The new microfluid PCR-based target enrichment technology has been considered to be less expensive option for large scale multi locus plant DNA barcoding.

DNA BARCODE WEBSITES

With the advancement of information technology, various databases and projects have been established for proper documentation of biodiversity, some of the public websites are as follows.

Barcode Related Websites

- Barcodes of Life (http://www.barcodeoflife.org)
- Canadian Centre for DNA Barcoding (http://www.dnabarcoding.ca/),
- ECBOL - European Consortium for the Barcode of Life (http://www.ecbol.org/)
- BOLNET.ca- Canadian Barcode of Life Network (http://www.bolnet.ca/)
- iBOL-International Barcode of Life Project (http://www.dnabarcoding.org/)
- All Birds Barcoding Initiative (http://www.barcodingbirds.org/)
- FISH-BOL: Fish Barcode of Life Initiative (http://www.fishbol.org/)
- All Leps - Barcode of Life (http://www.lepbarcoding.org/)
- BOLD- Barcode of Life Database (http://www.barcodinglife.org/)
- MarBOL - Marine Barcode of Life (http://www.marinebarcoding.org/)

- Sponge Barcoding Project (https://www.spongebarcoding.org/)
- Polar Barcode of Life (http://www.polarbarcoding.ca/)
- Belgian Network for DNA Barcoding (http://www.bebol.myspecies.info)
- CBOL- Consortium for the Barcode of Life (http://www.barcoding.si.edu)
- NorBOL (http://dnabarcoding.no/en/)
- MexBOL (http://www.mexbol.org/)
- JBOLI (https://www.jboli.org/)

CONCLUSION

DNA barcodes have provided a new tool for the organismal biologist to enhance their understanding of the natural resources. Over the last decade several DNA barcode markers have been identified, tested and used to address several basic questions in systematics, evolutionary biology, ecology and conservation. The important plant DNA barcode markers used are *rbc*L, *mat*K, *trn-psb*A, and *ITS*2. Forensic investigators have also used these plant DNA barcodes to identify endangered species and monitoring presence of banned plant origin material in commercial products, such as herbal supplements and foods. However, it is important to build the global plant DNA barcode library for its universal adoptability. Further, development and adoption of more cost effective genomic sequencing technologies shall encourage the application of DNA barcoding as genetic identification markers to additional fields of biology and commercial endeavors.

REFERENCES

Arnot, D. E., Roper, C., & Bayoumi, R. A. (1993). Digital codes from hypervariable tandemly repeated DNA sequences in the plasmodium *Falciparum circumsporozoite* gene can genetically barcode isolates. *Molecular and Biochemical Parasitology*, *61*(1), 15–24. doi:10.1016/0166-6851(93)90154-P PMID:8259128

Baldwin, B. G. (1992). Phylogenetic utility of the internal transcribed spacers of nuclear ribosomal DNA in plants: An example from the Compositae. *Molecular Phylogenetics and Evolution*, *1*(1), 3–16. doi:10.1016/1055-7903(92)90030-K PMID:1342921

Chase, M. W., Cowan, R. S., Hollingsworth, P. M., van den Berg, C., Madrinan, S., Petersen, G., Seberg, O., Jørgsensen, T., Cameron, K. M., Carine, M., Pedersen, N., Hedderson, T. A. J., Conrad, F., Salazar, G. A., Richardson, J. E., Hollingsworth, M. L., Barraclough, T. G., Kelly, L., & Wilkins.on, M. (2007). A propo.sal for a standardized protocol to barcode all land plants. *Taxonomy*, *56*(2), 295–299. doi:10.1002/tax.562004

Erickson, D. L., Spouge, J., Resch, A., Weigt, L. A., & Kress, W. J. (2008). DNA barcoding in land plants: Developing standards to qualify and maximize. *Taxonomy*, *57*(4), 1304–1316. doi:10.1002/tax.574020 PMID:19779570

Fazekas, A.J., Burgess, K.S., Kesanakurti, P.R., Graham, S.W., Newmaster, S.G., Husband, B.C., … Barrett, C.H. (2008). Multiple multilocus DNA barcodes from the plastid genome discriminate plant species equally well. *Public Library of Science (PLoS) ONE, 3,* e2802-e2819.

Hebert, P. D. N., Cywinska, A., Ball, S. L., & de Waard, J. R. (2003). Biological identifications through DNA barcodes. *Proceedings Royal Society of London Biological Science Series B*, *270*(1512), 313–321. doi:10.1098/rspb.2002.2218 PMID:12614582

Hollingsworth, M. L., Clark, A., Forrest, L. L., Richardson, J., Pennington, R. T., Long, D. G., ... Hollingsworth, P. M. (2009). Selecting barcoding loci for plants; evaluation of several candidate loci with species-level sampling in three divergent groups of land plants. *Ecological Restoration*, *9*, 439–457. PMID:21564673

Hollingsworth, M. L., Sean, M., Graham, W., & Damon, P. L. (2011). Choosing and using a plant DNA barcode. *Public Library of Science (PLoS). ONE*, *6*(5), 1–13. doi:10.1371/journal.pone.0019254

Kress, W. J., & Erickson, D. L. (2007). A two-locus global DNA barcode for land plants: Coding rbcL gene compliments the non-coding trnH-psbA spacer region. *Public Library of Science (PLoS). ONE*, *2*, e508–e522. doi:10.1371/journal.pone.0000508 PMID:17551588

Kress, W. J., Wurdack, K. J., Zimmer, E. A., Weight, L. A., & Lanzen, D. H. (2005). Use of DNA barcodes to identify flowering plants. *Proceedings of the National Academy of Sciences of the United States of America*, *102*(23), 8369–8374. doi:10.1073/pnas.0503123102 PMID:15928076

Lahaya, R. (2008). DNA barcoding the flores of biodiversity hotspots. *Proceedings of the National Academy of Sciences of the United States of America*, *105*(8), 2923–2928. doi:10.1073/pnas.0709936105 PMID:18258745

Liang, H., & Hilu, K. W. (1996). Application of the matK gene sequences to grass systematics. *Canadian Journal of Botany*, *74*(1), 125–134. doi:10.1139/b96-017

Magnacca, K. N., & Brown, M. J. F. (2010). Mitochondrial heteroplasmy and DNA barcoding in *Hawaiian hylaeus* (Nesoprosopis) bees (Hymenoptera: Colletidae). *BMC Evolutionary Biology*, *10*(1), 174–186. doi:10.1186/1471-2148-10-174 PMID:20540728

Nock, C. J., Waters, D. L. E., Edwards, M. A., Brown, S. G., Rice, M., Cordeoro, G. M., & Henry, R. J. (2011). Chloroplast genome sequences from total DNA for plant identification. *Plant Biotechnology Journal*, *9*(3), 328–333. doi:10.1111/j.1467-7652.2010.00558.x PMID:20796245

Prendini, L. (2005). Identifying spider through DNA barcodes. *Canadian Journal of Zoology*, *83*(3), 481–491. doi:10.1139/z05-025

Small, R. L., Cronn, R. C., & Wendel, A. S. (2004). Use of nuclear genes for phylogeny reconstruction in plants. *Australian Systematic Botany*, *17*(2), 145–170. doi:10.1071/SB03015

Souza, H. A. V., Muller, L. A. C., Brandao, R. L., & Lovato, M. B. (2012). Isolation of high quality and polysaccharide-free DNA from leaves of *Dimmorphandra mollis* (Leguminosae), a tree from the Brazalian Cerrado. *Genetics and Molecular Research*, *11*(1), 756–764. doi:10.4238/2012.March.22.6 PMID:22576834

Tringe, S. G., von Mering, C., Kobayashi, A., Salamov, A. A., Chen, K., Chang, H. W., ... Rubin, E. M. (2005). Compara.tive metagenomics of microbial communities. *Science*, *308*(5721), 554–557. doi:10.1126cience.1107851 PMID:15845853

Wiemers, M., & Fiedler, K. (2007). Does the DNA barcoding gap exist? – a case study in blue butterflies (Lapidoptera: Lycaenidae). *Frontiers in Zoology*, *4*(1), 126–134. doi:10.1186/1742-9994-4-8 PMID:17343734

Zhang, J. (2010). Exploiting formalin preserved fish specimens for resources of DNA barcoding. *Molecular Ecology Resources*, *10*(6), 935–941. doi:10.1111/j.1755-0998.2010.2838.x PMID:21565102

ADDITIONAL READING

Baker, D. A., Stevenson, D. W., & Little, D. P. (2012). DNA barcode identification of black cohosh herbal dietary supplements. *Journal of Association of Official Agricultural Chemists (AOAC). International*, *95*, 1023–1034.

Bolson, M., Smidt, E. C., Brotto, M. L., & Pereire, V. S. (2015). ITS and trnH-psbA as efficient DNA barcode to identify threated commercial woody angiosperms from Southern Brazilian Atlantic rain forest. *Public Library of Science (PLoS). ONE*, *10*, e0143049. doi:10.1371/journal.pone.0143049 PMID:26630282

Braukmann, T. W. A., Kuzmina, M. L., Sills, J., Zakarov, E. V., & Hebert, D. N. (2017). Testing the efficacy of DNA barcodes for identifying the vascular plants of Canada. *Public Library of Science (PLoS). ONE*, *12*(1), e0169515. doi:10.1371/journal.pone.0169515 PMID:28072819

Bruni, I., Mattia, F. D., Galimberti, A., Galasso, G., Banfi, E., Casiraghi, M., & Labra, M. (2010). Identification of poisonous plants by DNA barcoding approach. *International Journal of Legal Medicine*, *124*(6), 595–603. doi:10.100700414-010-0447-3 PMID:20354712

Burgess, K. S., Fazekas, A. J., Kesanakurti, P. R., Graham, S. W., Husband, B. C., Newmaster, S. G., Percy, D. M., Hajibabaei, M., & Barrett, S. C. H. (2011). Discriminating plant species in a local temperate flora using the rbcL+matK DNA barcode. *Methods in Ecology and Evolution*, *2*(4), 333–340. doi:10.1111/j.2041-210X.2011.00092.x

Chase, M.W., Salamin, N., Wilkinson, M., Dunwell, J.M., Kesanakurthi, R.P., Haider, N., & Savolainen, V. (2005). Land plants and DNA barcodes: short-term and long-term goals. *Philosophical Transaction of Royal Society B Biology Science. 360:* 1889-1895.

Chen, S., Pang, X., Song, J., Shi, L., Yao, H., Han, J., & Leon, C. (2014). A renaissance in herbal medicine identification: From morphology to DNA. *Biotechnology Advances, 32*(7), 1237–1244. doi:10.1016/j.biotechadv.2014.07.004 PMID:25087935

De Boer, H. J., Ichim, M. C., & Newmaster, S. G. (2015). DNA Barcoding and Pharmacovigilance of Herbal Medicines. *Drug Safety, 38*(7), 611–620. doi:10.100740264-015-0306-8 PMID:26076652

De Boer, H. J., Ouarghidi, A., Martin, G., Abbad, A., & Kool, A. (2014). DNA barcoding reveals limited accuracy of identifications based on folk taxonomy. *Public Library of Science (PLoS). ONE, 9*(1), e84291. doi:10.1371/journal.pone.0084291 PMID:24416210

De Vere, N., Rich, T. C., Ford, C. R., Trinder, S. A., Long, C., Moore, C. W., & Wilkinson, M. J. (2012). DNA barcoding the native flowering plants and conifers of Wales. *Public Library of Science (PLoS). ONE, 7*(6), e37945. doi:10.1371/journal.pone.0037945 PMID:22701588

Fahner, N. A., Shokralla, S., Baird, D. J., & Hajibabaei, M. (2016). Large scale monitoring of plants through environmental DNA metabarcoding of soil: Recovery, resolution, and annotation of four DNA markers. *Public Library of Science (PLoS). ONE, 11*(6), e01575505. doi:10.1371/journal.pone.0157505

Fatima, T., Srivastava, A., Somashekar, P. V., Hanur, V. S., & Rao, M. S. (2019). Development of DNA-based species identification and barcoding of three important timbers. *Bulletin of the National Research Center, 43*(1), 76–88. doi:10.118642269-019-0116-8

Fazekas, A. J., Fuzmina, K. L., Newmaster, S. G., & Hollingsworth, P. M. (2012). DNA barcoding methods for land plants. In W. J. Kress & D. L. Erickson (Eds.), *DNA barcodes: methods and protocols* (pp. 223–252). Springer Science. doi:10.1007/978-1-61779-591-6_11

Ferri, G., Alù, M., Corradini, B., & Beduschi, G. (2009). Forensic botany: Species identification of botanical trace evidence using a multigene barcoding approach. *International Journal of Legal Medicine, 123*(5), 395–401. doi:10.100700414-009-0356-5 PMID:19504263

Ghorbani, A., Saeedi, Y., & de Boer, H. J. (2017). Unidentifiable by morphology: DNA barcoding of plant material in local markets in Iran. *Public Library of Science (PLoS). ONE*, *12*(4), e0175722. doi:10.1371/journal.pone.0175722 PMID:28419161

Hajibabaei, M., Gregory, A. C., Herbert, P. D. N., & Hickey, D. A. (2007). DNA barcoding: How it complements taxonomy, molecular phylogenetics and population genetics. *Trends in Genetics*, *23*(4), 167–172. doi:10.1016/j.tig.2007.02.001 PMID:17316886

Hajibabaei, M., Singer, G. A., Hebert, P. D., & Hickey, D. A. (2007). DNA barcoding: How it complements taxonomy, molecular phylogenetics and population genetics. *Trends in Genetics*, *23*(4), 167–172. doi:10.1016/j.tig.2007.02.001 PMID:17316886

Hejibabaei, M., & McKenna, C. (2012). DNA mini-barcodes. In W.J. Kress and E. D. Ericksion (Ed.) DNA barcodes: methods and protocols (pp. 339-353), New York, Human Press, Springer Science+ Publishing Media, LLC. doi:10.1007/978-1-61779-591-6_15

Hollingsworth, P. M., Forrest, L. L., Spouge, J. L., Hajibabaei, M., Ratnasingham, S., van der Bank, M., ... Little, D. P.CBOL Plant Working Group. (2009). A DNA barcode for land plants. *Proceedings of the National Academy of Sciences of the United States of America*, *106*(31), 12794–12797. doi:10.1073/pnas.0905845106 PMID:19666622

Jurado-Rivera, J. A., Vogler, A. P., Reid, C. A. M., Petitpierre, E., & Gomez-Zurita, J. (2009). DNA barcoding insect-host plant associations. *Proceedings. Biological Sciences*, *276*(1657), 639–648. doi:10.1098/rspb.2008.1264 PMID:19004756

Kool, A., de Boer, H.J., KruÈger, Å., Rydberg, A., Abbad, A., BjoÈrk, L., & Martin, G. (2012). Molecular identification of commercialized medicinal plants in Southern Morocco. *Public Library of Science (PLoS) ONE, 7*, e39459. PMID: 22761800. doi:10.1371/journal.pone.0039459

Kress, W. J. (2017). Plant DNA barcodes: Applications today and in the future. *Journal of Systematics and Evolution*, *55*(4), 291–307. doi:10.1111/jse.12254

Kress, W. J., & Erickson, D. L. (Eds.). (2012) DNA barcoding: methods and protocols. New York, Humana Press, Springer Science+ Publishing Media, LLC. doi:10.1007/978-1-61779-591-6

Kress, W. J., Garcia-Robledo, C., Uriarte, M., & Erickson, D. L. (2014). DNA barcodes for ecology, evolution, and conservation. *Trends in Ecology & Evolution*, *30*(1), 25–35. doi:10.1016/j.tree.2014.10.008 PMID:25468359

Kress, W. J., Wurdack, K. J., Zimmer, E. A., Wright, L. A., & Janzen, D. H. (2005). Use of DNA barcodes to identify flowering plants. *Proceedings of the National Academy of Sciences of the United States of America*, *102*(23), 8369–8374. doi:10.1073/pnas.0503123102 PMID:15928076

Li, D. Z., Gao, L. M., Li, H. T., Wang, H., Ge, X. J., Liu, J. Q., Chen, Z.-D., Zhou, S.-L., Chen, S.-L., Yang, J.-B., Fu, C.-X., Zeng, C.-X., Yan, H.-F., Zhu, Y.-J., Sun, Y.-S., Chen, S.-Y., Zhao, L., Wang, K., Yang, T., & Duan, G. W. (2011). Comparative analysis of a large dataset indicates that internal transcribed spacer (ITS) should be incorporated into the core barcode for seed plants. *Proceedings of the National Academy of Sciences of the United States of America*, *108*(49), 19641–19646. doi:10.1073/pnas.1104551108 PMID:22100737

Liu, J., Yan, H.F., Newmaster, S.G., Pei, N., Ragupathy, S., & Ge X. J. (2015). The use of DNA barcoding as a tool for the conservation biogeography of subtropical forests in *Diversity and Distribution, 21*, 188-199.

Morello, L., Braglia, L., Gavazzi, F., Gianì, S., & Breviario, D. (2019). Tubulin-Based DNA Barcode: Principle and Applications to Complex Food Matrices. *Genes*, *10*(3), 229–235. doi:10.3390/genes10030229 PMID:30889932

Nadia, H. (2011). Identification of plant species using traditional and molecular-based methods. In R. E. Davis (Ed.), *Wild Plants: Identification, uses, and conservation* (pp. 1–66). Nova Science Publications, Inc.

Newmaster, S. G., Grguric, M., Shanmughanandhan, D., Ramalingam, S., & Ragupathy, S. (2013). DNA barcoding detects contamination and substitution in North American herbal products. *BioMed Medicine*, *11*(1), 222–234. doi:10.1186/1741-7015-11-222 PMID:24120035

Nithaniyal, S., Newmaster, S. G., Ragupathy, S., Krishnamoorthy, D., Vassou, S. L., & Parani, M. (2014). DNA barcode authentication of wood samples of threatened and commercial timber trees within the tropical dry evergreen forest of India. *Public Library of Science (PLoS). ONE*, *9*(9), e107669. doi:10.1371/journal.pone.0107669 PMID:25259794

Palhares, R. M., Goncalves, D. M., Dos, M. A. F. B. B., Pereive, C. G., Das, G. L. B. M., & Oliveira, G. (2015). Medicinal plants recommended by the world health organization: DNA barcode identification associated with chemical analyses guarantees their quality. Public Library of Science (PLoS). *ONE*, *10*(5), e0127866. doi:10.1371/journal.pone.0127866 PMID:25978064

SaÈrkinen, T., Staats, M., Richardson, J. E., Cowan, R. S., & Bakker, F. T. (2012). How to open the treasure chest? Optimising DNA extraction from herbarium specimens. *Public Library of Science (PLoS). ONE*, *7*(8), e43808. doi:10.1371/journal.pone.0043808

Taberlet, P., Coissac, E., Pompanon, F., Brochmann, C., & Willerslev, E. (2012). Towards next generation biodiversity assessment using DNA metabarcoding. *Molecular Ecology*, *21*(8), 2045–2050. doi:10.1111/j.1365-294X.2012.05470.x PMID:22486824

Thompson, K. A., & Newmaster, S. G. (2014). Molecular taxonomic tools provide more accurate estimates of species richness at less cost than traditional morphology-based taxonomic practices in a vegetation survey. *Biodiversity and Conservation*, *23*(6), 1411–1424. doi:10.100710531-014-0672-z

Van Velzen, R., Weitschek, E., Felici, G., & Bakker, F. T. (2012). DNA barcoding of recently diverged species: Relative performance of matching methods. *Public Library of Science (PLoS). ONE*, *7*(1), e30490. doi:10.1371/journal.pone.0030490 PMID:22272356

Webb, C. O., & Donoghue, M. J. (2005). Phylomatic: Tree assembly for applied phylogenetics. *Molecular Ecology Notes*, *51*(1), 181–183. doi:10.1111/j.1471-8286.2004.00829.x

Xu, S., Li, D., Li, J., Xiang, X., Jin, W., Huang, W., Jin, X., & Huang, L. (2015). Evaluation of the DNA Barcodes in *Dendrobium* (Orchidaceae) from Mainland Asia. *Public Library of Science (PLoS). ONE*, *10*(1), e0115168. doi:10.1371/journal.pone.0115168 PMID:25602282

Zhang, D., Mo, X., Xiang, J., & Zhou, N. (2016). Molecular identification of original plants of *Fritillariae cirrhosae* bulbus, a traditional Chinese medicine (TCM) using plant DNA barcoding. *African Journal of Traditional, Complementary, and Alternative Medicines*, *13*(6), 74–82. doi:10.21010/ajtcam.v13i6.12 PMID:28480363

APPENDIX

1. What are the basic components of DNA barcoding? Explain the purpose of each component.
2. Describe the step-wise procedure to be followed for DNA barcoding.
3. What are the advantages of DNA barcoding?
4. What are the factors to be considered for selection of markers for DNA barcoding?
5. What are the different markers used for plant DNA barcoding? Explain the merits and demerits of each one of them.
6. Which international bodies are involved in working on DNA barcoding? Explain their activities.
7. Explain different applications of DNA barcoding.
8. Explain the limitations of DNA barcoding.

Chapter 7
Gene Cloning

ABSTRACT

The discovery of two naturally occurring biological molecules, plasmid DNA and restriction enzymes, with remarkable properties have made possible the development of methods to isolate and manipulate specific DNA fragments. Through this technology, a DNA fragment, even an entire gene and its controlling elements, can be isolated and rejoined with a plasmid or phage DNA, and the hybrid DNA molecule can be inserted into a bacterium. The foreign DNA insert can be multiplied inside the bacterial host and induced to express or synthesize the protein product of the foreign DNA. The entire process through which this can be achieved is called recombinant DNA technology or genetic engineering. The recombinant DNA technology has been extended to animal and plant cells. In this chapter, methods for isolation, modification, rejoining and replication of genomic DNA, and production of new or enhanced protein products within a host cell have been described.

INTRODUCTION

During early 1970s several new methodologies were developed which have revolutionized the science of modern genetics. The discovery of two naturally occurring biological molecules with remarkable properties have made possible the development of methods to isolate and manipulate specific DNA fragments. These two molecules are plasmid DNA and restriction enzymes. Through this technology, a DNA fragment, even an entire gene and its controlling elements, can be isolated and rejoined with a plasmid or phage DNA, and

DOI: 10.4018/978-1-7998-4312-2.ch007

the hybrid DNA molecule can be inserted into a bacterium. The foreign DNA insert can be multiplied inside the bacterial host and induced to express or synthesize the protein product of the foreign DNA. The entire process through which this can be achieved is called recombinant DNA technology or genetic engineering. A fragment of DNA, representing a gene, when inserted into a vector, leading to the production of a recombinant DNA molecule, is inserted into a host cell, where it is multiplied and passed to its progeny, the inserted gene in the recombinant molecule is said to be cloned. This chapter describes how genomic DNA can be cut, modified, rejoined and replicated, ultimately to produce new or enhanced protein products, within a host cell.

TOOLS FOR GENE CLONING

For cloning and manipulation of genes, the genetic material (DNA or RNA) requires to be cut, modify and rejoin in the desired manner. Enzymes play a major role in these activities, which are described below.

Enzymes for Cutting DNA Molecules

In a DNA molecule, the phosphodiester bonds that link the nucleotides can be broken by treatment with nucleases. Two different types of nucleases are found: exonucleases (Figure 1a), which remove nucleotides from the end of a DNA strand, one at a time and endonucleases (Figure 1b), which breaks the phosphodiester bonds within a strand of DNA. Exonuclease like Ba131(derived from the bacterium *Alteromonas espejiana*) removes nucleotides from both strands of double-stranded molecule, in contrast enzymes exonuclease III (derived from *E. coli*) degrade only one strand of a double-stranded DNA molecule.

Similarly, S1 endonuclease (derived from fungus *Aspergillus oryzae*) cleaves single strand, both single and double stranded molecules can be cut by DNAse I (derived from pancreas of cow). DNase I is non-specific *i.e* it attack phosphodiester bond of DNA at any site. However, a special group of enzymes called restriction endonucleases cleave double stranded DNA at a specific recognition sites. Thus, this group of enzymes has been utilized in gene cloning.

Discovery of restriction enzymes played a major role in gene cloning. Restriction enzymes were discovered while investigating the mechanism of host-specific restriction in bacteriophages. The first duplex DNA cutting enzyme, called restriction enzyme, was discovered from *Haemophilus influenza* in 1970. It was observed that the bacteria are protected by restriction enzymes against virus infections and appear to serve a host-defense role. When virus particles are used to infect a strain of *E. coil* lacking a restriction enzyme, infection will always be successful. However, if the same strain contains a restriction enzyme, the probability of successful infection is reduced. The presence of additional restriction enzymes has multiple effects, and could make the bacteria virtually impregnable (Brown, 2015). The question that immediately comes to mind is why the restriction enzyme does not chew up the genomic DNA of their host? The answer lies on the fact that in almost all cases, a bacterium also synthesizes another enzyme called DNA methyltransferase, which protects the DNA target sequence, involved in restriction digestion, by methylation. Such interaction between restriction-endonuclease and DNA-methylase is called restriction-modification systems.

Restriction enzymes are found only in prokaryotes and few viruses. Their presence has been reported in thousands of bacteria and archaea. Many types of restriction endonucleases have been isolated from a wide variety of bacteria. Each enzyme is represented by a three-letter code in italics. For example, *Hin* for *Haemophilus influenza*, *Hae* for *Haemophilus aegypticus*, *Eco* for *Escherichia coli* etc. Sometimes a four-letter code is used when the enzymes are isolated from different serotypes of a species. For example, *Hin*f for *Haemophilus inflenzae* serotype f. In those case where more than one restriction enzyme is isolated from a single source (bacterium), they are denoted by Roman numerals *e.g* I, II, III etc. For example *Hin*dII, for the second enzyme of *Haemophilus influenza* serotype d.

Restriction enzymes vary in size from 157 (diminutive) to 1250 amino acids (giant *Cje*I). Over 3000 restriction enzymes have been purified and characterized, and about 250 of them show different sequence-specificities activities. Restriction enzymes with new specificities are found regularly.

Types of Restriction Enzymes

On the basis of their composition of the subunits, requirement of cofactors, position of the cleavage, and sequence specificity, restriction enzymes are classified into four types. However, on the basis of amino acid sequencing, large variations have been observed within restriction enzymes. Thus at molecular level there exist many more than four different types (Faraday 2018).

Figure 1. Cleavage by: (a) exonuclease and (b) endonuclease

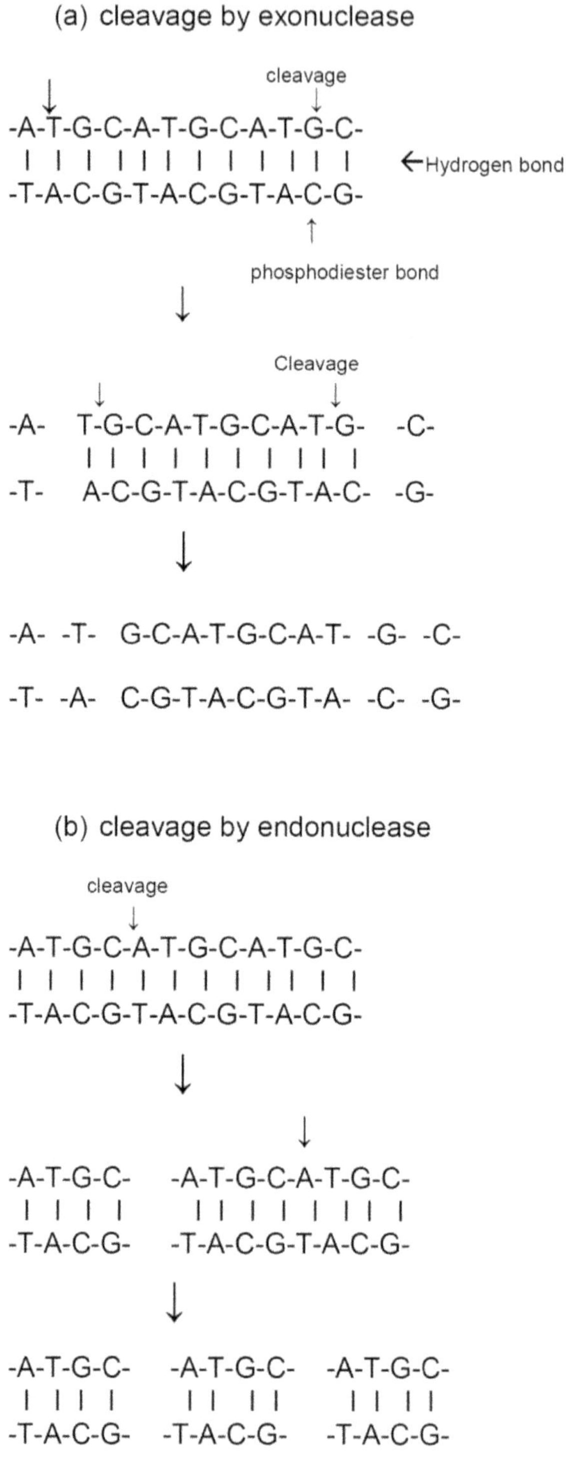

Type I enzymes cuts DNA far away from the site of their recognition randomly. They are complex molecule having multiple subunits, and have both restriction-and-modification characteristics. Originally thought to be rare, later found to be common. Although they are interesting biochemical molecules, they do not carry any practical value, as discrete restriction fragments are not produced.

Type II enzymes cuts DNA at defined positions, either within their recognition sequences or at nearby position. Discrete restriction fragments are produced which can be separated to distinct bands through gel electrophoresis. Therefore, they are used for DNA analysis and gene cloning.

Type II enzymes are composed of several unrelated proteins which differ in their amino acid sequences. They are considered to be a group of rapidly evolving proteins and are mainly involved in host-parasite interaction.

Most common Type II enzymes are those that cleave DNA within their recognition sequences e.g. *Hind*II, *Not*I. The next most common Type II enzymes referred to as 'Type IIS' are those that cleave at a site which is near to their recognized sequence e.g *Pok*I, *Atw*I. The third kind of Type II enzyme is 'Type IIG' restriction enzymes, that has both restriction and modification properties. They are composed of about 850-1250 amino acids, and the two enzymatic activities are performed by the same protein chain. They cut outside of their recognition sequences. Some of them recognizes continuous sequences (e.g. *Acu*I: CTGAAG) and cut on one side of the sequence. While others recognizes discontinuous sequences (e.g. *Bca*I: CGANNNNNNTGC) and cut on both the sides of the sequence, and a small fragments containing the recognition sequence are released. These enzymes may have varied amino acid sequences, but their physical organization remains consistent.

Type III enzymes are also comprised of large molecules having both restriction and modification functions. They also cut at a site away from their recognition sequences, but two such sequences has to be present in opposite orientations within the same DNA molecule for cleavage to be effective. Complete digestion can rarely be achieved by this type of enzymes.

Type IV enzymes can recognize modified nucleotides, for example methylated DNA. Typical examples are *Mcr BC* and *Mrr* systems of *E. coli*.

Restriction Enzyme Recognition Sequences

Specific sequences of double-stranded DNA known as recognition sequences act as the substrates for restriction enzymes. For different restriction enzymes, the length of the recognition sequence varies. The enzyme *Sau*3AI recognizes only 4-base-pairs (bp), whereas *Eco*RI, *Sac*I, *Sst*I, each recognizes a 6-bp and *Not*I recognizes an 8-bp sequence of DNA. The frequency at which the DNA molecule will be cut randomly is dependent upon the length of the recognition sequence. Restriction enzymes which recognizes a 4-bp sequence site will cut, on average every 256 bp (4^4), whereas enzyme that recognizes a 6-bp sequence site will cut on an average every 4^6 or 4096-bp.

The sequence shown is of one strand in the 5'®3' direction. Almost all recognition sequences are palindromes: when both strands are considered they read the same in each direction Py=pyrimidine, Pu=purine, N=any nucleotide. ^=indicate the site of enzyme cut within the specified nucleotide sequence.

Restriction enzymes that has the same recognition site, is called isoschizomers (Table 1). For example, recognition sites of *Sac*I and *Sst*I are identical. Some isoschizomers cut DNA at the identical site within their recognition sequence, however some others not. Most isoschizomers require different reaction conditions and stability parameters, which influence the cutting reactions.

Restriction recognition sites can be either ambiguous or unambiguous. For example, the enzyme *Bam*HI recognizes the sequence GGATCC only, i.e unambiguous. Whereas, *Hinf*I recognizes a 5 base-pair sequence starting with GA and ending in TC, having any base in-between. Thus *Hin*fI has an ambiguous recognition site. Similarly, *Xho*II also has an ambiguous recognition site (Table 1). The sequence may start with a purine (A or G) and end with a pyrimidine (T or C) or vice versa. Thus, *Xho*II can recognize and cut DNA sequences at GGATCC, AGATCT, GGATCT, and AGATCC.

The restriction recognition sequence of one enzyme may contain the restriction recognition sequence of another enzyme. For example, recognition site of *Sau*3AI is found within the recognition site of *Bam*HI. Similarly, four possible recognition sites of *Xho*II can also be cut with *Sau*3AI, and one site by *Bam*HI.

Most, but not all, recognition sequences is palindromes i.e they read from 5' to 3' ends in both the strands (forward and backward) of DNA double helix (Table 1). The restriction enzymes usually bind as dimers (pairs) to their recognition site.

Table 1. The recognition sequences for some of the most frequently used type II restriction endonucleases

Enzymes	Organism	Recognition sequence	Nature of cut ends
*Alu*I	*Arthrobacter luteus*	AG^CT	Blunt
*Apa*I	*Acetobacter pasteurianus*	GGGCC^C	Sticky
*Bam*HI	*Bacillus amyloliquefaciens*	G^GATCC	Sticky
*Bgl*II	*Bacillus globigii*	A^GATCT	Sticky
*BspI*20I	*Bacillus*	G^GGCCC	Sticky
*Dpn*I	*Diplococcus Pneumonia*	GA^TC	Blunt
*Dra*I (*Aha*III)	*Deinoccus radiophlus*	TTT^AAA	Blunt
*Eco*RI	*Escherichia coli*	G^AATTC	Sticky
*Hae*III	*Haemophilus aegyptius*	GG^CC	Blunt
*Hinc*II	*Haemophilus influnzae*	GTPy^PuAC	Blunt
*Hind*III	*Haemophilus influenzea* R_d	A^AGCTT	Sticky
*Hinf*I	*Haemophilus influenza* R_f	G^ANTC	Sticky
*Hpa*I	*Haemophilus parainfluenzar*	GTT^AAC	Blunt
*Hpa*II	*Haemophilus parainfluenzae*	C^CGG	Sticky
*Mae*III	*Methanococcus aerolicus*	^GTNAC	Sticky
*Not*I	*Nocardia otitidis-caviarum*	GC^GGCCGC	Sticky
*Pst*I	*Providencia stuartii*	CTGCA^G	Sticky
*Pvu*I	*Proteus vulgaris*	^CGCGAT	Sticky
*Pvu*II	*Proteus vulgaris*	CAG^CTG	Blunt
*Sal*I	*Streptomyces albus G*	G^TCGAC	Sticky
*Sau*3A	*Staphylococcus aureus 3A*	^GATC	Sticky
*Sfi*I	*Streptomyces fimbriatus*	GGCCNNNN^NGGCC	Sticky
*Sma*I	*Serratia marcescens*	CCC^GGG	Blunt
*Sph*I	*Streptomyces phaeochromogens*	GCATG^C	Sticky
*Taq*I	*Thermus aquaticus*	T^CGA	Sticky
*Xba*I	*Xanthomonas badrii*	T^CTAGA	Sticky

Pattern of DNA Cuttings by Restriction Enzymes

The bonds between deoxyribose and phosphate groups of DNA molecules are hydrolyzed by the restriction enzymes. There will be a phosphate group at the 5' ends and a hydroxyl at the 3' ends in both the strands. Some restriction enzymes cleave only one strand of the DNA molecule. Restriction enzymes cut within their recognition sites in the DNA molecule and according to the pattern of the cuts can produce different types of cut ends (Figure 2). (a) 5' overhang: Asymmetric cuts within the recognition site leads to production of a short single stranded overhang segment from the 5' ends. Example, *Bam*HI (Figure 2a), (b) 3' overhang: Asymmetric cuts within the recognition site leads to the production of short single-stranded overhang segment from the

Figure 2. Different types of cut ends generated by restriction enzymes. (a) A 5' overhang sticky end generated by BamHI, (b) A 3' overhang sticky end generated by KpnI, A blunt end generated by SamI

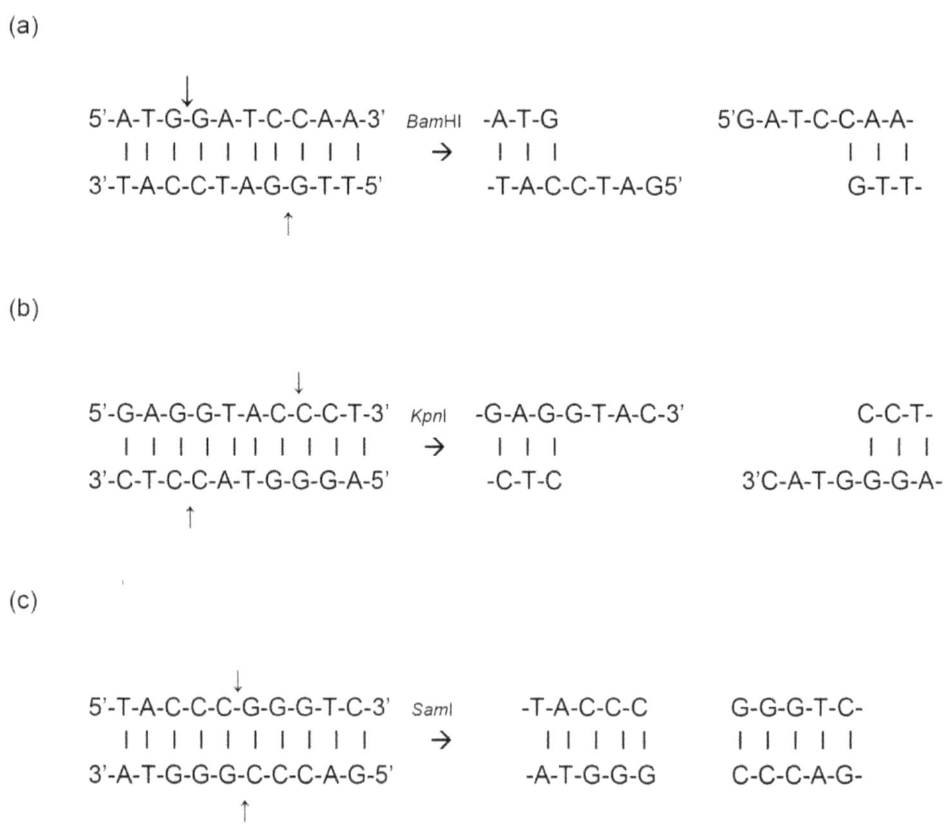

3' ends. Example, *Kpn*I (Figure 2b), and (c) Blunts: Symmetrical cuts within the recognition site in both the strands at precisely opposite locations leads to production of blunt ends without overhangs. Example, *Sma*I (Figure 2c). The 5' and 3' overhangs produced due to asymmetric cuts can readily stick or anneal with the complimentary bases and therefore called sticky ends or cohesive ends.

About 200 different target sequences have been identified to which the 3000 known restriction enzymes bind specifically. Although all restriction enzymes bind DNA specifically, they may also bind to the next best site (single base substitution) at a high rate under optimal conditions. For example, *Eco*RI and *Rco*RVcan can bind to their next best sites 5'-TAATTC-3' and 5'-GTTATC-3' respectively, at a high rate. However under non-optimal conditions binding with next best site is considerably reduced. This loss of specificity or increase in the frequency of binding at next best site is called star activity.

High-Fidelity Restriction Enzymes

Several advances in restriction enzyme research have been made to improve their efficiency. One such advancement is the development of high-fidelity restriction enzymes. These engineered enzymes have the same specificity as their established counterpart with the benefit of reduced star activity. These HF enzymes has fast digestion time (5 minutes) and increased buffer compatibility.

Cutting by DNase

In cases where DNA has very abnormal base composition, restriction endonucleases are unsuitable for cutting DNA. However, availability of wide range restriction enzymes with different recognition site, this should not be a problem. But the problem arises when it becomes necessary to randomly cut the DNA molecule into fragments with a mean size of few hundred base-pairs. Partial digestion with four-nucleotide-recognizing enzyme is not suitable, as fragments would be smaller or larger than the size required. The problem can be solved by digesting with a nuclease enzyme such as DNaseI, which has no sequence specificity. However, DNaseI will not produce sticky ends. Therefore cloning becomes difficult. This problem can also be solved by using appropriate DNA polymerase.

Cutting by Physical Stress

DNA molecules can also be cleaved at random by using physical shearing. This can be achieved by forcing the solution through a narrow opening such as a syringe needle or a pipette tip, or through sonication (which provide high-frequency vibrations). Sonicator is available in different forms. The simplest form consists of a metal probe dipped into the solution, which vibrates at high frequency. In cup-horn sonicator, the solution is put in a tube that floats in a small volume of water. The probe is dipped into water and vibrations generated by the probe are transmitted to the sample through the water. The fragments with varying size will be produced randomly.

Enzymes for Modifying DNA

Numerous enzymes are known that can modify DNA molecules by adding or removing specific chemical groups. Important features of some of the modifying enzymes are described (Wong 2006).

Alkaline phosphatase removes the phosphate group present in 5' terminus of a DNA molecule (Figure 3a). By treatment with alkaline phosphatase, both recircularization and plasmid dimer formation are prevented because ligase cannot join the ends.

Polynucleotide kinase acts by adding phosphate groups onto free 5' termini (Figure 3b). This enzyme is used in radiolabelling the 5' termini for sequencing by Maxam-Gilbert technique (see Chapter 12) and for other uses requiring terminally labeled DNA. It is also used in phosphorylating synthetic linkers and other fragments of DNA that lack termini 5' phosphates in preparation of ligation.

Terminal deoxynucleotidyl transferase adds one or more deoxyribonucleotde onto the 3' terminus of a DNA molecule (Figure 3c)

Methylases enzyme protects DNA molecules from endonuclease through methylation. Sometimes in cloning it is also necessary to protect DNA against cleavage by a particular enzyme. This can be done by treating with an appropriate methylase enzyme, which transfers methyl group onto the DNA from S-adenosyl methionine. For example, protection against digestion by *Eco*RI could be conferred by treatment with *Eco*RI methylase.

Figure 3. The reactions catalyzed by (a) alkaline phosphatase, which removes 5' –phosphate group, (b) polynucleotide kinase, which attaches 5' – phosphate groups and (c) terminal deoxynucleotidly transferage, which attaches deoxyribonucleotides to the 3' termini of polynucleotides in either single stranded or double stranded DNA molecules

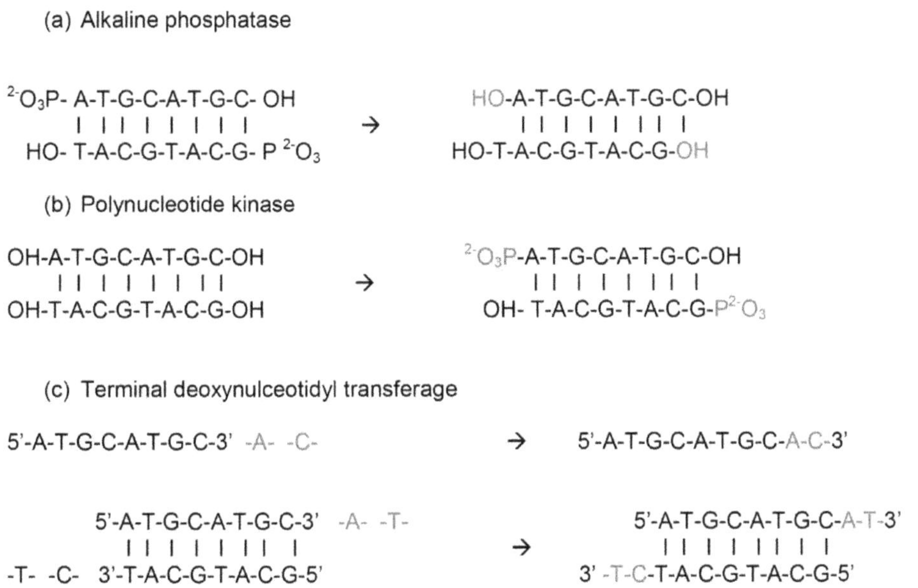

Enzymes for Joining DNA Molecules

Enzymes involved in joining DNA molecules are described in the following section.

DNA Ligases

The nicks produced in the phosphodiester backbone of DNA can be closed by DNA ligase (Figure 4). During replication of DNA, Okazaki fragments are produced, which are required to be joined to restore the continuity of the molecule. This is done by DNA ligase. DNA ligase is also involved in DNA repair process. Two classes of DNA ligases are found. The first, found only in bacteria uses NAD^+ as a cofactor. The other found in bacteriophages, viruses, and eukaryotes, uses ATP as a cofactor. The smallest known DNA ligase (41KDa) has been derived from bacteriophage T7. Eukaryotic DNA ligases are much larger in size. For example, human DNA ligase I is more than 100KDa in size. But most DNA ligases share some sequences and structural motifs.

The ligation of two DNA molecules through DNA ligase occurs in three stages (Figure 4).

Figure 4. Ligation of DNA with complimentary cohesive termini (a) blunt ends, (b) sticky ends

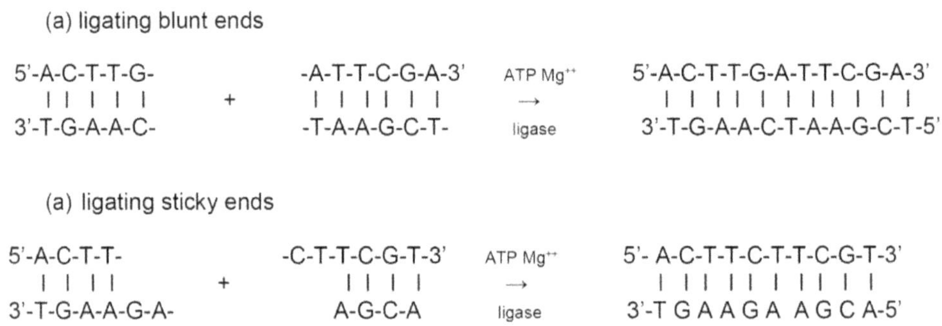

1. Formation of a link between covalent enzyme-AMP intermediate and a lysine side-chain in the enzyme.
2. Transfer of the AMP nucleotide to the nicked DNA strand at the 5' phosphate site.
3. Sealing the phosphate backbone and release of the AMP by attacking the AMP-DNA bond by the 3'-OH of the nicked DNA.

Prior to ligation, the DNA molecule that lack the 5' phosphate termini, has to be phosphorylated. Activation of phosphorylation can be done through T4 polynucleotide kinase and ATP. DNA fragments obtained after digesting with different endonucleases may contain protruding 5' termini that are not compatible (e.g. restriction digest of DNA with *Xba*I and *Hind*III). Such fragments can be made compatible by partial filling of the recessed 3' termini using the Klenow fragment of DNA polymerase from *E. coli* .

Bacteriophage T4 DNA ligase is often used in cloning because it can join DNA molecules of annealed complimentary cohesive DNA termini or nicks, as well as join blunt ended double- stranded DNA molecule, whereas *E. coli* DNA ligase is ineffective in ligating blunt-ended DNA fragments.

When the vector DNA and foreign DNA are both cut with the same restriction enzyme, the overlapping ends of the vector and foreign DNA are compatible and complementary. When these DNA fragments and vector DNA molecules are mixed together, they form complementary base pairs between overlapping terminal single-stranded DNA sequences. Ligases act on DNA substrates with 5' terminal phosphate groups and form the phosphodiester bond between the two adjacent DNA nucleotides and join them. This process is called ligation (Figure 4).

Topoisomerases

This is another enzyme with DNA ligase activity. The normal function of these enzymes is to alter the degree of supercoiling of DNA molecules. They do this by cleaving one or both strands, rotating the duplex, and resealing it. Given a linear DNA molecule with topoisomerase attached to the end and a suitable target molecule, the enzyme will ligate the two. This allows ligation to be accomplished more rapidly than with conventional DNA ligase.

Transposase

Transposable genetic elements are able to move from one place of chromosome (DNA molecule) to another, under the action of a transposase enzyme. This can be used for inserting features like say, origins of replication or antibiotic-resistance genes etc. into a molecule. But its applications are highly specialized.

Recombinase

Several phage-based recombination systems that catalyze breakage and rejoining of molecules at specific sites are known. In bacteriophage lambda and the infected bacterium has such a system of direct recombination between phage and bacterial genome. Integration of the phage DNA takes place by recombination between a site on the phage genome (*att*P) and a site on the bacterial genome (*att*B), to generate a phage-bacterial DNA chimera. The *crc-lox* recombinase system of bacteriophase P1 is frequently used for the detection of regions of DNA flanked by *loxP* sequences, and has applications in the modification of gene expression in transgenic animals.

Enzymes for Replication of DNA Molecules

Different enzymes involved in DNA replication are described in the following section.

Polymerases

DNA polymerases are involved in the synthesis of new strands of DNA from a complimentary strand of an existing DNA or RNA template. These enzymes can normally function when the template has a double-stranded region that acts as the initiation point for polymerization.

Four types of DNA polymerases are used in genetic engineering. First is DNA polymerase I (from *E. coli*) also known as DNA-dependent DNA polymerase. In a double-stranded DNA molecule, the enzyme attaches to a short single-stranded region (or a nick) and synthesizes a new strand. The existing strand is degraded as the synthesis of the new strand proceeds. Thus DNA polymerase I performs duel activity: DNA polymerization and DNA degradation. Different parts of the enzyme control this dual activity. The nuclease activity is performed by the first 323 amino acids of the polypeptide. After removal of this segment, a modified enzyme can be obtained that retains the polymerase function, but is not capable to carry out nuclease activity. This truncated enzyme, called the Klenow fragment. This fragment is still capable of synthesizing a new complimentary DNA strand on a single-stranded template, but cannot proceed further after the nick is synthesized. Klenow fragment is mainly useful during DNA sequencing.

Thermostable DNA polymerases (*Taq* DNA polymerase) are mostly used for amplification of DNA through polymerase chain reaction (PCR). They are isolated from those bacteria which can grow under extremely hot conditions, called thermophilic bacteria (e.g. *Thermus aquaticus*, *Thermococcus litoralis* and *Pyrococcus furiosus*). Some of these bacteria grow at temperature of over 100^0 C. The DNA polymerases obtained from them can function effectively at high temperature under *in vitro* conditions. Thermostable polymerases have their applications in DNA sequencing.

Reverse transcriptase is another type of polymerase enzyme having important in genetic engineering. These enzymes are also known as RNA-dependent DNA polymerase, often abbreviated as RTases, owing to their ability to reverse the usual flow of information by transcription in the 'central dogma'. This enzyme uses RNA as a template and not DNA to synthesize a DNA strand complimentary to the RNA template. The unique ability of this enzyme is central to the technique called complimentary DNA (cDNA) cloning.

Some enzymes can add one or more nucleotides to a molecule without depending on a template. In this category, two enzymes are particularly important. One is terminal transferase (from calf thymus), which can attach a series of deoxyribonucleotides one by one to the 3' end of a DNA molecule. This is used to add trails of a single nucleotide to existing DNA molecules. Another example is *Taq* polymerase used in the PCR. This adds a single dA-residue to the end of a PCR product, and this is exploited during cloning of PCR products. Such enzymes are temperature-independent polymerases.

CONSTRUCTION OF A RECOMBINED DNA MOLECULE

Steps involved in the construction of recombined DNA molecule are described in the following section (Watson 2007).

Frequency of Recognition Sequence in a DNA Molecule

For a particular restriction endonuclease, it is possible to calculate the number of recognition sequences within a DNA molecule of known length. For example, a tetranucleotide sequence (e.g. TACG) should occur once every 4^4 =256 nucleotides and a hexanucleotide sequence (e.g. GCTTAC) once every 4^6 =4096 nucleotides. The assumption of these calculations is that all the four nucleotides are available in equal proportions and that the nucleotides are placed in the DNA strand in a random fashion. In practice, however, these assumptions are not found to be valid. For example, the lambda DNA of 49kb should have 12 restriction sites for a hexanucleotide recognition sequence. But there exist six restriction sites for *Bgl*II, five for *Bam*HI and two for *Sal*I, indicating that GC content in lambda is less than 50%.

Further, restriction sites are not evenly distributed throughout a DNA molecule. Thus when digested with a particular restriction endonuclease, it will not produce fragments of equal size. Therefore, it should be remembered that although it may be possible to calculate the number of restriction sites in a DNA molecule of known size, for a given endonuclease, only experiments will provide the true picture. For construction of recombinant DNA, the DNA molecule to be cloned has to be joined to the vector molecule. This process is called ligation, and the enzyme DNA ligase catalyzes the reaction. The ligation may involve either blunt-end or sticky-end DNA molecules, depending on the restriction enzymes used for cutting the DNA molecules. T4 DNA ligase can carry out both blunt-ended and sticky ended ligations. However *E. coli* DNA ligase can carry out blunt-ended ligations efficiently.

Ligation reaction for joining two blunt-ended fragments is shown in Figure 7.4. Although this reaction can be carried out under *in vitro* conditions, it is not very efficiently. This is due to the fact that ligase enzyme normally unable to find the right molecular ends to be ligated, as it all depends on the chance event. Therefore, to increase the chances of right ends to come together, it is recommended that the for ligation blunt-ends high concentration of DNA should be used.

On the contrary, ligation of complimentary sticky ends is much more efficient. Since compatible sticky ends can pair with one another by hydrogen bonding, it provides a relatively stable structure for the enzymes to work.

Putting Sticky Ends onto a Blunt-Ended Molecule

It is possible to create sticky ends in a blunt-ended DNA molecule, by digesting both the fragment of the DNA to be cloned and the vector. Normally digestion is done by using the same restriction endonuclease, but different enzymes can also be used if the generate the same sticky ends. However, if one of the DNA molecules has sticky-end and the other has a blunt-end, this method is not applicable. In such a situation, one of the following methods may be followed.

Linkers

First of these methods involves use of linkers. These are short piece of synthetic double-stranded DNA molecules, of known nucleotide sequence. A typical linker sequence is blunt-ended and contains a restriction site (Figure 5a). Linkers can be attached to the ends of larger blunt-ended DNA strands by DNA ligase. Usually more than one linker will get attached to each end of the DNA strand, and a chain structure will be produced. But stick ends can be produced by digestion with the restriction endonuclease (Figure 5b). Therefore this modified fragment can be used for ligation into cloning vector, followed by restriction digestion with the same restriction enzyme.

Figure 5. Linkers and their use (a) structure of a typical linker, (b) attachment of linkers to the blunt ends of the DNA molecule and production of sticky ends through restriction digestion

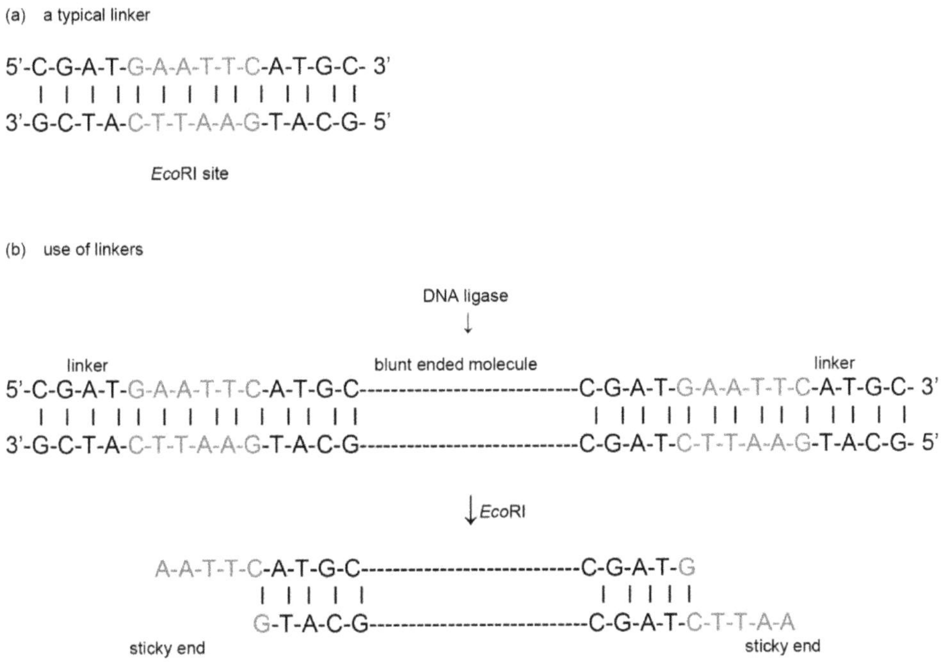

Adapters

Adapters are short synthetic oligonucleotides, having one sticky end (Figure 6a). The purpose is to ligate the blunt ends of the adopter and the DNA fragment, thereby construct a molecule having sticky ends. However, dimmers can be formed due to ligation of sticky ends of individual adapter molecules (Figure 6b), leaving the new DNA molecules blunt-ended (Figure 6c). Recreation of the sticky ends could be done by restriction endonuclease digestion (Figure 6d). But this situation has made no distinction between the linkers and adapters.

To overcome this problem, adapter molecules are synthesized in such a manner that the 5'-P terminus is modified, whereas the 3'-OH terminus of the sticky end remains undisturbed. The phosphate group and the 5'-OH terminus shall be missing (Figure 7). As a result the phosphodiester bridge shall not be formed between 5'-OH and 3'-OH ends, by DNA ligase. Thus the association shall never be stabilized. By this way adapters shall not be

Figure 6. Adapters and their use (a) structure of a typical adapter, (b) two adapters can ligate to produce a molecule similar to a linker, (c) attachment of adapters to the blunt ends of the DNA molecule and (d) production of sticky end through restriction digestion

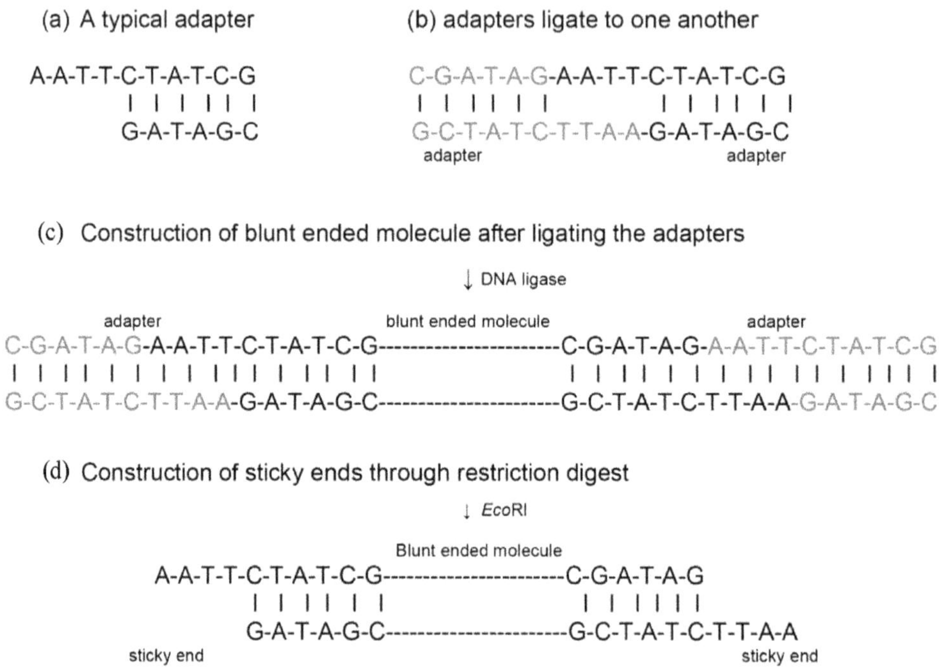

able to ligate themselves but shall be able to ligate to a DNA molecule. After joining the adapters, the 5'-OH end is modified to the natural 5'-P form with the help of polynucleotide kinase enzyme. The sticky-ended fragments thus produced can be incorporated to a vector.

Homopolymer Tailing

A polymer in which all the subunits are identical is called homopolymer. For example, a DNA strand made up of entirely of say, deoxyguanosine, is

Figure 7. An adopter having modified 5' terminus

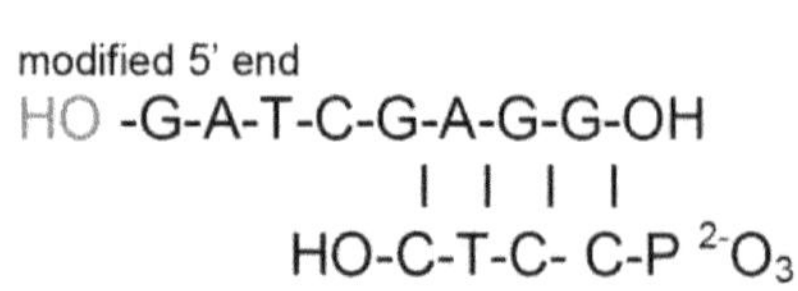

called polydeoxyguanosine or poly(dG). Addition of a series of nucleotides onto the 3'-OH termini of a double-stranded DNA molecule with the help of the enzyme terminal deoxynucleotidyl transferase is called tailing. A homopolymer tail can be produced when this reaction is carried out with just one deoxyribonucleotide (Figure 8a). Attachment of poly dC tails to the vector and poly d(G) to the DNA to be cloned, makes the homopolymers complimentary. When the DNA molecules are mixed, base pairing between the two occurs (Figure 8b).

Since the poly (dG) and poly (dC) tails are not exactly of same size, the recombinant molecules thus produced have nicks and discontinuities (Figure 8c). Therefore, the nicks have to be filled-in by Klenow polymerase followed by sealing with DNA ligase.

MULTIPLICATION OF RECOMBINANT DNA MOLECULE

In gene cloning experiments after creation of the novel recombinant DNA molecule, they have to be introduced into living cells (e.g. bacteria) to multiply and produce clones. Two important purposes are served by cloning. First, from a limited amount, large number of recombinant DNA molecules is produced. Inside bacterium the plasmid (recombinant DNA) divides several times to produce a colony. The bacterial cell itself multiplies and each cell of the colony contains multiple copies of the recombinant molecule. Thereby, large amount of recombinant DNA can be generated to study gene structure and expression (Wong 2006).

Another function of cloning is purification. The ligation mixture may contain, may contain several types of contaminants which has to be removed. The contaminants include, unligated DNA fragments, unligated vector molecule, 'self-ligated' vector without new DNA, and incorrect recombinant plasmid. Unligated molecules usually create no problem, as they will be degrades by the host enzyme and shall be able to replicate under exceptional circumstances. But the self-ligated vector molecules and incorrect recombinant plasmids shall be able to replicate within the host cell. But since usually one cell take up one DNA molecule, it is possible to purify the desired molecule. Since colonies are produced from a single cell, they will contain the same recombinant molecule. Different colonies should contain different molecules. Thus colonies that contain the correct recombinant plasmid have to be identified. Therefore, it is important to know about the different vectors, construction of recombinant molecules, introduction into the bacterial cells

Figure 8. Homopolymer tailing, (a) synthesis of homopolymer tail, (b) construction of a recombinant DNA molecule with a tailed vector and a tailed insert DNA and (c) repair of the recombinant DNA molecule with klenow polymerase and ligase

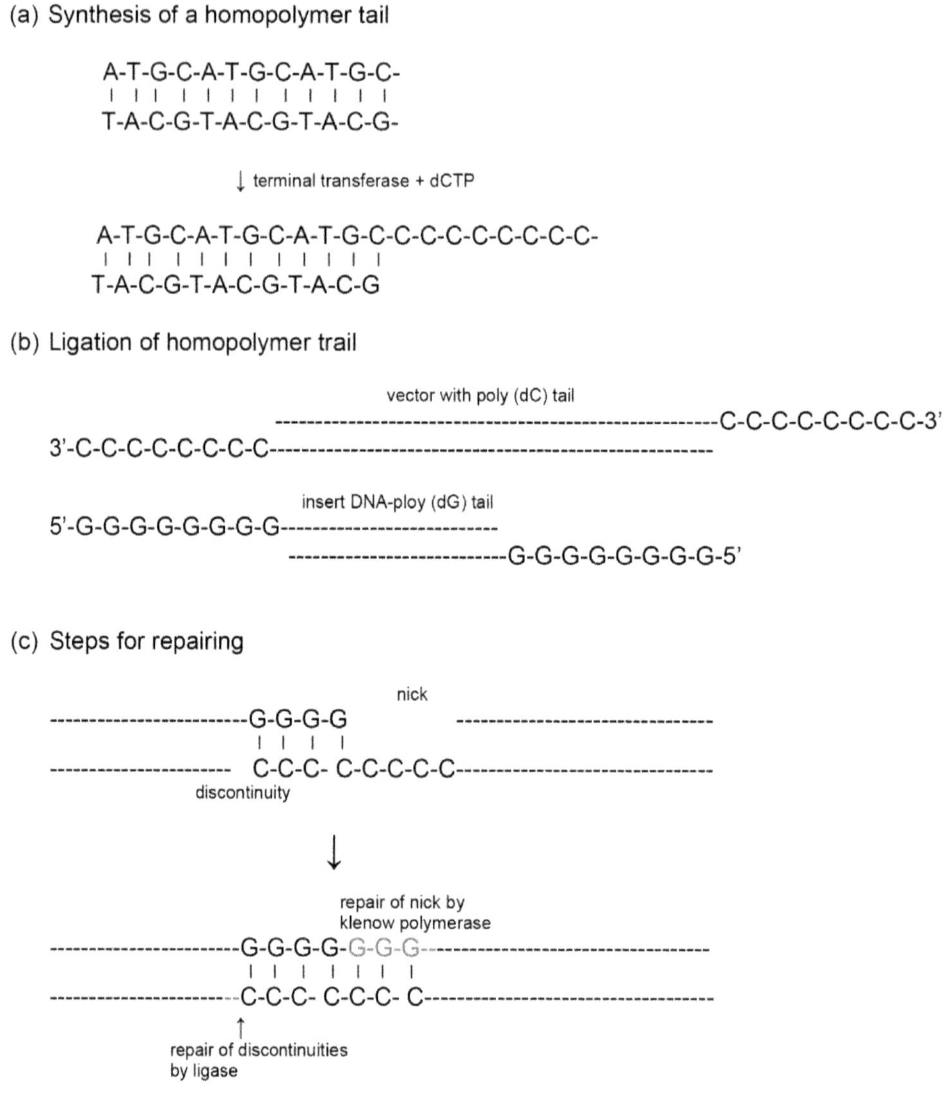

and selection of clones containing the correct recombinant DNA molecule (Wong 2006).

IMOPRTANT FEATURES OF A VECTOR

Important features of a vector are as follows:

- **An Origin of Replication**: The vector must have a replicon that enables it to replicate in host cell, so that the daughter cells have recombinant DNA copies.
- **A Suitable Marker**: The vector should have several marker genes. This is required to distinguish cells that have taken up the vector from those that have not.
- **Suitable Single Restriction Site**: The vector should have a single cleavage site within one of the marker genes so that insertion of foreign DNA into the marker gene leads to its inactivation and identification of recombinant DNA molecule.
- **Suitable Size**: The plasmids should have optimum size. A restriction enzyme cleavage site that comprises six nucleotides will occur on average approximately once every 4^6 bp (i.e. every 4 kbp or so). Therefore a vector that was much larger than 4 kbp might be expected to have several sites for a given enzyme, and cutting the vector would reduce it into several pieces that would be unlikely to be correctly rejoin in a ligation reaction. Although it is possible to remove excess restriction sites, it may not eliminate the problem. Large DNA molecules are very susceptible to physical shearing, and are difficult to handle.
- **Markers for DNA Insertion**: Ideally, the plasmids must have markers where insertion of foreign DNA is possible, so that insertion of foreign DNA can be detected by inactivation of the marker gene. However if the cloning site is not within a functional marker gene, detection of insertion of foreign DNA has to be done by digesting the isolated vector DNA and analyzing the fragments electrophoretically, which is very tedious.
- **Control Elements**: For the expression of cloned DNA, the vector DNA must contain suitable control elements, such as promoters, terminators and ribosome binding sites.
- **High Copy Number**: To maximize the yield of plasmid from transformed cells, the copy number in each cell should be as high as possible. Different plasmids have different copy numbers. When the replication of plasmids tied to replication of the chromosome, the copy numbers are small (1 to 5), and is known as stringent. For plasmids with a different origin of replication have less tightly controlled replication,

may have copy number as high as 50-700, and are call relaxed. The copy number of many low copy-number plasmids can be increased by chloramphenicol amplification. In this method, the host cells are treated with chloramphenicol, which inhibit bacterial protein synthesis. This in turn blocks chromosomal DNA replication. Plasmid replication, however, continues as they require proteins that are more long-lived. Eventually, plasmid DNA replication will stop too, as the supply of general replication proteins, such as DNA polymerase, runs out. In the meantime, the average copy number will increase substantially.

- **Disablement**: It has been a concern about the possibility of recombinant DNA molecules escaping into other bacteria and then spread in the environment. The best possible way to stop spreading is to make the recombinant plasmids disabled, so that they cannot spread to other bacteria through conjugation. Plasmids like pBR322 have been disabled by removing the '*mob*' gene, which is required to mobilize themselves by conjugation. But such plasmids can still be transmitted from cells containing other plasmids that can provide the necessary functions for mobilization. Such transmissions can be stopped by removing the sites called '*nic*' and '*bom*' from the plasmid which are involved in immobilization of the plasmid. Proteins provided by other plasmids can act on these regions. The pUC vector is one such vector having these regions removed.

CLONING PCR PRODUCTS

There may be some applications where the products of polymerase chain reaction are required to be ligated into a cloning vector. Thereafter, the ligated DNA sequence can be analyzed by following any of the standard procedures. However, it will be necessary to resolve certain issues before it will be possible to analyze the inserted DNA. Normally the fragments amplified by PCR are expected to be blunt-ended. In such case insertion into a cloning vector can be done by blunt-end ligation, or by converting them to sticky end with the help of linkers and adapters. But in reality, during synthesis Taq polymerase adds an additional nucleotide (mostly adenosine), to the end of each strand. This implies that a double stranded PCR product will not be blunt-ended, and instead most 3' termini have a single nucleotide overhang. The overhang can be removed by exonuclease enzyme treatment, making the dsDNA a blunt-ended molecule. But it is difficult to restrict with the exonuclease enzyme to cause further damage to the end of the molecules and therefore has limited use. The alternative solution could be to use special cloning vectors, which carries thymidine overhangs. These vectors can easily ligate to the PCR products. Such vectors can be made by digesting the blunt end of a standard vector, followed by *Taq* polymerase treatment in the presence of 2'-deoxythymidine 5'-triphosphate (dTTP). Since no primer is present, the polymerase will add thymidine nucleotide to the 3' end of the blunt-ended vector molecule. This will result into formation of T-tailed vector into which PCR products can easily be inserted (Figure 9).

Figure 9. Cloning of PCR product to a vector. Polymerase will add thymidine nucleotide to the 3' end of the blunt-ended vector molecule. This will result into formation of T-tailed vector into which PCR products can easily be inserted.

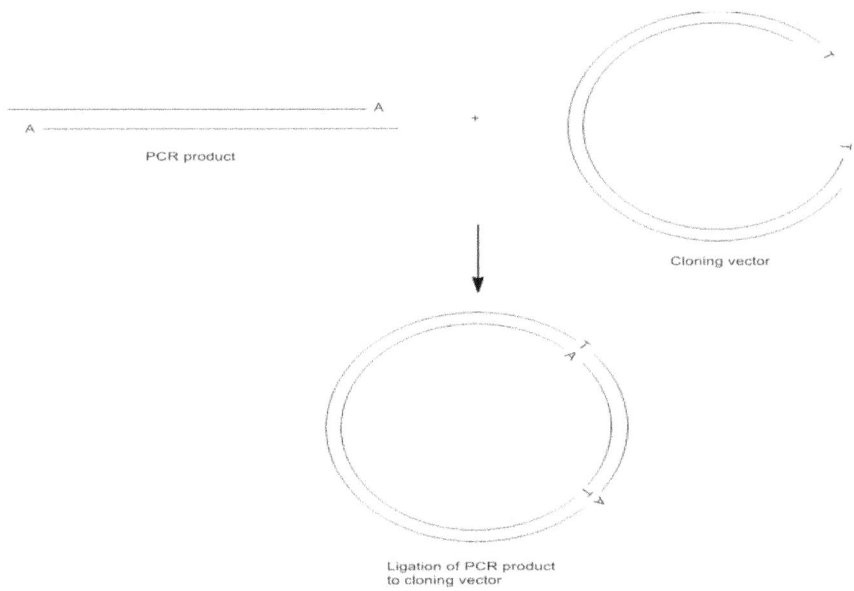

The above problem can also be resolved by another method wherein the primers for the PCR are designed to contain restriction sites. When the PCR products are treated with the specific restriction endonuclease, the molecule will be cut within the primer sequence, producing fragments with sticky-ends that can be efficiently joined to a standard cloning vector (Figure 10a,b). Restriction site should be included within a short extension at the 5' end of each primer, thereby restricting these extensions to hybridize with the template

Figure 10. (a) Use of a primer having a restriction sequence for obtaining a PCR product with sticky end. (b) A primer with a restriction site which can be used to be hybridize to the template molecule.

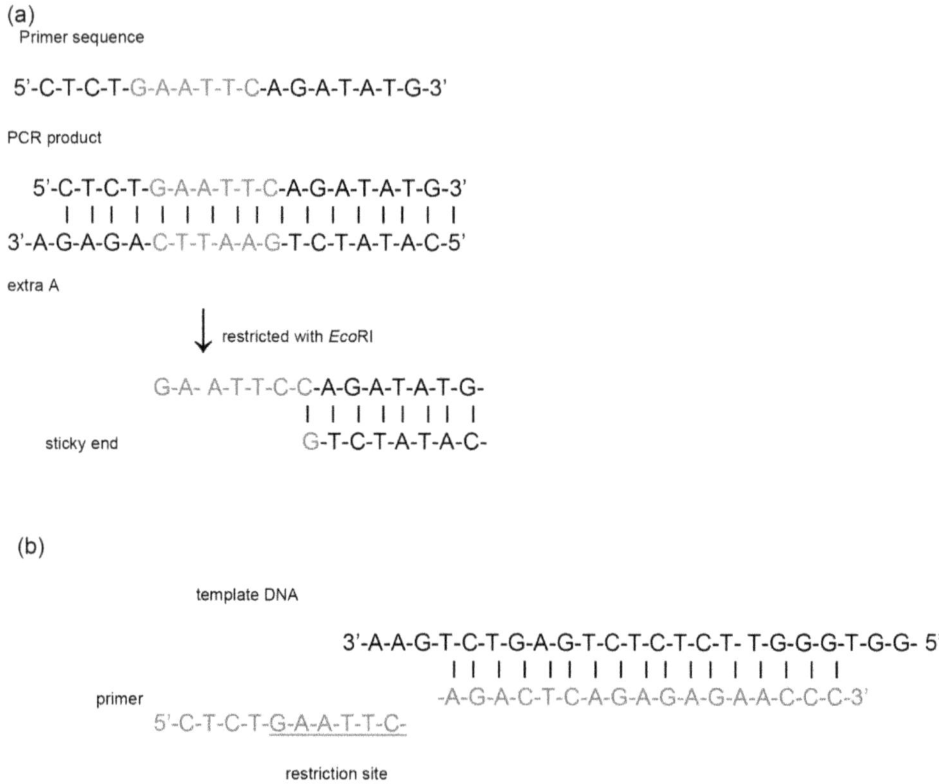

molecule, but can be copied during the PCR. This results into PCR products that carry terminal restriction sites (Watson 2007).

Disablement

It is also important to take precautions to ensure that the host cell carrying recombinant plasmids does not escape and propagate outside the laboratory. To overcome this problem, the host cells usually carry one or more mutant genes conferring auxotrophy (i.e. requirement of a particular metabolite to be supplied in the medium for its growth).

Markers

In the host cells, it is always useful to have genetically distinguishable markers. Several markers like nutritionally deficient, antibiotic resistance, DNA-specific endonuclease inhibitor etc are used in different bacterial hosts.

GENETIC TRANSFORMATION IN PLANTS

Transformation in crop plants is basically to undertaken to produce fertile transgenic plants with the integration of the transgenes at reasonable frequency. After isolation and cloning of the gene to be used for transformation, it must undergo several modifications to be enabled for its insertion into the plant. Successful system for plant transformation include: delivery of DNA to the plant genome without losing viability of the cells, selection of transformed cells, regeneration, production of fertile plants, and transmission of the transgene to the subsequent generations. For successful integration and expression of the transgene the following requirements should be fulfilled (Howe 2007).

1. A promoter which acts as the on/off switch to control the gene expression at different developmental stages in response to certain environmental factors. The most commonly used promoter is cauliflower mosaic virus (CaMV) 35S, which constitutive in nature.
2. Modification of the gene of interest to achieve higher expression in plants. For example, the bacterial origin *Bt* gene for insect resistance has a higher proportion of A-T nucleotide pairs. Since the plants prefer G-C nucleotides, the A-T nucleotides were substituted with G-C nucleotides, without significantly changing the amino acid sequence, in the *Bt* gene, thereby enhancing the production of the gene product in plant cells.
3. Presence of the termination sequence at end the gene.
4. Presences of a selectable marker in the gene construct to identify the integrated transgene.

Some of the other approaches of plant transformation include: (1) high throughput transformation, through which all candidate genes can be used for transformation, (2) plastid transformation, as in many plant species plastid DNA is not inherited, thus preventing gene flow from transgenic plants, (3) construction of chromosome and transformation, through which multiple genes with molecular weights can be delivered into plant cells.

Cloning Vectors for Higher Plants

In higher plants three types of vector systems are used: plasmids that occur naturally in bacteria *Agrobacterium*, plant viruses and direct gene transfer (Jones et al. 2005, Lodge et al. 2007).

Plasmid of *Agrobacterium tumefaciens*

Agrobacterium tumefaciens is a soil-borne bacterium that induces crown gall disease in many dicotyledonous plant species. The disease crown gall occurs due to infection of *A. tumefacians* through a wound in the plant. The bacteria induce proliferation of the stem tissue in the crown region of the infected plants. The bacteria contain a plasmid called Ti (tumor inducing) plasmid which is responsible for inducing crown gall. This is a large (>200kbp) plasmid that carries several genes involved in the infection process. A part of the plasmid DNA, called T-DNA of 15-30 kbp in size, gets integrated into the plant chromosomal DNA. Eight genes present in the T-DNA are responsible for the induction of cancerous properties in the plant cells that are transformed. These genes are also responsible for directing synthesis of compounds, called opines, by the plant cells that are used by the bacterium as nutrients. This is a typical example of naturally occurring genetic engineering where a bacterium uses the plant cell for its own purpose (Figure 11a).

While experimenting on utilization of Ti plasmid as a vector, it was realized that it was impossible to have a unique restriction site with a plasmid of 200 Kbp in size. Thus two different strategies were adopted for inserting new DNA into the plasmid. It was observed that physical attachment of T-DNA with the rest of the Ti plasmid is not required to exert its effect. It was also observed that T-DNA, a relatively small molecule, can complement with the rest of the plasmid in normal form, when present together. Thus although they are present as different entities they can act as an effective transforming system for plant cells. Since the T-DNA is a small molecule it is not possible to have a unique restriction site to be manipulated using standard technique. Therefore a technique called binary vector system is used (Figure 11b).

In the second strategy a new plasmid, developed on the basis of pBR322 or similar *E. coli* vector, but carrying small portion of the T-DNA is used. When both the new molecule and Ti plasmid are present in the same bacteria, recombination in the homologous segment can integrate the pBR plasmid into the T-DNA. In the unique restriction site of the small pBR plasmid, the

Figure 11. (a) The Ti plasmid integration into plant chromosome DNA (b) Binary vector; plasmid A and B complement each other when present together in the bacterium (c) Co-integration strategy.

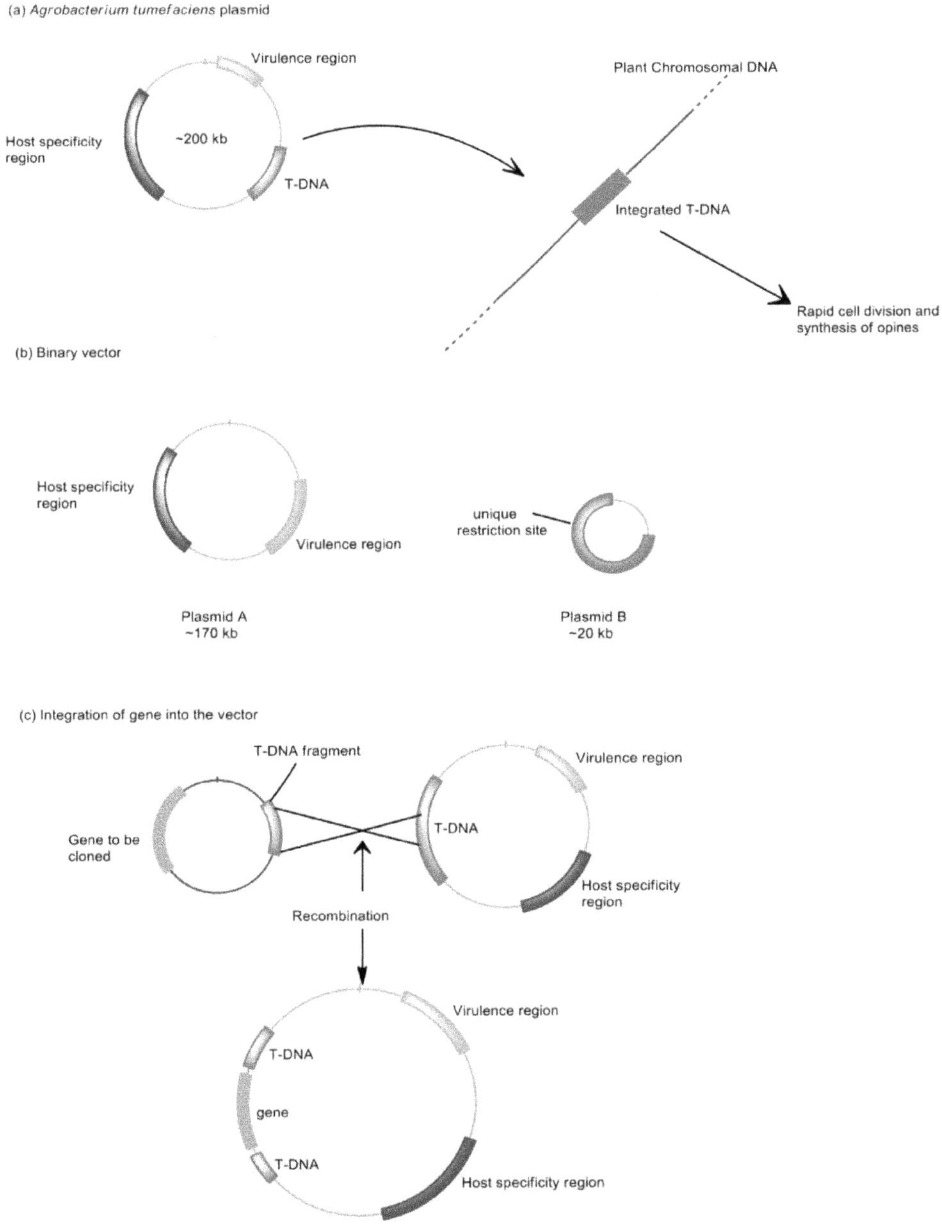

gene to be cloned is inserted. It is then put inside the cells of *A. tumefaciens* carrying a Ti plasmid, and left to the natural recombination process to integrate the new gene into T-DNA. The new gene, along with the rest of the T-DNA, can be introduced into the plant chromosome by infecting the plant cells by this bacterium. This is known as co-integration system (Figure 11c).

Before the Ti plasmid can be utilized as a cloning vector, they have to be modified in such a way that they lose the capacity to display cancerous property (crown gall formation) in the transformed cells. It was discovered that only parts of the T-DNA (25 bp repeat sequence found at the left and right boarders of the region) are integrated into plant genome (DNA). If a new DNA fragment is placed between these two repeat sequences it will be easily transferred to the plant. Therefore, it is possible to remove all the cancer (crown gall) causing genes from standard T-DNA, and insert new genes in that site, keeping the infection process undisturbed. Several Ti cloning vectors without the cancer causing genes have been developed, a common binary vector is pBIN19.

The Ri plasmid found in *Agrobacterium rhizogenes* is similar to Ti plasmid, but unlike Ti plasmid the Ri plasmid induces hairy root disease (massive induction and growth of a roots). This feature has been exploited for obtaining large amount of protein from genes cloned in plants by growing transformed roots in liquid culture, at high density.

Although in nature *Agrobacterium* sp. infect only dicotyledonous plants, it has now been possible to use them for gene transfer in monocotyledonous plants also, through artificial techniques.

Direct Gene Transfer

Direct gene transfer is based on the observation that supercoiled bacterial plasmid DNA, when introduced to plant cell, can integrate into the plant chromosome by recombination and express its genes. The recombination event is poorly understood, as it was observed that integration of the foreign DNA take place randomly in any chromosome at any position. Direct gene transfer makes use of supercoiled plasmid DNA (e.g. pBR322), into which the gene to be cloned and a selectable marker (e.g. resistant to kanamycine) are inserted. The modified plasmid is then introduced directly into plant cells, protoplasts, embryos etc. through biolistic-bombardment with microprojectiles. The modified plasmids (naked or fused with liposomes) can also be introduced into the protoplasts by treating with polyethylene glycol, through a process known as endocytosis. Intact plant cells can be vigorously shaken with DNA-coated silica needle, which penetrate the cell wall and transfer the DNA inside the cell.

Particle (microprojectile) bombardment method is carried out, to invade the plant cells/tissue, by using fine metal particles (tungsten or gold) coated with DNA, which is accelerated with helium gas under pressure. The process is also known as 'gene gun' or biolistic. Parameters of the particle bombardment include pressure (0.6-1.1um), tungsten or gold as material, target distance 97.5-10cm, and cell suspension, callus, meristem, protoplast, immature embryo as target material.

DNA is coated on the surface of the tungsten or gold particles by precipitation with $CaCl_2$ and spermidine. The transgene must incorporate stably into the host chromosome after hitting the nucleus. The foreign gene should also express in the host cell/tissue and perpetuate to the progeny cells to the whole plant. Application of particle bombardment has been particularly useful in monocot species, as it has no host limitations. It is also useful in for transferring organelle genome. One of the drawbacks is that in certain cases it produces multi-copy transformation which may result into instability and silencing.

Particle bombardment can be used any kind of host cell/tissue having potential to regenerate to whole plant. The exogenous DNA used plant transformation comprises a expression cassette inserted into a vector based on a bacterial cloning plasmid having high-copy number. In particle bombardment only the expression cassette is required for transgene expression. Unlike *Agrobacterium*-mediated transformation, cloning vectors are used in particular bombardment for convenience rather than necessity. Accordingly a clean DNA strategy was adopted wherein all vector sequences were removed prior to particle loading.

Particle bombardment is the most suitable method for organelle transformation. Introduction of transgenes into chloroplast genome offers several advantages. These include: very high level of expression of transgenes, uniparental (through eggs) inheritance of plastid gene in most plants (which prevent transmission of transgene through transgene), absence of gene silencing and position effects, integration through a homologous recombination process which facilitates targeted transgene insertion, precise transgene control, elimination of vector sequence, and sequestration of foreign proteins in the organelle (thereby preventing adverse interactions within the cytoplasmic environment).

Overall, *Agrobacterium*-mediated transformation offers several advantages over biolistic technique, such as simple integration pattern that results lower mutational consequences, and low transgene silencing. However, depending upon the objective(s) and the plant species the choice of the methodology to be used shall vary.

Plant Viruses as Vectors

Use of plant viruses as cloning vectors has mostly remained unsuccessful. The replicating genomes of plant viruses are nonintegrative as compared to *A. tumefaciens*, which is integrative in nature. One of the problems is that most of the plant viruses have RNA as genetic material and not DNA. Manipulation with RNA is rather difficult and therefore RNA viruses are not so useful as cloning vector. Only two classes of DNA virus, namely caulimoviruses and graminiviruses, are known to infect higher plants. But none of them are ideally suitable for gene cloning. In caulimovirus the virus genome capable of carrying inserted DNA, after deletion of non-essential sections, is very limited. The problem was solved by using a helper virus. In this strategy, the cloning vector genome lacks several essential genes, and therefore a large gene can be inserted but cannot direct infection by itself. The vector DNA along with normal virus genome is used to inoculate the plants. The normal viral genome will help in entering and packaging of the recombined DNA into the viral head and then spread to different parts of the plant. Although this approach is effective, it does not address another problem, i.e that caulimoviruses have very narrow host range, mainly brassicas. Caulimoviruses have, however, been is a source of highly active promoters that are used to express genes transferred through Ti plasmids or in genes transferred directly in all kinds of plants. They can easily be transmitted by mechanical means (by rubbing on to a leaf surface).

Graminiviruses can infect many agriculturally important plants like maize and wheat and thus could be potential vectors for these plants. These viruses contain single stranded DNA that replicate via a double stranded intermediate and thus makes *in vivo* manipulation in bacterial plasmids more convenient. But during the infection cycle these viruses undergo rearrangements and deletions. When additional DNA molecules are inserted they usually scrambled up and thus unsuitable for a cloning vector. They do not readily be transmitted by mechanical means. With the recent breakthrough in solving these problems, graminiviruses are starting to find special applications in plant DNA cloning.

Binary Vectors

The plasmid that carries an artificial T-DNA is usually called a binary vector. A binary vector consists of artificial T-DNA and the vector backbone. T-DNA is delimited by the boarder sequences, the right boarder (RB) and the left boarder (LB), and may contain a selectable marker for plants, a reporter gene and other gene of interest. The vector backbone contains plasmid replication functions for *A. tumefacians* and *E. coli*, selectable marker for bacteria, and a function for mobilization of the plasmid between the bacteria and other necessary components. Insertion of gene of interest into the binary vector is carried out by standard subcloning techniques.

Till 1990s, *Agrobacterium*-mediated transformation could be applied in dicot plant species. Subsequently, the finding that some of the virulence genes exhibit gene dosage effect, resulted into development of a superbinary vector, with incorporation of additional virulence genes. This superbinary vector was found to be highly effective in transforming various plants, including monocots such as cereals. Thereafter, an improved version of the superbinary vector was developed, having 14.8 kb KpnI fragment that contains the *vir*B, *vir*C and *vir*G genes. The improved version was found to be highly efficient for transformation of the monocot plants, including rice and maize.

Cloning Without a Vector

It was discovered in early 1990's that new genes can be transferred into mammalian cells by microinjection most effectively. When copies of the linear DNA molecules representing specific genes and bacterial plasmids are microinjected into mammalian nuclei, they can get integrated into the chromosomes, often as multiple copies in a tandem. This procedure is preferred over viral vector, as it avoids the possibility of causing any adverse effect by the viral DNA.

Gateway Cloning Technology

Gateway technology originally developed by researchers at Life Technologies, Inc. and then commercialized by Invitrogen, is a molecular biology method that enables transfer of DNA fragments between plasmids in a fast and efficient manner. According to this novel technology, the gene of interest becomes ready for cloning needs after entering the 'gateway'. Gateway cloning is an extremely fast and efficient technology. In this technique all categories of DNA molecules including genomic DNA, cDNA, and PCR products can be cloned. Gateway technology is also applicable for diverse organisms like *E. coli*, insects and mammals.

The Gateway system take advantage of the site-specific recombination reactions, provided by the *att* site, enabling the bacteriophage λ to integrate and excise itself in and out of a bacterial chromosome in combination with enzyme clonase mixes (Curtis and Grossniklaus 2003). The first step to Gateway cloning is to insert the gene of interest into an entry clone. A plasmid having the Gateway *att*L recombination site to which the gene of interest can be inserted is called an entry clone. This clone can be developed by four different ways. The first is the application of enzymes such as restriction endonucleases and ligases. However, instead of normal vector the Gateway vector with the *att*L site should be used. In the second method, a PCR product having terminal *att*B site is created, using primers carrying a 25 bp *att*B sequence, and four G's at the terminal ends. In the entry clone this product will then be inserted through the BP method (discussed later). The entry clones can also be constructed by using cDNA library or an expression clone having the gene of interest.

Once the entry clone is made, one can insert the DNA using LR reaction (Figure 12) or use previously cloned vectors to construct new entry clones through BP reaction (Figure 13). The L's, R's, B's and P's are the key components of the Gateway system, and are restriction sites abbreviated from *att*X, where X is replaced with L,R,B or P. The attachment sequence *att*L are present on both sides their gene of interest in all the entry clones. In the Gateway system these L's are cleaved to generate sticky ends. These sticky ends can bind with the sticky ends of the destination vector containing *att*R restriction site. This process is called a LR reaction. This is how the expression clones is formed. From these expression clones, proteins can be analyzed.

The expression clone thus formed contains parts of the L and R restriction sites. The new site present in the expression clone called B site, and the one present in the byproduct is called P site. The Gateway system is completely reversible, i.e. it is possible to make more entry vectors from the expression clone through BP reaction. In this system when B sites recombine with P sites they generate L's and R's (Earley et al. 2006).

Gateway-based binary vectors developed so far are applicable for dicotyledonous plants. These are not useful for monocot plants primarily because of the limited functionality of promoters that are used to drive either the specific selectable marker or the gene of interest.

Figure 12. GATEWAY cloning reactions: the LR reaction. An entry clone, containing a gene flanked by recombination sites, recombines with a destination vector to yield an expression clone and a by-product plasmid. The result is that a gene sequence in the entry clone is transferred into an expression vector, donated by the destination vector. The by-product plasmid contains the ccdB gene, and hence gives rise to no colonies when using standard strains of E. coli.

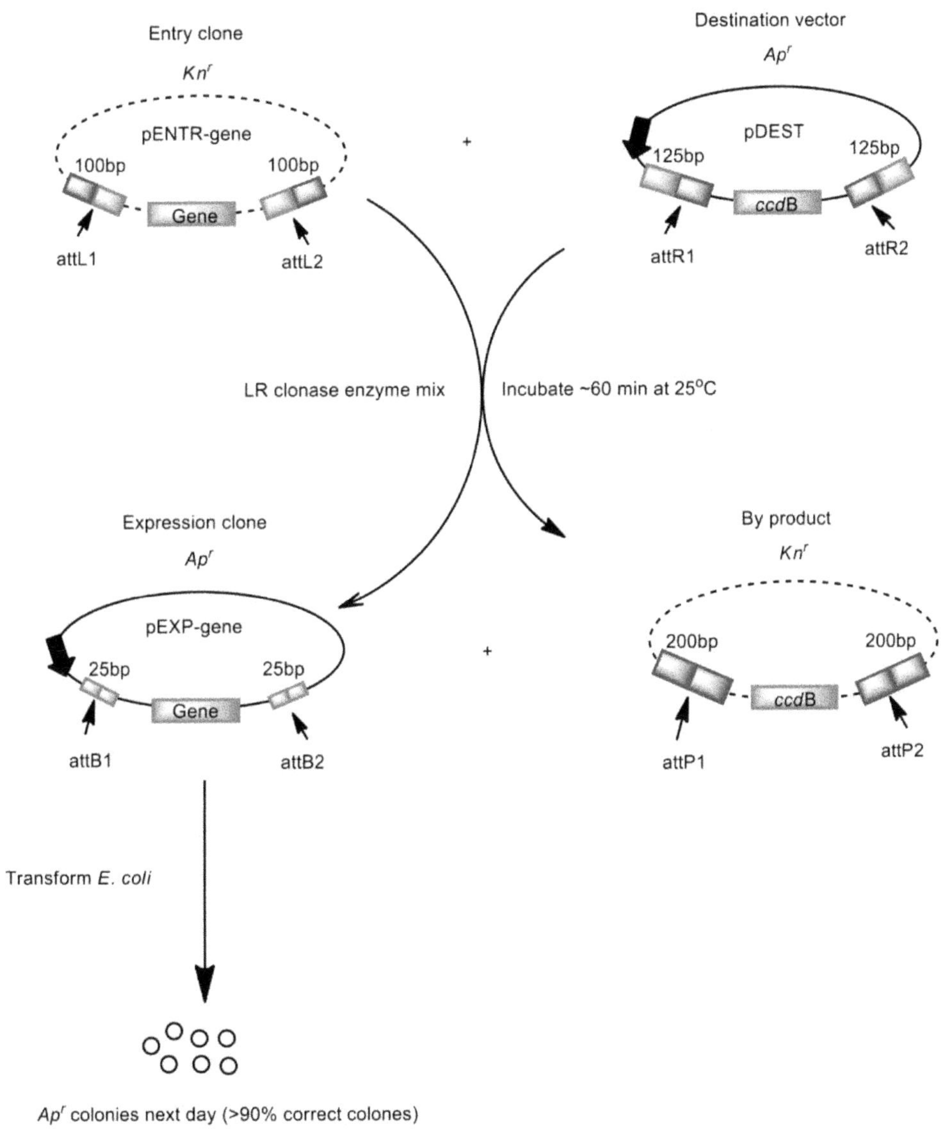

Figure 13. GATEWAY cloning reactions: the BP reaction. A gene in an expression clone can be transferred into an Entry Vector by the BP reaction. Only plasmids without the ccdB gene that are also kanamycin resistant (Kn^r) will yield colonies.

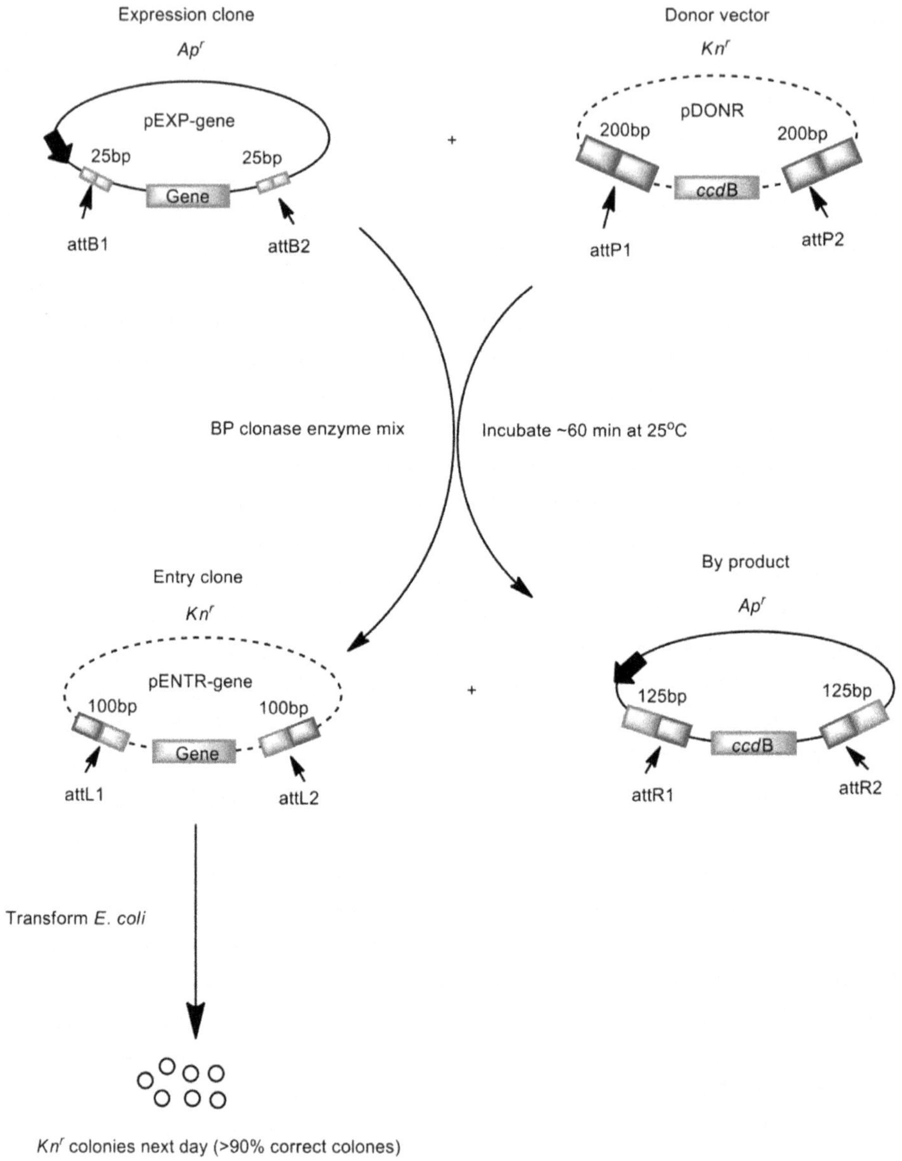

Selectable Marker Gene

Selectable marker gene facilitates the process of transformation process and allows easy recovery of the transgenic crop plants. As only few plant cells shall take-up and stably integrate the foreign DNA, they will be lost in the background of large wild-type cells. Under certain conditions, it may not be required to have selectable markers, as it becomes possible to identify and select transgenic without the support of selectable marker(s).

After integration of the foreign gene into the host chromosome, the next step is to regenerate the transformed cells. This step is very critical, as the frequency of regeneration of the transformed cells vary in different plant species, or in different genotypes of the same species. The second task is to distinguish the transgenic cells from the bulk of non-transgenic cells. To overcome this issue selectable marker genes are used. Genes conferring resistance to various antibiotics and herbicides are normally used as marker genes. In the absence of selectable markers, the transformation efficiency is low in all systems. On the contrary, transformation frequency are extremely high in most crops, in the presence of selectable markers. With high co-transformation frequencies, the transgenic plants can be easily identified with the help of the selectable marker. Following insertion of the foreign gene into the host genome, plant tissues are cultured in selective medium containing either antibiotic or herbicide, depending upon which selectable marker is used. In such a medium, plant tissue having the resistant marker gene shall grow and regenerate to whole plants. It is assumed that these plants will always carry the transgene of interest.

Elimination of Selectable Markers from Transgenic Plants

One of the major reasons as to why transgenic plants are not approved for commercial cultivation is the presence of marker genes, which are usually of bacterial origin. There is a strong apprehension that these markers genes shall cause seriously harm the environment and human health, if consumed for long time. In fact, after obtaining the transgenic plants, there is no need for the marker gene to be present in the transformed plant. Therefore, several methods have been developed to eliminate the marker gene(s) from the transgenic plants. Alternatively, plant breeders have now shifted their attention to used marker genes of plant origin. Considerable progress has been made in this direction. On resolved, there should not be any difficulty in approving the transgenic plants for large scale cultivation. Some of the methods used to eliminate the selectable markers from transgenic plants are as follows (Wong 2006).

Through Co-Transformation

In co-transformation, plant cells are co-transformed with two separate pieces of T-DNA, one with the selectable marker and the other with the gene of interest and selecting the marker-free progeny from the segregating population. Co-transformation can be carried out using either single strain or two strains of *A. tumefaciens*. Selection of the method shall depend on the suitability of the analytical methods required.

Marker-free transgenic plants were produced by another method, by using dual binary vector system pGreen/pSoup. pGreen is a small Ti binary vector which cannot replicate in *Agrobacterium* in the absence of another binary plasmid, pSoup. When pGreen carrying the gene of interest was co-transduced with pSoup, it is possible to produce marker-free transgenic plants in subsequent progeny.

Through Recombination

Recombinases derived from yeast and phages, such as *FLP*, *R* and *cre* which recombines specific sites such as *FRT*, *RS* and *loxP*, respectively, can be used to remove selectable markers. A DNA segment placed between any two of the recombination sites mentioned above can be excised from the plant chromosome, if the corresponding recombinases can be expressed in the plant cells. However, application of this system is yet to be popularized due to non-availability of a dependable system.

It is possible to remove selectable markers by auto-excision vector, which was constructed by placing the promoter, which was specifically functional during morphogenesis/ in pollen/ or in seed, upstream of a site-specific recombinase gene. It has now been clearly shown that germline-specific auto-excision is an efficient, versatile and flexible system for removing selectable marker gene from the transgenic plants.

Through Transposons

Novel T-DNA vectors have been created by using maize Ac/Ds transposable elements, for separating linked genes from T-DNA after insertion into plants. Expression of the Ac transposon can led to transposition of the gene of interest from T-DNA to another location of the same or some other chromosome, resulting separation of selectable marker from the gene of interest.

Through Homologous Recombination

Homologous recombination between direct repeats can be used for removing marker genes from the transgenic cells/shoots. In this method marker genes are flanked by engineered direct repeats. The size and number of the direct repeats used to flank the marker shall determine the rate of excision. Excision is executed automatically, and the loss of marker can be determined by selection. Excision is unidirectional, which leads to accumulation of marker-free genome. Marker-free plants can be obtained from the progeny of appropriate cross, or through vegetative propagation.

Through Positive Markers

The marker removal system developed with a positive marker is called MAT vector. The MAT vector system includes oncogenes (such as *ipt, rol, iaaM/H*) of *Agrobacterium* that control the level of plant hormones produced endogenously and response of the cells to plant growth regulators for differentiation of transgenic cells. The site-specific recombination system (R/RS) is combined with the oncogenes. After transformation, the oncogenes help to regenerate transgenic plants and R/RS system removes the selectable marker. The choice of the promoter for the oncogenes and recombinase (R) gene, and the conditions of the plant material/tissue culture greatly influence the regeneration potential of the transgenic plants and production of marker-free plants.

Although several methods are now available for elimination of marker gene from the transgenic plants, there exist limitations in all these techniques. Therefor choice of the technique shall depend on the plant species, availability of vector system, availability of reproducible tissue culture protocol for regeneration and selection system adopted.

TRANSGENE INTEGRATION, EXPRESSION AND LOCALIZATION

After regeneration of whole plants and production of seeds, progenies of the transgenic plants are required to be evaluated for integration, expression and localization of the transgene.

Transgene Integration

Transgenic lines with complex integration patterns are considered to be undesirable. *Agrobacterium*-mediated integration of DNA is a defined process that results into integration of low copy number of T-DNA. Integration usually takes place in the transcriptionally active sites. Through gene targeting it is possible to place foreign gene sequences at predetermined region of the host genome. Through this it is possible to nullify the so-called position effects on transgene expression.

The integration of transgenes in a pre-determined genomic locus of the host plant can be achieved through site-specific recombinase systems, such as *cre/lox* and *FLP/frt*. Usually homologous recombination follow a simple integration pattern and allow insertion of transgene into a known and stable site of the host genome. Transposons can also be used for site-specific integration through recombination process.

Transgene Expression

Characterization of expression of the transgenes can be done by using reporter gene, transgenes having novel phenotypic traits, transgenes which modify endogenous metabolic activities, inactivation of genes through anti-sense or co-suppression techniques, and identification of genes through complementation. Use of other parameters such as constitutive and non-constitutive promoters, transcription termination, transcript stability, post-transcriptional modifications, efficiency of translation and protein targeting are also equally exploited to determine transgene expression.

The methods used to confirm putative transgenic plant are as follows: Southern blotting, Northern blotting, Western blotting, ELISA, functional assay (test for the presence of selectable marker and the target gene), PCR, *in-situ* hybridization, and progeny analysis.

For agriculture, it is important to have stable inheritance and expression of the transgenes. A perfect transgenic plant should have a single copy of the foreign gene that would segregate in a Mendelian fashion, and show stable expression from in all subsequent generations. One of the concerns with the plant breeders is the aberrant segregation pattern shown by the transgenic cereals plants. Several factors may contribute towards such variations in transgenic plants, including variations induced through tissue culture, position effect, copy number of the transgene, mutation in the transgene, and gene silencing. Gene silencing, loss or decline in gene expression in the progeny of primary transformants, can occur both at the transcriptional or post-transcriptional levels. It is also linked with high copy number of the transgene. Considering the seriousness of the problem, it is recommended that transgenic lines having economically important gene(s), should be tested for expression of the levels of the gene(s) over many generations.

Vectors developed on the backbone of plant viruses can be used efficiently for high level of transient expression of foreign protein in plant cells. Since the virus can replicate within the plant cells automatically, they face no problem in replication and expression of the genes. Viral vectors are currently built on the backbone of plus-sense RNA viruses, such as potato virus or tobacco mosaic virus (TMV) and used for transformation and expression of transgenes in various plants species.

Reporter Genes

Expression of reporter genes can be easily monitored and thus very useful in many ways in plant transformation. Regulation of promoters (strength, temporal, spatial) and other elements can be assayed by connecting these elements to the reporter gene. Constitutive promoters connected to reporter genes can be used to monitor the process of transformation. Expression of reporter genes in plant cells, immediately after the inoculation, is called 'transient expression'. Expression of reporter genes later in a aggregate of plant cells growing in selection media indicate integration of the T-DNA into the genome (chromosome) of the host cells. A good reporter gene should have the following properties (Watson 2007).

1. Express in plant cells
2. Low background activity in transgenic plants
3. No detrimental effect on metabolism of the plant

4. Moderately stable to detect down regulation of gene expression and gene activation
5. Amenable to non-destructive, quantitative, versatile and sensitive assay system

One of the reporter systems, having all the properties, is the Coral-derived red fluorescent protein DsRed, currently used in cereal transformation. Some of the reporter gene systems used in plant transformation is described in the following section.

- **β-Glucuronidase (GUS)**: GUS catalyzes the hydrolysis and cleavage of fluorometric and histochemical β-Glucuronidase substrates. Because of stability and sensitivity of the enzyme it has become the most widely used marker system in plant transformation. Expression of GUS gene can be quantified by fluorometry, and local gene activity can be analyzed by histochemical analysis. However, there exist several problems in using GUS as the reporter gene. First, assay of GUS gene expression is destructive, and GUS protein show high in vivo stability which create problem while monitoring gene deactivation. Further, GUS enzyme activity can be 'leaky' and create error during histological analysis.
- **Luciferase**: The luciferase gene (luc) catalyzes the process of oxidation of D(-)-luciferin in the presence of ATP, thereby generating oxyluciferin and yellow-green light. This property of the luc gene can be assayed in transformed plant cells non-destructively. However, there exist several problems in using luc gene as reporter gene. The luciferin substrate cannot penetrate easily in whole plants. On the other hand, luc genes are widely used as an internal standard with fusion constructs of gus gene for assessing transient expression in transgenic plants. The equipment required to detect and monitor luciferase gene expression is relatively expensive. Because of the above facts use of luc gene as reporter gene has not become popular.
- **Anthocyanin Biosynthetic Pathway Genes**: In maize, the C1, B and R genes codes for trans-acting substances that regulate the anthocyanin biosynthetic pathway. When these genes, with constitutive promoters, are introduced into cereal cells, they induce autonomous pigmentation in non-seed tissue. For this reporter system application of external substrate is not required for its detection.
- **Green Fluorescent Protein**: In 1992, genes expressing green fluorescent protein (GFP) were isolated from jellyfish and have been modified and

transformed into many organisms. The procedure for detection and monitoring of the transgenes expressing GFP has been standardized in many organisms. Further external substrates are not required for detection of GFP and the process is non-destructive. Application of GEP in plant transformation has been restricted due to its relatively weak activity in transformed plant cells. Several modifications have been made to overcome the problem. The variant mgfp5-er gene has been identified to be a potential gene for monitoring transgenes in plants under field conditions. GFP has also been shown as a qualitative marker for detection of linked synthetic Bt cry1Ac endotoxic transgene. Thus GFP has good potential to be used as a reporter system in transgenic plants.

- **Promoters**: Promoters are used for both constitutive and non-constitutive transgene expression in plants. Promoters used for the above mentioned applications are described in the following section.
- **Promoters for Constitutive Transgene Expression:** The cauliflower mosaic virus (CaMV) promoter 35S was initially used for constitutive expression of the transgenes in cereal transformation. CaMV 35S promoter was extensively used in dicot transformation also. However, application of CaMV 35S promoter in cereals showed relatively low activity in transient assays and was not found to be completely constitutive. A number of strategies ware adopted to overcome these problems, which include introduction of an intron or other enhancers within the promoter.
- **Promoters for Non-Constitutive Transgene Expression**s: Several promoters derived from monocot (rice rcS, maize Adh1, wheat His3), dicot (potato pinII, tomato rcS), bacteria (Agrobacterium rhizogenes rolC), and virus (rice tungro bacilliform virus major transcript) are being used as non-constitutive promoters. Choice of the non-constitutive promoter shall depend on the plant species and the breeding objective(s).

INACTIVATION OF TRANSGENE

The reasons for inactivation of transgenes are not conclusive. It might be associated with high copy number, complex integration process, interaction between multiple copies of homologous DNA sequence etc. Inverted repeats of multiple copies and dis-integrity of the inserted DNA segment shall cause gene silencing.

Inactivation of transgenes can happen at various steps of its expression. These include: transcription inactivation due to de novo methylation of the promoter regions or heterochromatin formation, post-transcriptional inactivation due to increased RNA turnover or antisense and defective transcript effects.

Several steps may be taken to minimize transgene inactivation. These are: (1) elimination of repeated elements from transgene, (2) development of recombination system for targeting transgene to suitable location in the genome, (3) regulation of the rate of transcription of the transgene, (4) incorporation of the transgene at specific location of the chromosome by using site-directed gene targeting system such as *cre/lox, FLP/frp*, (5) using double-haploid system to evaluate stability of the transgene in homozygous plants, (6) induction of stress-mediated hypermethylation during tissue culture by applying stress mimics such as butyric acid or propionic acid, and (7) evaluation of expression of transgene under different environmental conditions and different genetic backgrounds. Application of RNAi to understand gene silencing in plant has great potential (Howe 2007).

STACKING OF TRANSGENE

Most important agronomical characters are multigenically controlled. Accordingly for genetic improvement of important traits shall require manipulation of complex metabolic and regulatory pathways involving many genes. Thus it is desirable to put all the genes in one plant to make it most productive. However, this task is difficult to achieve through conventional breeding methods. Advances in molecular biology and biotechnology has generated hope to achieve this goal by integrating multiple transgenes into the plant genome and coordinated expression of these transgenes in transformed plants. Such an approach is called multi-transgene pyramiding or stacking.

Some of the approaches used to produce transgenic plants having multiple transgenes are as follows: (1) co-transformation, in which several transgenes are delivered together in a single transformation experiment, (2) re-transformation, in which several transgenes are stacked into one plant through several successive delivery system, and (iii) sexual crossing, in which crosses are made between transgenic plants carrying different transgenes (Russel and Sambrook 2001).

- **Co-Transformation**: In co-transformation transgenes are present either in one plasmid or in different plasmids. The advantage of co-transformation is that by one single transformation event it is possible to deliver multiple transgenes. One of the difficulties in co-transformation using single plasmid is the construction of a complex plasmid with multiple transgenes. Even after construction of the transformation vector with multiple transgenes, it is often not possible to insert the transgenes in single step, due to presence of restriction sites at undesirable locations.

Co-transformation with multiple vectors has the obvious advantage. The success of this method depends on the co-transformation frequency, i.e. the frequency at which two or more independent transgenes are transferred together and integrated into the plant genome. Results on co-transformation of more than one transgenes using multiple vectors in rice are very encouraging.

Particle bombardment has been found to be the most convenient method for multiple gene transfer to plants. In this method, there is no need for the construction of complex cloning vector, as DNA mixers having any number of different transformed constructs can be used. Successful integration of two/three different transgenes along with selectable marker through particle bombardment have been achieved in several plants. However, the technique has its own limitations such as, undesirable incorporation of complex T-DNA integration pattern, integration of transgenes at different locations in the plant genome which may create various expression patterns, segregation of the transgene in the offspring etc. Therefore, it is important to evaluate the pros and cons of various methods before selecting adopting a method for co0transformation.

- **Re-Transformation:** In this method, the same plant is used for successive transformation for different transgenes. The method has several limitations and therefore should be used only when the repeated transformation is successful.
- **Sexual Crossing**: For sexual crossing, one gene of interest is introduced in one parent and the second gene to another. The progeny of such a cross it is expected that 25 percent of the offspring should have both the transgenes, provided both the parents are hemizygous or homozygous for the transgenes.

The method is technically simple and transgenic populations of each parent can be screened for the presence of the transgenes. However, the method is time consuming, particular when more than two transgenes need to be combined. If the transgenes reside in different chromosomes, the breeding procedure becomes more complicated. However it has been possible to exploit the potential of this method and develop multi-staked maize by Monsanto. This super trait maize carries five transgenes, four of which are synthetic genes linked to regulatory elements from bacteria, viruses and unrelated plants. Expression of two synthetic genes does not inactivated by glyphosate, the herbicide. The third synthetic gene encodes cry3Bb1 protein, which acts against Coleoptera, and the forth gene encodes cry1Ab protein, which provide tolerance to certain Lepidopteran insects. The fifth gene is kanamycin resistance gene which encodes neomycin phosphotetransferase (nptII). Cultivation of this cultivar in millions of hectare in USA has been able to reduce pesticide use substantially.

ANALYSIS OF THE PROGENIES OF THE TRANSFORMED CELLS

Restriction digests of the total genomic DNA of the simplest organism like *E. coli* leads to production of fragments carrying the desired gene as well as fragments carrying all the other genes. During ligation reaction selection of desired fragments is not possible. Consequently after transformation, different types of recombinant clones will be produced. To make the transformation experiment successful, the desired recombinant must be identified. Two different strategies are applied to select the desired clone: direct selection for the desired gene and identification of the clone from a gene library.

In direct selection method the transformants are cultured onto an agar medium, which promote growth of only the desired recombinants, and not others. The colonies that developed from this transformed cell(s) will contain the desired DNA molecule. An example of direct selection is resistance to an antibiotic say, kanamycine. The R6-5 plasmid carries genes resistance to four antibiotics: kanamycine, chloramphenicol, streptomycin and sulphonamide. The gene for kanamycine resistance is present within one of the 13 *Eco*RI cleaved fragments of the plasmid. For cloning this gene, the fragment produced by *Eco*RI in the plasmid R6-5 should be inserted into the *Eco*RI site of the vector (say for example pBR322). Out of the 13 different recombinant DNA molecule produced, one will carry the gene for kanamycine resistance. Transformants are then plated onto agar medium containing kanamycine, on which the cells containing the recombinant DNA with kanamycine resistance gene will only survive (Figure 14). Apart from using antibiotic resistant genes, auxotroph mutant strains of *E. coli* can also be used for direct selection of transformants. For example, a mutant *E. coli* strain has non-functional *trp*A gene (*trp*A$^-$) and is unable to survive in the absence of tryptophan in the medium, is transformed with a vector carrying *trp*A$^+$ gene (derived from the wild type strain), the transformed cells will be able to grow in the absence of tryptophan. Selection of the transformants is done by plating them onto minimal medium, containing the basic nutrients without any supplements (e.g tryptophan) (Lodge et al. 2007).

In the second, technique clones are identified from the gene library. Collection of clones from a particular organism representing every single gene sufficient in number is called a genomic library. Genomic libraries are made by purifying of the total genomic DNA, and then partially digesting with restriction enzyme and cloning the resultant fragments into a suitable vector (Figure 15). For prokaryotes like bacteria and lower eukaryotes like yeast and fungi, due to smaller size of the genome, number of clones required is comparatively small and thus manageable. However, for higher eukaryotes, like plants and animals, a complete library contains large number of clones which makes it difficult to identify the desired clone easily. Therefore, development of a second type of library, specific to a cell type and to the whole organism, may be useful. In multicellular organisms each cells contains the same complements of genes, but in different cell types (tissue – brain, heart, liver, blood etc.), different sets of genes are switch on, while rest are silent. Since in any one type of cells only a few genes will be expressed, the library can be constructed by RNA (mRNA) that is expressed and not by DNA. In this process the resultant clones will contain only a few selected genes from the total genome of the cell. Availability of a cloning method for mRNA would be of great help for those genes which expresses at high rate in a particular cell type. For example, the gliadin gene, a nutritionally important protein in wheat, expresses at high level (>30%) in the cells of germinating wheat seeds. If clones of the mRNA from wheat seeds are made they are expected to contain large number of clones specific to gliadin. Messenger RNA cannot be ligated into a cloning vector. But complimentary DNA (cDNA) derived from mRNA through reverse transcriptase can be used to ligate a cloning vector.

Once the library has been prepared, several procedures can be adopted to identify the desired clone. The procedures involved include, either through detection of translation product for the cloned gene or identification of correct recombinant DNA molecule through hybridization probing, which is an easier technique. Hybridization probing technique can be applied to identify recombinant DNA molecule within bacterial colonies and plaques formed by bacteriophage. The colonies or plaques should be transferred to a nitrocellulose

Figure 14. Procedure for direct selection for the cloned kanamycin resistance (Kan^r) gene

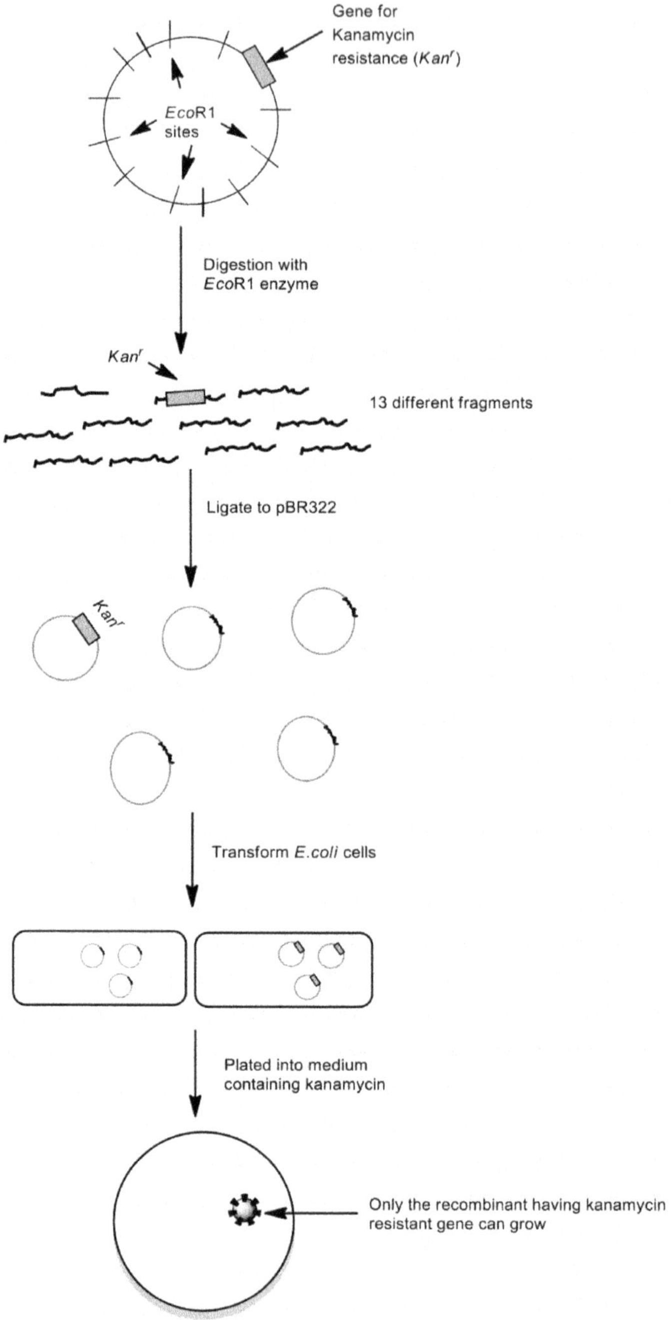

Figure 15. Procedure for preparation of a gene library in a cosmid vector

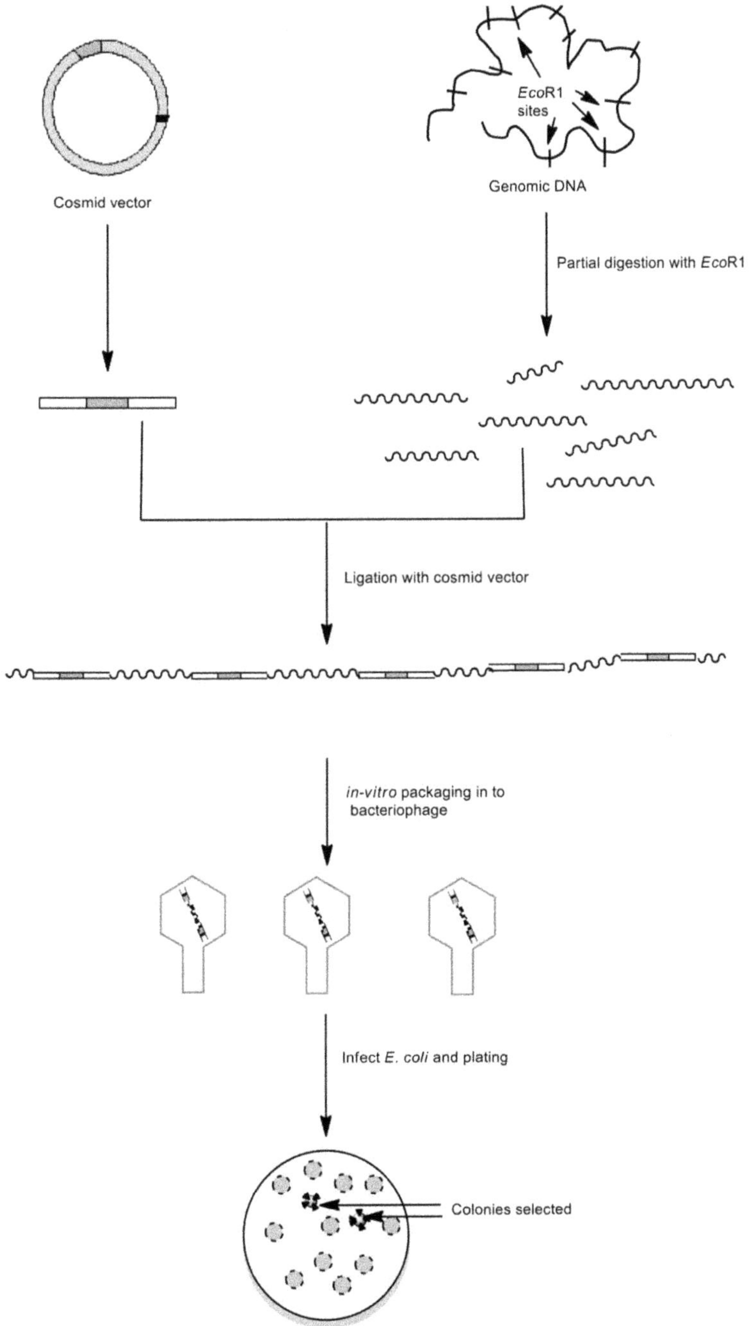

or nylon membrane and the contaminating materials are removed by washing. The DNA is then denatured by chemical treatment. The single stranded molecules thus formed are then fixed tightly to the membrane by exposing them for short period at 80^0 C (for nitrocellulose membrane) or ultraviolet light (for nylon membrane). The molecules bind to the membrane with the help of their sugar-phosphate backbone, and thus the bases are free to pair with complimentary nucleic acid molecules. The labeled probe is denatured by heating and poured as a solution to the membrane and incubated to hybridize. After a period to allow hybridization, the membrane is washed thoroughly to remove unbound material, dried and the position of the bound probe detected. Labeling of the probe can be done through radioactive nucleotide, or through non-radioactive fluorescent marker (Lodge et al. 2007).

Labeling with radioactive nucleotide can be done either through nick translation, end filling or random priming techniques. In non-radioactive labeling, deoxyuridine triphosphate (dUTP) nucleotides are modified by treating with biotin (an organic molecule), that has strong affinity for avidin (a protein). The site of the hybridized biotinylated probe can be determined by washing with fluorescent marker attached avidin. In the second procedure of non-radioactive hybridization probing, a complex is made between the probes DNA with horseradish peroxidase enzyme. Detection is made based on the ability of the enzyme to degrade luminol and emission of chemiiluminescence. Normal photographic films can be used for recording of the signal.

The method of identification of cloned genes through detection of the transient product of the gene can be carried out for those genes, where amino acid sequences of the protein coded by the gene(s) are available. Based on the genetic codes of the amino acid sequences, the nucleotide sequence of the relevant gene can be predicted. However, these predictions will always be approximation, as only methionine and tryptophane can be assigned to triplet codon unambiguously. At least two codons are assigned to all other amino acids. How this problem can be resolved as described in the following example. The amino acid sequence of cytochrome c protein from yeast is available and a sequence amino acids starting from 59 to 64 runs as Trp-Asp-Glu-Asn-Asn-Met. The corresponding genetic code for this region will be TGG-GAT(C)-GAA(T)-AAT(C)-AAT(C)-ATG. Under laboratory conditions oligonucleotides of up to 50 nucleotides in length can be synthesized easily. Thus based on the predicted nucleotide sequence, a probe can be constructed and can be used to identify a gene coding for the protein. In the present example of yeast cytochrome c, all the 16 possible ologonucleotides that can code for the amino acids of this particular segment can be synthesized and used as a probe for the yeast genome or cDNA library. The probe that has the correct oligonucleotide sequence for this region can be identified through its hybridization signals. The results can be checked by constructing a second probe, the sequence of which has been predicted from a different portion of the cytochrome c protein. However, the segment of protein selected for prediction of the second probe must be chosen carefully, so that it does not lead to requirement of production of several thousand different nucleotide sequences.

CONCLUSION

Gene cloning is a common practice in molecular biology laboratories that is used by researchers to create copies of a particular gene for downstream applications, such as sequencing, mutagenesis, genotyping or heterologous expression of a protein. The traditional technique for gene cloning involves the transfer of a DNA fragment of interest from one organism to a self-replicating genetic element, such as a bacterial plasmid. This technique is commonly used today for isolating long or unstudied genes and protein expression. A more recent technique is the use of polymerase chain reaction (PCR) for amplifying a gene of interest. The advantage of using PCR over traditional gene cloning is the decreased time needed for generating a pure sample of the gene of interest. However, gene isolation by PCR can only amplify genes with predetermined sequences. For this reason, many unstudied genes require initial gene cloning and sequencing before PCR can be performed for further analysis.

Gene cloning also involves in the production *in vitro* of new DNA molecules which contain novel combinations of genes or oligonucleotides and the propagation of such recombinant DNA molecules by the exploitation *in vivo* of the replicative mechanisms of bacteria and other organisms. The developments of genetic engineering techniques have permitted the alteration of the genome of microorganisms so that it produces substances of little intrinsic value but of great medical or economic value to mankind.

Foreign genes have been implanted into the DNA of *E. coli* to enable the production of useful proteins. In agriculture, techniques have been developed which permit the transfer of the characteristics of one plant to another through bacterial infection. Such techniques may create new varieties of plants with desirable characteristics, e.g. resistance to infection, the ability to withstand adverse weather conditions or the capability of nitrogen fixation.

REFERENCES

Brown, T. A. (2015). *Gene cloning and DNA analysis: an introduction* (7th ed.). Wiley.

Curtis, M. D., & Grossniklaus, U. (2003). A Gateway cloning vector set for high-throughput functional analysis of genes in plants. *Plant Physiology*, *133*(2), 462–469. doi:10.1104/pp.103.027979 PMID:14555774

Earley, K. W., Haaag, J. R., Pontes, O., Opper, K., Juehne, T., Song, K., & Pikaard, C. S. (2006). GATEWAY-compatible vector for plant functional genomics and proteomics. *The Plant Journal*, *45*(4), 616–629. doi:10.1111/j.1365-313X.2005.02617.x PMID:16441352

Faraday, P. (2018). *Gene cloning*. Syrawood Publishing House.

Howe, C. J. (2007). *Gene cloning and manipulation* (2nd ed.). Cambridge University Press., Retrieved from www.cambridge.org/9780521817936 doi:10.1017/CBO9780511807343

Lodge, J., Lund, P., & Minchin, S. (2007). *Gene cloning: principles and applications*. Taylor and Francis. doi:10.4324/9780203967287

Russel, D. W., & Sambrook, J. (2001). *Molecular cloning: a laboratory manual*. Cold Spring Harbor Laboratory Press.

Watson, I. D. (2007). Recombinant DNA: genes and genomics: a short course. San-Francisco: WH Freeman.

Wong, D. (2006). *The ABC of gene cloning*. Springer.

ADDITIONAL READING

Cheng, M., Lowe, B. A., Spencer, T. M., Ye, X., & Armstrong, C. L. (2004). Factors influencing Agrobacterium-mediated transformation in monocotyledonous species. *In Vitro Cellular & Developmental Biology. Plant*, *40*(1), 31–45. doi:10.1079/IVP2003501

Chung, S. M., Frankman, E. L., & Tzfira, T. (2005). A versatile vector system for multiple gene expression in plants. *Trends in Plant Science*, *10*(8), 357–361. doi:10.1016/j.tplants.2005.06.001 PMID:15993643

Chung, S. M., Vidya, K., & Tzfira, T. (2006). *Agrobacterium* is not alone: Gene transfer to plants by viruses. *Trends in Plant Science*, *11*(1), 1–4. doi:10.1016/j.tplants.2005.11.001 PMID:16297655

Conner, A. J., Barrell, P. J., Baldwin, S. J., Lokerse, A. S., Cooper, P. A., Erasmuson, A. K., Nap, J.-P., & Jacobs, J. M. E. (2007). Intragenic vectors for gene transfer without foreign DNA. *Euphytica*, *154*(3), 341–353. doi:10.100710681-006-9316-z

Coutu, C., Brandle, J., Brown, D., Brown, K., Miki, B., Simmonds, J., & Hegedus, D. D. (2007). pORE: A modular binary vector series suited for both monocot and dicot plant transformation. *Transgenic Research*, *16*(6), 771–781. doi:10.100711248-007-9066-2 PMID:17273915

Himmelbach, A., Zierold, U., Hensel, G., Riechen, Y., Douchkov, D., Schweizer, P., & Kumiehn, J. (2007). A set of modular binary vectors for transformation of cereals. *Plant Physiology*, *145*(4), 1192–1200. doi:10.1104/pp.107.111575 PMID:17981986

Jones, H. D., Doherty, A., & Wu, H. (2005). Review of methodologies and a protocol for the *Agrobacterium*-mediated transformation of wheat. *Plant Methods*, *6*, 1–5. PMID:16270934

Karimi, M., Depicker, A., & Hilson, P. (2007). Recombinational cloning with plant gateway vectors. *Plant Physiology*, *145*(4), 1144–1154. doi:10.1104/pp.107.106989 PMID:18056864

Komari, T., Imayama, T., Kato, N., Ishida, Y., Ueki, J., & Komari, T. (2007). Current status of binary vectors and super-binary vectors. *Plant Physiology*, *145*(4), 1155–1160. doi:10.1104/pp.107.105734 PMID:18056865

Komari, T., Takakura, Y., Ueki, J., Koto, N., Ishida, Y., & Hiei, Y. (2006). Binary vectors and super-binary vectors. In K. Wang (Ed.), *Methods in molecular biology: Agrobacterium protocols* (2nd ed., Vol. 1, pp. 245–256). Humana Press.

Old, R. W., & Primerose, S. B. (1985). *Principles of gene manipulation: an introduction to genetic engineering*. Blackwell Scientific Publications.

Pattern, C. L., Glick, B. R., & Postmak, J. (2008). *Molecular chemistry: principles and applications of recombinant DNA*. ASM Press.

Shrawat, A. K., & Lorz, H. (2006). *Agrobacterium*-mediated transformation of cereals: A promising approach crossing barriers. *Plant Biotechnology Journal*, *4*(6), 575–603. doi:10.1111/j.1467-7652.2006.00209.x PMID:17309731

Tzfir, T., Kozlovsky, S. V., & Vitaly Citrovsky, V. (2007). Advanced expression vector systems: New weapons for plant research and biotechnology. *Plant Physiology*, *145*(4), 1087–1089. doi:10.1104/pp.107.111724 PMID:18056858

Tzfira, T., & Citovsky, V. (2006). *Agrobacterium*-mediated genetic transformation of plants: Biology and biotechnology. *Current Opinion in Biotechnology*, *17*(2), 147–154. doi:10.1016/j.copbio.2006.01.009 PMID:16459071

Xu, Y. (2010). *Molecular plant breeding*. CABI International. doi:10.1079/9781845933920.0000

APPENDIX

1. What do the following terms signify?

Endonuclease, exonuclease, cloning vector, polymerases, topoisomerases, isoschizomers, high fidelity restriction enzymes, homopolymers, bianary vector, star activity, competent cells, retroviruses.

2. is the biological function of a restriction enzyme?
3. What common feature is present in most base sequences recognized by a restriction enzyme?
4. What type of cuts is made by restriction enzymes in dsDNA molecule? Describe them.
5. Under what conditions two different restriction enzymes yield identical pattern of cuts?
6. What are the ideal properties of a cloning vector?
7. What is insertion inactivation? How this function can be used in cloning?
8. What is Gateway cloning technology? In what ways, Gateway technology is superior over the conventional cloning technology?
9. What is the difference between a genomic library and a cDNA library? What are the major differences in the structure of a gene cloned from genomic and cDNA library?
10. Why pBR322 is the most popular plasmid cloning vector?
11. What is the difference between phagmid and cosmid vectors? What are the benefits of using cosmids as a vector?
12. What strategy was adopted to construct Yeast Artificial Chromosome (YAC) as a cloning vector?
13. What are the different vectors available for cloning genes in higher plants? Explain the importance of each one of them.
14. What strategy was adopted to use *Agrobacterium tumefaciens* as a cloning vector in plants?
15. How heterologous probing can be used to identify related genes?
16. The restriction enzyme *Hind*III was used to cut (a) linear DNA molecule containing six *Hind*III restriction sites and (b) a circular DNA molecule also containing six *Hind*III restriction sites.
    a. How many fragments are produced from the linear molecule and from circular molecule?

b. Show how many of these fragments produced from the linear molecule and circular molecule can circularized and how many cannot and why?

17. Restriction enzyme X was used to cut phage T5 DNA and Y to cut phage T7 DNA. A particular fragment of T5 of 2 kbp size was mixed with a particular fragment of T7 of 1.5 kbp size and treated with low concentrations of DNA ligase. This treatment has resulted into three major circular forms. All of them can be cut with both X and Y restriction enzymes, resulting either one or two fragments. Diagram the most likely structures of these circles. Can you guess about the restriction enzymes X and Y?
18. One of the strategies adopted for selection of cloned gene is through antibiotic resistance genes. Is there any other method(s) by which cloned genes can be selected?
19. How RNA viruses can be used as cloning vectors?

Chapter 8
Hairy Roots

ABSTRACT

Agrobacterium rhizogenes induces hairy root disease in plants. The neoplastic (cancerous) roots produced by A. rhizogenes infection, when cultured in hormone free medium, show high growth rate and genetic stability. These genetically transformed root cultures can produce levels of secondary metabolites comparable to that of intact plants. Several elicitation methods can be used to further enhance the production and accumulation of secondary metabolites. Thus, hairy root culture offer promise for high production and productivity of valuable secondary metabolites in many plants. Hairy roots can also produce recombinant proteins from transgenic roots, and thereby hold immense potential for pharmaceutical industry. Hairy root cultures can be used to elucidate the intermediates and key enzymes involved in the biosynthesis of secondary metabolites, and for phytoremediation due to their abundant neoplastic root proliferation property. Various applications of hairy root cultures and potential problems associated with them are discussed in this chapter.

INTRODUCTION

Hairy root is a disease of higher plants caused by the bacterium *Agrobacterium rhizogenes*, a gram negative soil born bacterium. When the bacterium infects the plant, the pathogen transfers a DNA segment (T-DNA region bounded by 25 bp direct oligonucleotide repeats) from its large root-inducing (Ri) plasmid into the infected plant. The T-DNA integrates into the nuclear genome of the

DOI: 10.4018/978-1-7998-4312-2.ch008

host plant. The T-DNA carries a set of genes that code for enzymes for the phytohormone auxin control and cytokinin biosynthesis (*iaaM*, *iaaH*, *ipt*) and also for opines (an unusual amino acids). Expression of these new hormones induces the formation of proliferated roots. These roots emerge at the wound sites and are called hairy roots (HR).The mechanism of transfer of T-DNA from *A. rhizogenes* to the host plant is similar to the mechanism involved in *A. tumefaciens,* which causes crown gall disease in the infected plants.

The hairy roots have unique characteristics like fast hormone independent growths, lack of geotropism, lateral branching, and genetic stability. Further, hairy roots can be cultured under *in vitro* conditions indefinitely. Because of these characteristics hairy roots cultures were investigated for several decades to exploit the possibilities of their commercial use. One of the objectives was to exploit the production of high value secondary metabolites. Long term aseptic hairy root cultures have been established from more than 200 plant species, having the ability to synthesize a wide variety of secondary metabolites and to adjust their metabolic activities in response to biotic and abiotic stress. However, it is important to note that not every hairy root culture displays these characteristics. Further, many researchers have experienced problems with hairy root initiation and maintenance. But the overwhelming positive results have generated hope for its utilization.

In some species, regeneration of whole plants has been possible from the hairy roots. Transgenic plants have been obtained in 89 different plant taxa, representing 79 species from 55 genera and 21 families, through *A. rhizogenes* mediated transformation. The transgenic plants show a characteristic phenotype, called HR syndrome, which include reduced apical dominance in stems and roots, shortened internode, high growth rate of roots in culture, wrinkled leaves with increased width to length ratio, plagiotropic roots, with altered geotropism, altered flower morphology, late flowering, reduced fertility and reduced pollen and seed production.

HAIRY ROOT INDUCTION AND SELECTION

Procedures for induction and selection of hairy roots are described in the following section.

Establishment of Hairy Root Culture System

Establishment of hairy root culture system involves several requirements. These include: the bacterial strain of *Agrobacterium rhizogenes*, an appropriate explant, a proper antibiotic to eliminate redundant bacteria after cocultivation, and a suitable culture medium. On the basis of types of opines produced, *A. rhizogenes* strains can be divided into five lines: octopine, agropine, nopaline, mannopine, and cucumopine. Owing to their strongest hairy root induction ability, agropine strains are the most often used strains. Explants from leaf, stem, shoot tip, stalk, petiole, cotyledon, hypocotyls, tubers, and protoplast can be used to induce hairy roots. However, selection of right type of explants and its age are critical for the success of induction of hairy roots. Juvenile material is the preferred choice.

For induction of hairy roots, explants are separately wounded and cocultivation or inoculated with *A. rhizogenes*. After 2-3 days, the explants are transferred into solid media with antibiotics, such as cefotaxime sodium, carbencilli disodium, vancomycin, ampicilin sodium, claforan, streptomycin sulfate, or tetracycline. Antibiotics are used to kill or eliminate the redundant bacteria at a concentration ranging from 100 to 500 g/ml. Depending on the plant species and type of explants used, hairy roots are normally induce within a week to over a month. Hairy roots thus induced can be sub-cultured on phytohormone-free medium.

To activate the virulence genes of *A. rhizogenes* and to enhance the transfer of foreign genes into the plant genomes, acetosyringone (ranging from 10 to 150 μM) has been used (Kumar et al. 2006).

For high production of secondary metabolites, optimization of the nutrients in the culture medium and physical factors of the culture, are equally important. Factors such as carbon source and its concentration, the ionic concentration of the medium, phytohormones, light, temperature, pH of the medium and inoculums are known to influence growth and secondary metabolism. Supplementation of auxin and elicitors often increases the levels of secondary metabolites. In view of the above findings, it is important to determine the requirements of nutrient conditions for hairy root culture of each species and each clones.

A protocol for *A. rhizogenes*-mediated hairy root transformation was developed for *Phaseolus vulgaris*, a recalcitrant plant species (Estrada-Navarrete et al. 2007). However the efficiency of hairy roots transformation by this method is highly variable, and the transformation frequency never goes beyond 70 percent. By utilizing radical severed, cotyledon-bearing common bean (*P. vulgaris*) young seedlings as explants, 100% transformation was obtained through *A. rhizogenes*. The same protocol was reported to be equally effective for soybean (Khandual and Reddy 2014). A different technique called sonication-assisted *Agrobacterium*-mediated transformation (SAAT) was developed to induce hairy roots in those plant species which are difficult to transform (Trick and Finer 1997).

Reporter Gene

To monitor transfer of genes from *A. rhiziogenes* to the plants, a gene which expresses in the plant and can be easily identified is required. Accordingly, a gene named β-glucuronidase (*GUS*) which fulfills this requirement has been identified. This gene is transferred from *A. rhizogenes* into hairy roots, and it can be easily analyzed by histological assay. Thus this gene is called as reporter gene. It is the most common means of monitoring in most of the plant systems. In some cases, neomycin phosphotransferase II (*NPT-II*) encoding the kanamycine-resistance enzyme has been used. However, sometimes both *GUS* and *NPT-II* were used. Another gene named green fluorescent protein (*GFP*) was also used as reporter gene in *Catharanthus roseus* L.

Infection Condition

The infection condition and the choice of the *A. rhizogenes* strain are of paramount importance for the success of transformation. The strain virulence has strong repercussions on the properties (morphology, growth rate and metabolite level) of the transformed plant tissue. The strain LBA 9402 has showed stronger infective ability, while the strain R1601 generated a faster growing clone on *Rheum palmatum*

Selection of Hairy Root Line

The site of integration of T-DNA into the host plant genome is usually uncertain. Therefore, hairy roots obtained after integration of T-DNA show different accumulation pattern of secondary metabolites. After analyzing 45 hairy root clones of *Duboisia leichhardtii* F, it was observed that there was considerable variation in growth rate, alkaloid content, and productivity among the clones. Usually sub-culturing of hairy roots is easy and they remain stable over time. However, certain amount of heterogeneity has been reported, even in hairy roots obtained from the same root tips. Therefore, it is important to follow a stringent selection process to obtained high secondary metabolite producing hairy root line (Yukimune et al. 1994).

AGROBACTERIUM RHIZOGENES GENES AND INDUCTION OF HAIRY ROOTS

Development of hairy roots in higher plants by *A. rhizogenes* is caused by the transfer of one or two fragments of T-DNA from a root-inducing (R_i) plasmid to the host plant genome. Among the 18 ORFs located in the R_i T-DNA, four coincide with genetic loci (*rol*A, B, C, D). These genes correspond to ORFs 10, 11, 12, and 15 of TL-DNA. Two other ORFs, ORF13 and ORF14 also play significant roles in the induction of roots on carrot disks and tobacco leaf segments.

When inserted individually, *rol* genes affect plants growth and development, each with its distinctive features. Individually they do not show HR syndrome, which develops probably due to the synergistic action of these genes. The *rol*C gene has been most widely studied because its effects are the most advantageous for improving ornamental and horticultural traits. Effect of this gene include: reduced apical dominance, altered leaf morphology, reduced seed production, dwarfness, increase in lateral shoots, smaller flowers, advanced flowering, better rooting capacity. The *rol*B show flower heterostyly abundant adventitious rooting. Transgenic *rol*A plants show wrinkled leaves and reduced internode length. The *rol*D have been shown to consist primarily in a strong acceleration and stimulation of flowering, while *rol*D has been shown to promote flowering.

Although formation of hairy from infected cells is regulated by *rol*A, *rol*B and *rol*C genes, the role of rolB gene in thought to be the most important. Some possible mechanisms of action of the *rol* genes are discussed.

Mechanism of Action of rolA Gene

The rolA gene has been proposed to be involved in the metabolism of gibberellins in transgenic tobacco roots. This could explain the dwarfing of these plants, since a similar phenotype was observed by applying inhibitors of gibberellins synthesis. The rolA gene was also reported to be responsible for changes in polyamine metabolism by inhibiting their conjugation.

Mechanism of Action of rolB Gene

The rolB gene produces abundant lateral roots in plants. RolB protein was suggested to be β-glucosidase. This enzyme has the capacity to Increase the level of free indoleacetic acid (IAA) by releasing it from its inactive glucose conjugates. The function of RolB protein as a tyrosine phosphatase or as an auxin-binding protein has been proposed. However, no direct evidence has been provided between these functions and adventitious rooting. Therefore the auxin effects observed in rolB-transformed plants could be due to an altered perception of the hormone stimuli (Pistelli et al. 2010).

Moriuchi et al. (2004) reported that *rol*B might function as a transcriptional coactivator/mediator. Although the exact role of such activity on adventitious root induction is not known, protein of *rol*B family may alter developmental plasticity in higher plants by association with plant protein.

Increase in the anthraquinone (AQ) production in *rol*B and *rol*C-transformed plant cultures indicate that this might be involved in plants secondary metabolism. Similarly, involvement of tyrosine phosphorylation in plant secondary metabolism was reported (Kiselev et al. 2007).

The signaling pathway by which the rolB gene activates plant defense reactions does not depend on oxidative signals. In fact the stimulatory effect of the gene on stilbene production was abolished when the *rol*B-calli were cultivated in the presence of either phenylarsine oxide (PAO) and Na-orthovanadate inhibitors of Try phosphatases. These results indicate that Try phosphorylation indeed is involved in the stimulatory function of the *rol*B gene. Try phosphorylation has been identified to be involved in critical functions in plants, regulating activity of MAP kinase, transcription factors and reactive oxygen species (ROS) signaling.

Mechanism of Action of the rolC Gene

The gene *rol*C affects the plant size and architecture. These include apical dominance, decreased height, internode length, male fertility, increased number of flowers, changes in leaf size, color and shape. They produce essentially similar phenotypes in all the analyzed plant species. The reduction in height has been observed at different degrees of dwarfness among independent transformants carrying the same *rol*C gene construct. These differences are dependent on several factors like, site of integration, copy number, mutation, somaclonal variation and changes in expression level.

The effect of *rol*C gene on plant morphology may be due to cytokinin-beta-glucoside activity that increases cytokinin levels. Moreover, significant increase of some carbohydrate isoforms in *rol*C transformed cells of ginseng has been reported. Increase in the level of activity of β-and α-D-galactosidase and 1,3-β-D-glucanase were detected in *rol*C transformed cells compared to control cells (Bulgakov et al. 2002).

Stimulatory effect of *rol*C gene on secondary metabolism was demonstrated in tropane alkaloids, pyridine alkaloids, indole alkaloids, ginsennosides, and anthraquinone phytoalexins (Bulgakov et al 2002). Usually *rol*C gene-mediated signal did not interfere with general plant defense pathways. An inhibitory effect of *rol*C gene has been reported on rabdosiin and rosmarinic acid production in *Eritrichium scriccum* and *Lithospermum crythrorhizon* hairy roots. The inhibitory action may be due to interference of *rol*C gene product with different regulatory backgrounds existing in plants. One of the possible targets of *rol*C gene could be protein phosphatases enzyme.

Mechanism of Action of rolD Gene

The *rol*D gene stimulates the reproductive phase transition in plants. RolD protein has been suggested to exert its effect on increased flowering through changes in the concentration of plant hormones in transformed plants (Mauro et al. 1996). In transgenic tobacco, the *rol*D gene induces earliness in the flowering process and increase in the number of flowers speculated to be due to accumulation of proline or depletion of ornithine (Mouro et al. 1996). High proline concentration in tomato flowers led to believe that proline-mediated role of *rol*D in flowering. An increase in the amount of proline could affect the biosynthesis of hydroxyproline rich glycoproteins (HRGPs, extensions and arabinogalactan proteins). These proteins are structural constituents of the plant cell wall and are thought to play a key role in the regulation of cell division, cell wall assembly and cell extension.

Mechanism of Action of ORF13

The sequence of ORF13 is highly conserved in the agropine, mannopine, cucumopine and mikinopine- type R_i plasmids. Although it was thought to be involved in hormone signaling pathways, most subsequent studies excluded hormone-relate biochemical function for ORF13. ORF13 expression leads to the formation of spikes (protrusion between minor veins) on leaves and petals of tobacco. Increased cell division in the vegetative shoot apical meristems and accelerated formation of leaf primordial were observed in plants expressing ORF13. It has been proposed that ORF13 confers meristematic competence to cells infected by *A. rhizogenes* by inducing the expression of KNOX genes and promotes the transition of infected cells from the G_1 to the S phase by binding to RB (retinoblastoma) (Stieger et al. 2004).

ELICITATION

Application of biotic or abiotic stress on plant tissues in cultures has been shown to have an effect on the secondary metabolite accumulation. The process of elicitation involves treatment of the cultures with a physical and a chemical agent that will cause phytoalexin production leading to defense mechanisms in the plant cells. The eliciting agents are classified into two categories: abiotic elicitors (physical, mineral and chemical factors), and biotic elicitors (factors of plant or pathogen origin). In nature, secondary metabolites are generally produced as a defense mechanism against pathogenic and insect attack. Therefore, elicitors are used to enhance the production of secondary metabolites while the plant tissues are grown in culture. Elicitor is mainly used when the hairy roots cultures have reached its stationary phase, usually around 2-3 weeks after inoculation. Some examples of application of elicitors for increased production of secondary metabolites are shown in Table 1.

APPLICATIONS OF HAIRY ROOTS

Hairy roots have been identified to have several applications. Some of the important applications of hairy roots are described in the following section (Figure 1).

Functional Analysis of Genes

Lotus japonicas were transformed by infection with *A. rhizogenes* containing gene constructs for the expression of hairpin RNAs (hpRNAs) having sequences

Table 1. Elicitors (biotic and abiotic) used on hairy root cultures for enhanced production of metabolites

Plant species	Metabolites (medicinal properties)	Medicinal properties	Elicitor (type)	Fold Increase in metabolite content
Ammi majus	Coumarine, furocoumarine	Antioxidant	Benzo (1,2,3) – thiadiazole-7-carbothionic acid S—methyl easter (abiotic)	1.3
Aracheis hypogaea	Resveratrol and trans-resveratrol	Antioxidant and atheroscler-osis prevention	Sodium acetate, 10.2 mM (abiotic)	60.0
Artemisia annua	Artemisinin	Anti-malarial	Oligosaccharides from *Colletotricum gloeosporoides* (fungus), 0.4 mg/l (biotic)	1.5
			Polysaccharide fraction of the yeast, 2 mg/l (biotic)	3.0
			Chitosan, 150 mg/l (biotic)	6.0
Azadirachta indica	Azdirachtin	Pesticide	Salicylic acid, 100 mM (biotic)	6.0
Beta vulgaris	Peroxidase	Antioxidant	$CaCl_2$ (5 mM)	1.2
			Jasmonic acid, 100 mM (biotic)	9.0
Centella asiatica	Asiaticoside	Anti-inflammatory	MeJA, 0.1 μM (biotic)	*de novo* accumulation
Hyoscyamus niger	Polyamines, tropane alkaloids	Mydriatic, parasympa-tholytic, antiparkinsonian	MeJA, 50 μM (biotic)	2.0
Oxalis tuberosa	Harmaline, harmine	Stimulant	Phytosphtora cinnamon (biotic)	1.3
Pasax ginseng	Total saponin	Tonic, stimulant, adaptogenic	$NiSO_4$, 20μM (abiotic)	1.2
			Selenium 0.5 mM (abiotic)	1.3
			NaCl 1% (abiotic)	1.2
			Oligosaccharides from *Paris polyphylla* (plant), 30 mg/l (biotic)	1.5
			Methyl jasmonate (abiotic)	3.8
Pharbitis nil	Umbelliferone, scopoletin, skimmin	Antibacterial	$CuSO_4$ (abiotic), MeJA (biotic)	
Portulaca oleracea	Dopamine	Anti-parkinsonian (100 Mmol)	Methyl jasmonate	4.3
Salvia miltiorrhiza	Tanshinone	Antioxidant and anti-inflammatory	Sorbital, 50g/l (abiotic)	4.5
			Yeast elicitor (biotic), Ag (abiotic)	1.2
			Yeast extract + Ag + Methyl jasmonate (biotic)	2.2
Silybum marianum	Silymarin	Antibacterial	Ag, 2 mM (abiotic)	2.0
Scopolia parviflora	Scopolamine	Antioxidant	*Pseudomonas aeruginosa, Basillus ceeus, Staphylococcus aureus* (biotic)	1.5
Solanum tuberosum	Sesquiterpene (rishitin, lubimin, phytuberin, phytuberol), lypoxygenase	Anti-fertility	*Rhizoctonia bataticola,* MeJa (biotic), B cyclodextrin (abiotic)	2.0

Figure 1. Biotechnological prospects for hairy root research. Agrobacterium rhizogenes transfers the T-DNA segment from its large plasmid (Ri) into the plant genome after leaf infection. A few days later, roots emerge from the inoculation site of the leaf. Hairy roots develop in agitated liquid culture. To become an acceptable biotechnological process, hairy root cultures must be scaled-up in a bioreactor in order (a) to produce valuable metabolites from medicinal plants, (b) to increase the production of secondary metabolites by chemical means, (c) to introduce foreign genes into plant genomes to produce recombinant proteins or to overexpress proteins that are otherwise limiting to metabolic pathways, (d) to uptake heavy metals from phytoremediation systems, and (e) to conserve germplasm through artificial seeds, (f) clonal propagation of elite transgenic plants, and (g) clonal propagation of plants with altered phenotypes having ornamental value.

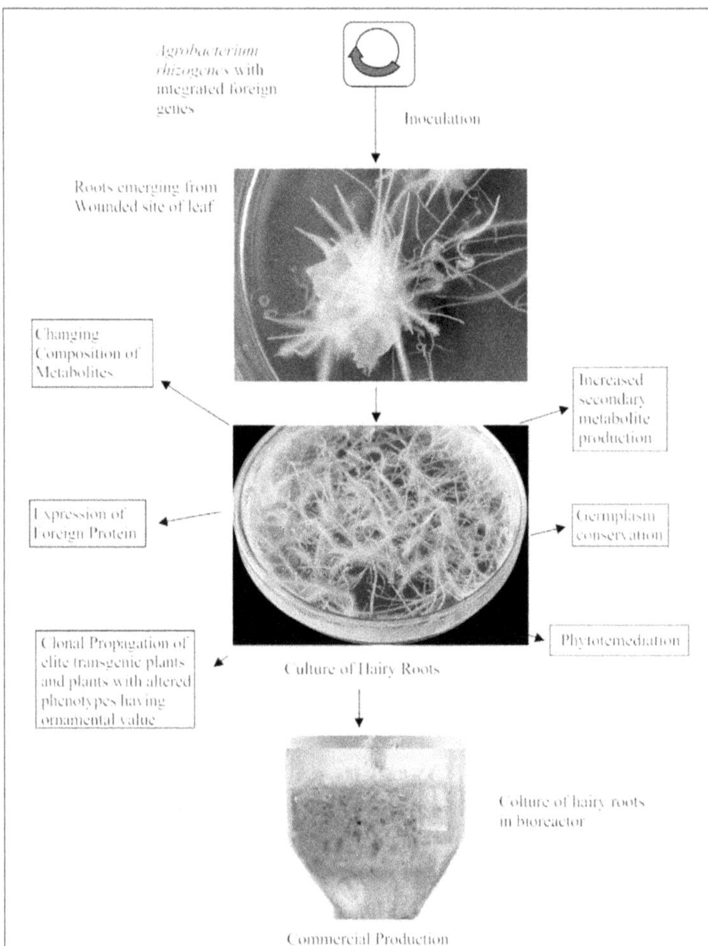

complimentary to the GUS coding region. In the transformed lines the GUS activity decreased by more than 60 percent. This result suggested that transient RNA silencing by hairy root transformation provides a powerful tool for loss-of-function analysis of genes that are expressed in roots. In another study, Hughes et al. (2002) transformed hairy roots of *Catharanthus roseus* with a *GFP* gene controlled by glucocorticoid-inducing promoter. The inducible promoter showed a tightly controlled, reversible, and dose-dependent response to the glucocorticoid dexamethasone in the hairy roots. Several experiments have demonstrated that when different promoters are fused with *GUS* gene, and transferred to hairy roots show different expression patterns. For example, when expression of *GUS* gene fused with alcohol dehydrogenase (Adh) promoter introduced into hairy roots of soybean and tested in different conditions (cold temperature, wounding anoxia and absicsic acid treatment), it showed differential expression (Preiszner et al. 2001). Similarly, an antisense dihydroflavonol reductase (DFR) gene was introduced into the hairy roots of *Lotus corniculatus* and effectively downgraded tannin biosynthesis.

Expression of Foreign Proteins

Possibility of production of industrial and therapeutic protein in plants has been studied intensely for its commercial value. Introduction and expression of pea lectin gene in white clover (*Trifolium repens*) hairy roots was reported (Diaz et al. 1995). Similarly, expression of murine-lgG1 in the hairy roots of tobacco and increase in the production of the antibody by increasing the dissolved oxygen tension to 150% air saturation. Three genes (responsible for poly 3-hydroxybutyrate, PHB, synthesis) from the bacteria *Ralstonia eutropha* were introduced into hairy roots of sugar beet. Twenty transgenic hairy root clones produced up to 55 mg high molecular PHB per gram of dry weight.

In vitriculture, antiviral traits are important economic factors. Plants transformed with virus protein often show resistance to virus infection. Transformation of grapevine hairy roots with coat protein of grapevine chrome mosaic nepovirus was reported. However, they could not regenerate the whole plants from the hairy roots. The fact that coat protein genes can be expressed in the hairy roots, indicate that this approach could be viable for the development of virus resistant plants in future.

Production of Secondary Metabolites

Hairy roots have several advantages for the production of secondary metabolites. These include: highly productive under hormone-free cultural conditions, fast growth, low doubling time, ease of maintenance, ability to synthesize a range of chemical compounds, and stability in culture. These roots can also synthesize more than one metabolite and therefore, prove economical for commercial production processes. Many medicinal plants have been transformed successfully by *A. rhizogenes* and hairy roots induced show a relatively high productivity of secondary metabolites (Sevon 2002). A list of important secondary metabolites produced in hairy root cultures of different plant species is presented in Table 2.

Sometimes, the efficiency of secondary metabolite production by hairy roots is not found to be of desirable level for commercial production. In such cases, new approaches have been made for improving the production of secondary metabolites. Such approaches are designed to increase the level of enzymes involved in the metabolism and which consequently result in the accumulation of the target product.

Based on the type of gene to transfer, two different transformation methods are used. The first method utilizes the foreign genes that encodes enzyme activities not normally present in a plant. This may cause the modification or diversion of plant metabolic pathways. Two direct repeats of a bacterial lysine decarboxylase gene expressed in the hairy roots on tobacco markedly increased cadaverine and anabasine. Similarly, production of anthraquinone and alizarin in hairy roots of *Rubia peregrine* was enhanced by the introduction of isochorismata synthase. The hairy roots of *Atropa belladonna* transformed with the rabbit P450 2E1 gene led to increased production of the metabolites. Increased production of ajmalicine and cantharanthine or serpentine and campesterol by the hairy roots of *Catharanthus roseus*, harboring hamstar 3-hydroxy 3-methylglutaryl coenzyme A (CoA) reductase (HMGR) cDNA without the membrane-binding domain was reported.

The second method enhances the overexpression of enzymes that are already present in a plant. When tobacco putrescine N-methyltransferase (PMT) gene was transformed into *Datura metel* and *Hyoscyamus muticus,* the enzyme catalyzed the first committed step in the tropane alkaloid pathway, stimulated the growth of transgenic roots and enhanced accumulation of tropane alkaloid. To overcome the deficiency of oxygen in the hairy root culture, two genes namely Adh and pyruvate decarboxylase, were transferred

Table 2. Plant species transformed by A. rhizogenes for the production of secondary metabolites

Plant species	Secondary metabolites	Medicinal property
Arachis hypigaea	Resveratrol	Anti-inflamatory, anti-oxidant, anti-cancerous
Arbus precatorium	Glycyrrhizin	Diuretic, tonic, alexitric, anti-fertility
Artemisia annua	Artemisinin	Anti-malarial
Atropa belladonna	Tropane alkaloids (hyoscamine, atropine, hyoscine)	Anti-parkinsonian
Beta vulgaris	Betalains	Aphrodisiac, laxative
	Peroxidase	--
Camptotheca acuminate	Camptothecin	Anti-cancer, anti-viral
Catharanthus roseus	Indole alkaloid (vinblastive, vincristine)	Anti-cancerous
Datura innoxia	Tropane alkaloids (scopolamine, hyoscyamine)	Nacrotic, anti-cholinergic, anti-spasmodic
Datura metal	-do-	-do-
Datura stromonium	-do-	-do-
Fagopyrum esculentum	Rutin	Anti-oxidant, anti-carcinogenic, anti-thromobotic
Ginko biloba	Ginkgolides	Aging disorders
Glycyrrhiza palliflora	Flavonoids	Anti-Gastric ulcer, anti-inflammatory, anti-tumor
Glycyrrhiza uralensis	-do-	-do-
Gmelina arborea	Verbascoside	Aging disorder, anti-inflammatory, wound healing
Gynostemma pentaphyllum	Gypenoside	Detergent
Panax ginseng	Phytosterols, ginsennosides	Tonic, stimulant, adaptogenic
Salvia broussonetii	Diterpens	Antioxidant, anti-inflammatory
S. cinnabarina	*p*-sitosterol, ursolic acid	--
S. inolucrata	Apigenin, total flavonoids	--
S. miltiorrhiza	Cryptotanshinone, tanshinone -I, IIA & IIB, lithospermic acid B, diterpenoid,tanshinones	Antioxidant, anti- inflammatory
S. sclarea	Ortonaphtoquinone diterpens	--
S. wagneriana	Rosmarinic acid	--
Solanum khasianum	Solasodin	Barth control

into the hairy roots of *Arabidopsis thaliana*. The transformed hairy roots lines could maintain similar growth rate under conditions of low oxygen to the rate achieved with full aeration. Thus the problem of oxygen deficiency in hairy root culture caused by poor mixing and mass transfer could be solved.

Production of Compounds not Found in Untransformed Roots

Sometime, hairy roots produce new compounds which are not found in the untransformed roots. Such compounds having industrial or pharmaceutical value can be exploited further for commercial production. Transformed hairy roots of *Scutellaria baicalensis* accumulated glucoside conjugates of flavonoids instead of glucose conjugates accumulated in untransformed roots was reported.

Changing Composition of Metabolites

Composition of several metabolites has been reported to have changed in the hairy roots culture lines. For example, expression of an *Antirrhinum* dihydroflavavonol reductase gene resulted in changes in condensed tannin structure and accumulation in hairy root culture of *L. conrniculatus*. In another study, selected hairy roots culture lines indicated alteration of monomerlevels during growth and development without changes in composition.

Phytoremediation

Phytoremediation takes advantage of the ability of plants and associated micro-organisms to remove, contain, or render harmless inorganic as well as organic contaminants. Phytoremediation studies carried out with whole plants provide useful information related to the removal capabilities of the plant species under study and also about the phytotoxic effects of the contaminant. In vitro cultures of roots are of particular interest for studying the interaction of contaminants with the plant system. Isolated organ culture permit the characterization of the uptake capability of the roots while avoiding the interference of translocation to other plant tissues. The drawback of root cultures is their low growth rate, which can impair experimental procedures. However, the in vitro culture of transformed roots (hairy roots) can overcome this drawback. Development of a protocol for application of hairy root culture of *A. lapathifolia* for studying the removal of organic compounds like phenol, a model organic contaminant was reported (Fiocco and Giulietti 2007). *A. lapathifolia* roots contain high levels of peroxidases enzyme that are known to be involved in the detoxification of plenols and other aromatic compounds. The cultures were exposed to different concentrations of the contaminant and the remaining amounts of it, and some physiological parameters were monitored at different time intervals. The experimental procedure permitted the estimation of the capability of the plant species to remove the contaminant and also the main variables that may affect the remediation process. Such information is essential for assessing the feasibility of a remediation process prior to its field application. Thus hairy root culture constitutes a valuable tool for phytoremediation research.

Germplasm Conservation

Artificial seed are considered to be good material for ex-situ conservation of germplasm. Hairy roots can be used for the production of artificial seeds. This provides an effective tool for ex-vitro conservation of germplasm. Hairy roots in the form of artificial seeds are reliable delivery system for clonal propagation of elite plants with genetic uniformity, high yield and low production cost. Artificial seed were produced from hairy roots of horseradish and *Ajuge reptans.* While cryopreserved root-tips derived from the hairy roots of *Panex gensing* and shoot tips of HRs regenerants of horseradish were regenerated.

Hairy root induced from rare medicinal plants can be used for regeneration of whole plants, making them an alternative and complimentary ex-site biodiversity conservation method. For germplasm conservation, development of long term preservation technique for hairy root cultures is essential. Hairy root cultures have been stored at ambient, low and –zero temperatures with success. Stable alkaloid production was observed in transgenic *Catharanthus roseus* hairy roots after five-year maintenance in liquid cultures (Peeebles et al. 2009).

Regeneration of Whole Plants

Regeneration of transgenic plants depends mostly on the *in vitro* cultural conditions for each species. But genotype and juvenility of the explants are equally important. Transformed roots can be regenerated into somatic embryos following the addition of the appropriate phytohormone. Regeneration of shoots from the hairy roots of *Robinia pseudoacacia* following the addition of 10 μmol/l α-NAA (naphthalene acetic acid) and 5 μmol/l 6-BAP (benzyl aminopurine) was reported. Induction of somatic embryos from the hairy roots of *Astragalus sinicus* was observed when the medium was supplemented with 2,4-dichlorophenoxy acetic acid (7.5 to 10.0 mg/l).

Transformed roots are able to regenerate genetically stable plants as transgenics or clones. This property of rapid growth and high plantlet regeneration frequency allow clonal propagation of elite plants. In addition, the altered phenotype of hairy root regenerants (hairy root syndrome) is useful in plant breeding programmes with plants of ornamental interest.

MEDICINAL PLANTS TRANSFORMED BY *A. rhizogenes*

Agrobacterium rhizogenes has been successfully used to transform medicinal plants and for production of secondary metabolites. Some important medicinal plants having potential for secondary metabolite production through hairy root transformation are discussed.

Basil (*Ocimum basilicum*)

Ocimum basilicum (sweet basil) plants are valued for their pharmaceutical properties. Among the different secondary metabolites found in *Ocimum* species, rosemarinic acid (RA), having antioxidant property, is one of the most abundant. Suspension hairy root cultures of sweet basil obtained from the leaves accumulated RA up to 10 mg/g dry weight, a value up to 11 times higher than in callus cultures or in the leaves of the donor plant.

To maximize the production of secondary metabolites, HR cultures of *O. basilicum* were exposed to salicylic acid, jasmonic acid, chitosan and fungal cell wall elicitors. Among the different elicitors tested, application of fungal cell wall elicitor from *Phytophthora cinnamoni* has increased the production of RA to 2.67 fold compared to untreated control (Bais et al. 2002).

Echinacea (*Echinacea* spp.)

The genus *Echinacea* has nine species and is widely grown and used for the immunostimulant property. Several caffeic acid derivatives (CADs) such as cichoric acid, caftaric acid, chlorogenic acid and caffeic acid have been identified in *Echinacea* species. They are believed to have immunostimulatory activity.

The method of transformed HR cultures of *E. purpurea* was established by infecting different types of explants with three strains of *A. rhizogenes*. Higher concentrations of polysaccharides were detected in transformed HR than in nontransformed roots. Accumulation of secondary metabolites could be increased by change and/or addition of nutritional ingredients, hormones or precursors. Production of significant amounts of CADs (especially cichoric acid and caftaric acid) was achieved by the HR of *E. purpurea* (Liu et al. 2006).

There exist a definite positive effect of light on root growth, cell viability and PAL (phenylalanine ammonia lysate) activity in relation to the biosynthesis of CADs in *E. purpurea*. The photoregulation of CADs biosynthesis in *E. purpurea* HR may offer additional advantages of quantitative and qualitative improvements of these medicinally important metabolites.

Ginseng (*Panax ginseng*)

Panax ginseng is of Chinese origin and is well known for its use in oriental medicine. Crude ginseng root extracts have tonic, stimulatory and adoptogenic properties, mainly due to the presence of numerous saponins and sapogenins.

Successful induction and establishment of *P. ginseng* rhizomes HR after *A. rhizogenes* infection has been reported. These transformed roots exhibited rapid growth and higher levels of ginsenosides than the normal cultured roots obtained by hormonal control. It was demonstrated that the growth of *P. ginseng* HR is strongly influenced by the inoculums size and its age.

The *P. ginseng* root lines can produce the highest levels of ginsenosides at the third day of culture, when they were cultured in the presence of methyl jasmonate and when the cultures were in the advanced progressive deceleration growth phase (Pavlov et al. 2002). To improve the productivity of useful metabolites, several elicitors were used in *P. ginseng* HR cultures. These include: salicylic acid (SA), acetylsalicylic acid (ASA), yeast elicitor and bacterial elicitor. In general, elicitor treatments were found to inhibit the growth of the HR, but enhance ginseng saponin biosynthesis (Jeong et al. 2003). Also, addition of 20 μM $NiSO_4$ increased ginseng saponine content and productivity. It was also reported that sodium chloride added into the media increased the amount of synthesized ginseng saponins. These results suggest that application of elicitors can reduce the processing time for the generation of ginseng saponin in HR culture.

Oligosaccharides, [such as, heptasaccharide (HS), octasaccharide (OS) etc.] from plant and microbial sources represent a class of the most widely recognized elicitors, acting as inducers of plant defense responses and playing regulatory roles in plant growth and development. Therefore, these compounds were tested as elicitors in HR of *P. ginseng* to see their effect in the induction of secondary metabolites. Separate addition of HS and OS to HR cultures of *P. ginseng* increased the root biomass dry weight by more than 70% and the total saponin content of roots by more than 1-fold (Zhou et al. 2007). These results suggest that HS and OS may have activities like plant growth regulators in plant cell cultures.

Mint (*Mentha* spp.)

The genus *Mentha* includes more than 25 species, and is widely cultivated for their essential oil. This valuable product is mainly composed of monoterpens and is largely used in the food, cosmetics and pharmaceutical industries. The quality of oil depends on the composition of the monoterpene, and therefore emphasis has been made to develop strains producing better quality oil.

Regeneration of HR and shoots from *M. piperita* after infection with *A. rhizogenes* was first reported in 2003. The regenerated plants showed HR syndrome. Hairy roots was also induced in *M. pulegium* (clone line MPH-4), a species containing high levels of endogenous phenolics, by *A. rhizogenes* infection. The inoculated explants exhibited higher levels of total phenolic components and guaiacol peroxidase activity. Enhancement of phenolics production was more apparent when explants were treated with polymeric dye R-478. Polymeric dyes are widely used in textile industry and are known environmental pollutants. Polymeric dyes have a structure analogous to common polycyclic aromatic hydrocarbons (PAHs). Plants that show an inherent tolerance to these dyes are currently being investigated for use in phytoremediation. The response of *M. pulegium* line MPH-4 to *A. rhizogens* and polydye R-478 has created interest to study the possible use of this clone in phytoremediation (Strycharz and Shetty 2002)..

Saga (*Salvia officinalis*)

The genus *Salvia* includes more than 900 species. Some of them are known for their medicinal potential and are commercially used. Only few are cultured *in vitro*. The possibility to manipulate the *in vitro* production of rosmarinic acid (RA) in cell cultures of *S. officinalis* and *S. fruticosa* was demonstrated. Production of RA and lithospermic acid B from transformed cells of *S. miltiorrhiza* was demonstrated. In 2003, a new sccoisopimarane diterpenoid in *S. cinnabarina* was identified, while in 2004, other new terpenoids in *S. wagneriana* having interesting biological activity were discovered.

Several HR lines were obtained from *S. wagneriana, S. cinnabarina,* and *S. jamensis* through single transformation event. Several elicitors were used to enhance the production of secondary metabolites from these HR lines grown in liquid medium. Screening of the secondary metabolites produced by *S. wagneriana* does not reflect any of the secondary metabolites produced by the matured plants. However, it showed presence of diterpenes, ursolic acid, p-sitosterol, caffeic acid and other flavonoids different from the available reference material.

Plume Poppy (*Macleaya cordata*)

Huang et al. (2018) established hairy root cultures of *Macleaya cordata* for the production of Sanguinarine, co-cultivating leaf and stem explants with *Agrobacterium rhizogenes*. Sanguinarine is used to replace antibiotic growth promoters in animal feeding and has anticancer properties. By comparing the metabolic profiles and gene expression of hairy roots and wild-type roots sampled, they found that the sanguinarine and dihydrosanguinarine contents of hairy roots were far higher than those of wild-type roots, and could reveal the molecular mechanism that causes these metabolites to increase. Consequently, this finding demonstrated that the hairy root system has the potential for bioengineering and sustainable production of sanguinarine on a commercial scale.

POTENTIAL PROBLEMS

Hairy roots have been recognized as a potential tool for various applications. However, there exist certain potential problems which need to be solved. Some of these problems are presented in the following section.

Different Regulation of Secondary Metabolism in Related Species

Normally related plant species share the same pathways for the production of secondary metabolites, and their regulation may differ according to different pattern. For example, the tropane alkaloid was increased in the hairy roots of two related species, *D. metel* and *H. muticus*. But different patterns of accumulation were observed in the two species mentioned above. Both hyoscyamine and scopolamine were accumulated in the hairy roots of *D. metel*, whereas only hyoscyamine levels increased in *H. muticus*. This indicates that the same pathway in two related species is regulated differently. Expression of 4- Hydroxycinnamoly-CoA hydratase (HCHL) in the hairy roots of *Datura stromonium* was demonstrated, although it was not expressed in the normal mature plants.

Depending on the genotype of the recipient HR, expression of the transferred gene may differ. The antisense DFR down regulated tannin biosynthesis in two genotypes (S33 and S50) of *L corniculatus*, but accumulated high levels of tannins in the third genotype (S41).

Key Enzyme's Overexpression may not Improve Secondary Metabolism

In the biosynthesis of shikonin, two key enzymes encoding chorismate pyruvate-lysate and HMGR, are assumed to be involved. Genes controlling these two enzymes were transferred to the hairy roots of *Lithospermum erthrorhizon*. However, accumulation of sikonin remained unchanged, even with high expression of both the enzymes.

Reduction of Chromosome Numbers during Sub-Culture

The number of chromosome may be reduced in certain HR cultures and thereby affect production of the secondary metabolites. For example, after four months of culture the chromosome number of *Onobrychis viciaefolia* was found to have reduced from 85 to 23.5 percent, and after eight months to only 4.1 percent cells had normal chromosome number.

Co-Suppression of Endogenous and Foreign Genes

Transfer of more copies of the gene(s) does not result in greater expression of the task enzyme and a corresponding increase in the product(s). For example, *Catharanthus roseus* hairy roots having hamster HMGR cDNA expressed a different alkaloid production pattern. One of the clone (236) having more copies of the gene HMGR, had the lowest HMGR activity but increased levels of ajmalicine and catharanthine. Another clone (19), with low copy number of HMGR, expressed more HMGR and produced more campesterol and serpentine, but had a low level of ajmalicine and showed no accumulation of catharanthine.

Loss of Expression of Foreign Gene after Long Period Culture

Hairy roots of *Cinchona officinalis* transformed with tryptophan decarboxylase (TDC) and strictosidine synthase (STR) produced high amounts of tryptamine and strictosidine at the beginning of the culture. However, they completely lost their capacity to accumulate alkaloids, after one year in culture, without any change in the growth and morphology.

Alternation in the Morphology of the Regenerated Plants

Plants regenerated from hairy roots often have altered morphological characters. The changes include an extremely abundant and plagiotropic root system, wrinkled leaves, reduced internode length and leaf size, reduced apical dominance, and an increased ability of leaf explants to differentiate roots in phytohormone-free medium. Apparently these changes have originated from either the insertion of foreign DNA or somaclonal variations, rather than from the expression of T-DNA genes in the transformants. Further, compared to untransformed plants, the transgenic plants show higher mortality.

Difficulties in Scaling-Up

The present established culture systems are based on flask or small-scale bioresctors. Many attempts have been made to develop economical bioreactors containing airlift, bubble column, mist, dual, and wave reactors. Except few, most of the existing culture systems could not resolve the cost benefit ratio, making them impractical for commercial use. To overcome this issue, future research should focus on the establishment of effective and economical scale-up culture system. While designing the bioreactor following points should be considered: it should permit the growth of interconnected tissues unevenly distributed throughout the vessel, rheological characteristics of heterogeneous system, oxygen consumption and excretion of products to the medium. Once such breakthrough is achieved, commercial application of hairy roots shall be a reality. Few examples on modifications made in the major types of bioreactors for increased production of biomass and secondary metabolite from hairy roots are discussed in the following section.

Stirred tank Bioreactor

In stirred tank bioreactors, the aeration and medium currency is regulated by morter-derived impeller or turbine blades. Usually the temperature, pH, amount of dissolved oxygen, and nutrient concentration can be better controlled within this reactor. In general, the impellers used in this reactor produce a high-shear stress compared to other type. For hairy root culture, the impeller must be operated with restricted power input and speed to minimize the shear stress. Improvement in the impeller performance by modifying internal reactor geometry has been reported. In *Catharanthus trichophyllus*, hairy roots cultured in stirred bioreactors showed similar alkaloid composition as found in normal roots. Success in hairy root culture of *Swertia chirata* and Panax ginseng in stirred-tank reactor was successful only when a stainless-steel mesh fitted inside the culture vessel.

Airlift Bioreactor

In airlift bioreactors, aeration and liquid currency are driven by externally supplied air. Therefore this type of reactor is useful for culturing plant cells and organs, which are sensitive to shear stress. However, this reactor is not suitable for high-density culture because of insufficient mixing process inside the reactor. It was reported 200 times increase in puerarin accumulation from *Pueraria phaseoloides* hairy root culture grown in 2.5L airlift reactor compared to 250ml shake flask culture. In *Asteagalas membranaceas*, hairy roots cultured in 30L airlift reactor accumulated more astragaleside, compared to when cultured in 10L bioreactor. In *Panax gensing*, growth of hairy roots was reported to have increased several times when cultured in stirred bioreactor compared to flask cultivation. Growth and production of hyescyamine and scopolamine in the culture of hairy roots of *Datura metel* was enhanced by the treatment of permeabilizing agent Tween 20 in an airlift bioreactor with root anchorage.

Bubble Column Bioreactor

The bubble column reactor creates less shear stress compared to other stirred type reactor, and thus useful for hairy root culture. In this case, the bubbling rate needs to be gradually increased with the growth of hairy roots. However, at a high tissue density level, the bubble column has been observed to reduce growth performance. It was reported that in bubble column hairy root culture of *Solanum tuberosum*, stagnation and channeling of gas through the bed of growing roots exists, however, the gas-liquid interface was not the dominant resistance factor to oxygen mass transfer and oxygen uptake of growing tips increase with the oxygen tension of the medium. Inclusion of polyurethane foam in the vessel of air-sparged bioreactor reduces the entrapping of gas by hairy roots, which in-turn improves biomass and alkaloid production in *Duboisia leichhardtii* hairy root culture. In *Artemisia annua* hairy root culture, the bubble column reactor was reported to be superior to mist reactor for biomass concentration, whereas mist reactors produce significantly more artemisinin. High density culture of red beet hairy roots was obtained by a radial flow reactor, which consists of a cylindrical vessel with a radial flow of medium.

Liquid-Dispersed Bioreactor

Bioreactors used for hairy root culture can be classified as either liquid-phase or gas-phase. The liquid-dispersed bioreactors provide sufficient oxygen supply to roots and low sheer stress environment compared to reactors in which the roots remained submersed in a liquid medium. In liquid-dispersed bioreactors, roots are exposed to ambient air or gas mixture, and the nutrient liquid, which is dispersed as spray or mist onto the top of the root bed. The sprayed liquid and mist are drained from the bottom of the bioreactor to a reservoir and is re-circulated. The degree of distribution of liquid varies according to the mechanism of liquid delivery at the top of the reactor chamber. Various types of liquid-dispersed bioreactors are developed for the hairy root culture. These include: nutrient mist, trickle-bed or tricking film, and drip tube. In all these bioreactors, certain types of support like glass beads, rasching rings, steel wire scaffolding, polyurethane foam, horizontal mesh trays, and cylindrical stainless steel mesh, to the roots are provided. It was reported production of twice as much aesculin in acoustic mist bioreactor grown hairy root cultures of *Cicborium intybus* compared to roots grown in bubble column and nutrient sprinkle reactors. *Artemisia annua* hairy roots grown in nutrient mist reactors produce nearly three times as much artemisinin as roots grown in bubble column reactor. Apparently a higher level of artemisinin was produced due to a response to the increased osmotic strength of the medium within the mist reactor, as the medium becomes concentrated due to water evaporation. However, in general, the mist reactor accumulates lower biomass than does the bubble column reactor due to insufficient nutrient availability.

Balloon Type Bubble Bioreactor

The balloon-type bubble reactor (BTBB) was found to be superior to the bubble column bioreactor and stirred tank bioreactor, for biomass growth of hairy root culture of *Taxas caspidota*, *Beta valgaris*, *Panax ginseng*, and adventitious roots of *Panex ginseng*. The fresh weight of ginseng hairy-like adventitious root culture in 20L BTBB was three-times higher than that of the stirred tank reactor. In a 20L bioreactor, the maximum biomass production of 2.2 kg fresh weight was obtained after 42 days of inoculation of 240 gm adventitious root cultures of ginseng. Growth of biomass in mountain ginseng cell line, maintained by the CBN Biotech Co., Korea, increased 30 fold after 42 days of culture. Pilot scale 500 and 1000 L stainless steel bioreactors was designed according to BTBB type for commercial production of saponin from hairy root culture of ginseng by the CBN Biotech Co., Korea. In the 1000L reactor, the saponine content increased by six times (33.6 mg/g).

Plants are rich source of different bioactive molecules having pharmaceutical and industrial application. Medicinal plants are mostly growing wild. Habitat loss and indiscriminate collection has let to severe loss of genetic diversity of important medicinal plants. Thus the use of tissue culture technology under controlled environment can overcome some of these problems. Further genetic engineering techniques can be applied on tissue culture grown plants to enhance the production of bioactive compounds for commercial production.

REFERENCES

Bais, H. P., Waker, T. S., Schweizer, H. B. T., & Vivanco, J. (2002). Rot specific elicitation and antimicrobilal activity of rosmarinic acid in hairy root culture of *Ocimum basilicum. Plant Physiology and Biochemistry*, *40*(11), 983–995. doi:10.1016/S0981-9428(02)01460-2

Bulgakov, V. P., Kusaykin, M., Tchernoded, G. K., Thomas, K., & Kim, J. (2002). Carbohydrate activities of the ro1C gene transformed and nontransformed ginsing cultures. *Fitoterapia*, *73*(7-8), 638–643. doi:10.1016/S0367-326X(02)00231-9 PMID:12490223

Diaz, C. L., Longman, T. J. J., Stam, H. C., & Kijne, J. W. (1995). Sugar-binding activity of pea lectin expressed in white clover hairy roots. *Plant Physiology*, *109*(4), 1167–1177. doi:10.1104/pp.109.4.1167 PMID:12228660

Estrada-Navarrete, G., Alvarado-Affantranger, X., Olivares, J. E., Guillen, G., Diaz-Camino, C., Campos, F., Quinto, C., Gresshoff, P. M., & Sanchez, F. (2007). Fast, efficient and reproducible genetic transformation of *Phaseolus* spp. by *A. rhizogenes. Nature Protocols*, *2*(7), 1819–1824. doi:10.1038/nprot.2007.259 PMID:17641650

Flocco, C. G., & Giulietti, A. M. (2007). In vitro hairy root cultures as a tool for phytoremediation research. *Methods in Biotechnology*, *23*, 161–173. doi:10.1007/978-1-59745-098-0_14

Huang, P., Xia, L., Liu, W., Jiang, R., Loin, X., Tang, Q., ... Zeng, J. (2018). Hairy root induction and benzylisoquinoline alkaloid production in *Macleaya cordata. Scientific Reports*, *8*(1), 11986–11997. doi:10.103841598-018-30560-0 PMID:30097605

Hughes, E. H., Hong, S. B., Shanks, J. V. S., San, K. Y., & Gibon, S. I. (2002). Characterization of an inducible promoter system in *Catharanthus roseus* hairy roots. *Biotechnology Progress*, *18*(6), 1183–1186. doi:10.1021/bp025603o PMID:12467449

Jeong, G. T., Purk, D. H., Hwang, B., & Woo, J. C. (2003). Comparison of growth characteristics of *Panex ginseng* hairy roots in various bioreactors. *Applied Biochemistry and Biotechnology*, *107*(1-3), 493–503. doi:10.1385/ABAB:107:1-3:493 PMID:12721430

Kiselev, K. V., Dubrovine, A. S., Veselova, M. V., & Dubrovina, A. S. (2007). The ro1B gene induced overproduction of resveratrol in *Vitis amurensis* transformed cells. *Journal of Biotechnology*, *123*, 618–692. PMID:17166613

Kumar, V., Sharma, A., Prasad, B. C. N., Gururaj, H. B., & Ravishankar, G. A. (2006). *A. rhizogenes* mediated genetic transformation resulting in hairy root formation is enhanced by ultrasonication and acetosyringone treatment. *Electronic Journal of Biotechnology*, *9*(4), 349–357. doi:10.2225/vol9-issue4-fulltext-4

Liu, C. Z., Abbasi, B. H., Gao, B., Murch, S. J., & Saxena, P. K. (2006). Caffeic acid derivatives production by hairy root cultures of *Echinacca purpurea*. *Journal of Agricultural and Food Chemistry*, *54*(22), 8456–8460. doi:10.1021/jf061940r PMID:17061821

Mauro, M. L., Trovato, M., De Paolis, A., Gallelli, A., & Altamura, M. M. (1996). The plant oncogenic ro1D stimulates flowering in transgenic tobacco plants. *Developmental Biology*, *180*(2), 693–700. doi:10.1006/dbio.1996.0338 PMID:8954737

Moriuchi, H., Okamoto, C., Nishihama, R., Yamashita, I., Machida, Y., & Tanaka, N. (2004). Nuclear localization and interaction of ro1B with plant 14-3-3 proteins correlates with induction of adventitious roots by the oncogene ro1B. *The Plant Journal*, *38*(2), 260–275. doi:10.1111/j.1365-313X.2004.02041.x PMID:15078329

Peebles, C. A., Sander, G. W., Li, M., Shanks, J. V., & San, K. Y. (2009). Five year maintenance of the inducubble expression of anthranilacte synthesis in *C. roseus* hairy roots. *Biotechnology and Bioengineering*, *102*, 1512–1525. doi:10.1002/bit.22173

Pistelli, L., Giovannini, A., Ruffoni, B., Bertoli, A., & Pistelli, L. (2010). Hairy Root Cultures for Secondary Metabolites Production. *Advances in Experimental Medicine and Biology*, *698*, 167–184. doi:10.1007/978-1-4419-7347-4_13 PMID:21520711

Preiszner, J., Van-Toai, T. T., Huynh, L., Bolla, R. I., & Yen, H. H. (2001). Structure and activity of a soybean Adh promoter in transgenic hairy roots. *Plant Cell Reports*, *20*(8), 763–769. doi:10.1007002990100385

Sevon, N., Oksman, C., & Kirsi, M. (2002). Agrobacterium rhizogenes-mediated transformation: Root cultures as a source of alkaloids. *Planta Medica*, *68*(10), 859–867. doi:10.1055-2002-34924 PMID:12391546

Zhou, L., Cao, X., Zhang, R., Peng, Y., Zhao, S., & Wu, J. (2007). Stimulation of saponin production in *Panax ginseng* hairy roots by two oligosaccharides from *Paris polyphylla* var. yunnanensis. *Biotechnology Letters, 29*(4), 631–634. doi:10.100710529-006-9273-6 PMID:17216538

ADDITIONAL READING

Abraham, J., & Thomas, T. D. (2017). Hairy root culture for the production of useful secondary metabolites. In S. Malik (Ed.), *Biotechnology and Production of Anti-Cancer Compounds* (pp. 201–230). Springer. doi:10.1007/978-3-319-53880-8_9

Banihashemi, O., Nejad, R. A. K., Yassa, N., & Najafi, F. (2015). Induction of hairy roots in *Atropa komarovii* using *Agrobacterium rhizogenes. Indian Journal of Fundamental and Applied Life Sciences, 5*, 2014–2020.

Bensaddel, L., Villarreal, M. L., & Fliniaux, M. A. (2008). Induction and growth of hairy roots for the production of medicinal compounds. *Journal of Integrative Bioscience, 3*, 2–9.

Borkataky, M., Kakoti, B. B., & Saikia, L. R. (2014). Analysis of primary and secondary metabolite profile of *Costus speciosus* (Koen Ex.Retz.) Sm. rhizome. *Journal of Natural Product and Plant Resources, 4*, 71–76.

Brijwal, L., & Tamta, S. (2015). *Agrobacterium rhizogenes* mediated hairy root induction in endangered *Berberis aristata* DC. *SpringerPlus, 4*(1), 443–454. doi:10.118640064-015-1222-1 PMID:26312208

Cardarelli, M., Mariotti, D., Pomponi, M., Spanò, L., Capone, I., & Costantino, P. (1987). *Agrobacterium rhizogenes* T-DNA genes capable of inducing hairy root phenotype. *Molecular & General Genetics, 209*(3), 475–480. doi:10.1007/BF00331152 PMID:17193709

Cardillo, A. B., Otalvaro, A. A. M., Busto, V. D., Talou, J. R., Velasquez, L. M. E., & Giulietti, A. M. (2010). Scopolamine, anisodamine and hyoscyamine production by *Brugmansia candida* hairy root cultures in bioreactors. *Process Biochemistry, 45*(9), 1577–1581. doi:10.1016/j.procbio.2010.06.002

Cardoso, J. C., Oliveira, M. E. B. S., & Cardoso, F. C. I. (2019). Advances and challenges on the *in vitro* production of secondary metabolites from medicinal plants. *Horticultura Brasileira*, *37*(2), 124–132. doi:10.15900102-053620190201

Chashmi, N. A., Sharifi, M., Karimi, F., & Rahnama, H. (2010). Differential production of tropane alkaloids in hairy roots and *in vitro* cultured two accessions of *Atropa belladonna* L under nitrate treatments. *Zeitschrift für Naturforschung. Section C*, *65*(5-6), 373–379. doi:10.1515/znc-2010-5-609 PMID:20653239

Chaudhury, A., & Pal, M. (2010). Induction of shikonin production in hairy root cultures of *Arnebia hispidissima* via *Agrobacterium rhizogenes*-mediated genetic transformation. *Journal of Crop Science and Biotechnology*, *13*(2), 99–106. doi:10.100712892-010-0007-x

Chen, L., Cai, Y., Liu, X., Guo, C., Sun, S., Wu, C., Jiang, B., Han, T., & Hou, W. (2018). Soybean hairy roots produced *in vitro* by *Agrobacterium rhizogenes-mediated transformation. The Crop Journal*, *6*(2), 162–171. doi:10.1016/j.cj.2017.08.006

Chen, S. L., Yu, H., Luo, H. M., Wu, Q., Li, C. F., & Steinmetz, A. (2016). Conservation and sustainable use of medicinal plants: Problems, progress, and prospects. *Chinese Medicine*, *11*(1), 37–43. doi:10.118613020-016-0108-7 PMID:27478496

Cheruvathur, M. K., Jose, B., & Thomas, T. D. (2015). Rhinacanthin production from hairy root cultures of *Rhinacanthus nasutus* (L.) Kurz. *In Vitro Cellular & Developmental Biology. Plant*, *51*(4), 420–427. doi:10.100711627-015-9694-9

Chio, Y. E., Kim, T. S., & Pake, K. Y. (2008). Types and designs of bioreactors for hairy root culture. In P. Dutta Gupta (Ed.), *Plant tissue culture engineering* (pp. 161–172). Springer.

Choi, S. M., Son, S. H., Yum, S. R., Kwon, W., Seon, J. H., & Pack, K. Y. (2000). Pilot scale culture of adventitious roots of ginseng in a bioreactor system. *Plant Cell, Tissue and Organ Culture*, *62*(3), 187–193. doi:10.1023/A:1006412203197

Christey, M. C. (1997). Transgenic crop plant using *Agrobacterium rhizogenes* mediated transformation. In P. M. Doran (Ed.), *Hairy roots: culture and applications* (pp. 99–111). Academic Publisher.

Christey, M. C. (2001). Use of Ri-mediated transformation for production of transgenic plants. *In Vitro Cellular & Developmental Biology. Plant*, *37*(6), 867–700. doi:10.100711627-001-0120-0

Fang, J., Reichelt, M., Hidalgo, W., Agnolet, S., & Schneider, B. (2012). Tissue-specific distribution of secondary metabolites in rapeseed (*Brassica napus* L.). *Public Library of Science (PLoS). ONE*, *7*(10), e48006. doi:10.1371/journal.pone.0048006 PMID:23133539

Fu, X., Yin, Z. P., Chen, J. G., Shangguan, X. C., Wang, X., Zhang, Q. F., & Peng, D. Y. (2015). Production of chlorogenic acid and its derivatives in hairy root cultures of *Stevia rebaudiana*. *Journal of Agricultural and Food Chemistry*, *63*(1), 262–268. doi:10.1021/jf504176r PMID:25548875

Gai, Q.Y., Jiao, J., Luo, M., Wei, Z.F., Zu, Y.G., Ma, W., & Fu, Y.J. (2015). Establishment of hairy root cultures by *Agrobacterium rhizogenes* mediated transformation of *Isatis tinctoria* L. for the efficient production of flavonoids and evaluation of antioxidant activities. *Public Library of Science (PLoS) ONE*. Retrieved from: .0119022 doi:10.1371/ journal.pone

Gelvin, S. B. (2000). Agroacterium and plant genes involved in T-DNA transfer and integration. *Annual Review of Plant Physiology and Plant Molecular Biology*, *51*(1), 223–256. doi:10.1146/annurev.arplant.51.1.223 PMID:15012192

Grzegorczyk, I., & Wysokinska, H. (2010). Antioxidant compounds in *Salvia officinalis* L. shoot and hairy root cultures in the nutrient sprinkle bioreactor. *Acta Societatis Botanicorum Poloniae*, *79*(1), 7–10. doi:10.5586/asbp.2010.001

Guillon, S., Tremnouillaux-Guiller, J., Patil, P. K., Rideau, M., & Gantel, P. (2006). Hairy root research: Recent scenario and exciting prospects. *Current Opinion in Plant Biology*, *9*(3), 341–436. doi:10.1016/j.pbi.2006.03.008 PMID:16616871

Gurusamy, P. D., Schaefer, H., Ramamoorthy, S., & Wink, M. (2017). *Biologically active recombinant human erythropoietin expressed in hairy root cultures and regenerated plantlets of Nicotiana tabacum L. Public Library of Science (PLoS) ONE*. doi:10.1371/journal.pone.018236

Ha, L. T., Pawlicki-Jullian, N., Pillon-Lequart, M., Boitel-Conti, M., Duong, H. X., & Gontir, E. (2016). Hairy root cultures of *Panax vietnamensis*, a promising approach for the production of ocotillol-type ginsenosides. *Plant Cell, Tissue and Organ Culture*, *126*(1), 93–103. doi:10.100711240-016-0980-y

Hu, Z. B., & Du, M. (2006). Hairy root and its application in plant genetic engineering. *Journal of Integrative Plant Biology*, *48*(2), 121–127. doi:10.1111/j.1744-7909.2006.00121.x

Jiao, J., Gai, Q. Y., Fu, Y. J., Ma, W., Yao, L. P., Feng, C., & Xia, X. X. (2015). Optimization of *Astragalus membranaceus* hairy root induction and culture conditions for augmentation production of astragalosides. *Plant Cell, Tissue and Organ Culture*, *120*(3), 1117–1130. doi:10.100711240-014-0668-0

Kevers, C., Jacques, P. H., Thonart, P. H., & Gaspar, H. (1999). *In vitro* root culture of *Panas ginseng* and *P quinquefolium. Plant Growth Regulation*, *27*(3), 173–178. doi:10.1023/A:1006266413919

Kochan, E., Szymczyk, P., Kuźma, L., Lipert, A., & Szymariska, G. (2017). Yeast extract stimulates ginsenoside production in hairy root cultures of American ginseng cultivated in shake flasks and nutrient sprinkle bioreactors. *Molecules (Basel, Switzerland)*, *22*(6), 880–897. doi:10.3390/molecules22060880 PMID:28587128

Kochan, E., Szymczyk, P., Kuźma, L., & Szymańska, G. (2016). Nitrogen and phosphorus as the factors affecting ginsenoside production in hairy root cultures of *Panax quinquefolium* cultivated in shake flasks and nutrient sprinkle bioreactor. *Acta Physiologiae Plantarum*, *38*(6), 149–165. doi:10.100711738-016-2168-9

Lokhande, V. H., Kudale, S., Nikalje, G., Desai, N., & Suprasanna, P. (2015). Hairy root induction and phytoremediation of textile dye, Reactive green 19A-HE4BD, in a halophyte, *Sesuvium portulacastrum* L. *Biotechnology Reports (Amsterdam, Netherlands)*, *28*, 56–63. doi:10.1016/j.btre.2015.08.002 PMID:28352573

Lonoce, C., Salem, R., Marusic, C., Jutros, P. V., Scaloni, A., Salzano, A. M., ... Donini, A. (2016). Production of a tumour-targeting antibody with a human-compatible glycosylation profile in *N. benthamiana* hairy root cultures. *Biotechnology Journal*, *11*(9), 1209–1220. doi:10.1002/biot.201500628 PMID:27313150

Mai, N. T. P., Boitel-Conti, M., & Guerineau, F. (2016). *Arabidopsis thaliana* hairy roots for the production of heterologous proteins. *Plant Cell, Tissue and Organ Culture, 127*(2), 489–496. doi:10.100711240-016-1073-7

Martin, K. P., Sabovljevic, A., & Madassery, J. (2011). High-frequency transgenic plant regeneration and plumbagin production through methyl jasmonate elicitation from hairy roots of *Plumbago indica* L. *Journal of Crop Science and Biotechnology, 14*(3), 205–212. doi:10.100712892-010-0123-7

Md Setamam, N., Sidik, N. J., Rahman, Z. A., & Zain, C. R. C. M. (2014). Induction of hairy roots by various strains of *Agrobacterium rhizogenes* in different types of *Capsicum* species explants. *BMC Research Notes, 7*(1), 414. doi:10.1186/1756-0500-7-414 PMID:24981787

Moghadam, Y. A., Piri, K. H., Bahramnejad, B. H., & Ghiasvand, T. (2014). Dopamine production in hairy root cultures of *Portulaca oleracea* (Purslane) using *Agrobacterium rhizogenes. Journal of Agricultural Science and Technology, 16*, 409–420.

Nakweti, K. R., Vaissayre, V., Ndofunsu, D. A., Ndiku, L. S., Bonneau, J., & Franche, C. (2015). Hairy roots production in *Phyllanthus odontadenius* Mull. Arg. by seedlings transformed with *Agrobacterium rhizogenes* A4RS/pHKN29. *African Journal of Plant Science, 9*(2), 50–55. doi:10.5897/AJPS09.024

Nuutila, A. M., Toivonen, L., & Kauppinen, V. (1994). Bioreactor studies on hairy root cultures of *Catharanthus roseus*: Comparison of three bioreactor types. *Biotechnology Techniques, 8*(1), 61–66. doi:10.1007/BF00207635

Pala, Z., Shukla, V., Alok, A., Kudale, S., & Desai, N. (2016). Enhanced production of an anti-malarial compound artesunate by hairy root cultures and phytochemical analysis of *Artemisia pallens* Wall. *Biotechnology, 3*(6), 182-198.

Pandey, R., Krishnasamy, V., Kumaravadivel, N., & Rajamani, K. (2014). Establishment of hairy root culture and production of secondary metabolites in Coleus (*Coleus forskohlii*). *Journal of Medicinal Plants Research, 8*(1), 58–62. doi:10.5897/JMPR12.1182

Park, Y. J., Thwe, A. A., Li, X., Kim, Y. J., Kim, J. K., Arasu, M. V., Al-Dhabi, N. A., & Park, S. (2015). Triterpene and flavonoid biosynthesis and metabolic profiling of hairy roots, adventitious roots, and seedling roots of *Astragalus membranaceus*. *Journal of Agricultural and Food Chemistry*, *63*(40), 8862–8869. doi:10.1021/acs.jafc.5b02525 PMID:26402168

Patra, N., & Srivastava, A. K. (2014). Enhanced production of artemisinin by hairy root cultivation of *Artemisia annua* in a modified stirred tank reactor. *Applied Biochemistry and Biotechnology*, *174*(6), 2209–2222. doi:10.100712010-014-1176-8 PMID:25172060

Patra, N., & Srivastava, A. K. (2015). Use of model-based nutrient feeding for improved production of artemisinin by hairy roots of *Artemisia annua* in a modified stirred tank bioreactor. *Applied Biochemistry and Biotechnology*, *177*(2), 373–388. doi:10.100712010-015-1750-8 PMID:26206459

Pillai, D. B., Jose, B., Satheeshkumar, K., & Krishnan, P. N. (2015). Optimization of inoculum density in hairy root culture of *Plumbago rosea* L. for enhanced growth and plumbagin production towards scaling-up in bioreactor. *Indian Journal of Biotechnology*, *14*, 264–269.

Singh, R. S., Chottopadhyay, T., Thakur, S., Kumar, N., Kumar, T., & Singh, N. (2018). Hairy root culture for *in vitro* production of secondary metabolites: a promising biotechnological approach. In S. Mishra (Ed.), *Biotechnological approaches for medicinal and aeromatic plants* (pp. 235–250). Springer Nature. doi:10.1007/978-981-13-0535-1_10

Sivakumar, G., Liu, C., Towler, M. J., & Weathers, P. J. (2010). Biomass production of hairy roots of *Artemisia annua* and *Arachis hypogaea* in a scaled-up mist bioreactor. *Biotechnology and Bioengineering*, *107*(5), 802–813. doi:10.1002/bit.22892 PMID:20687140

Srivastava, V., Malhotra, S., & Mishra, S. (Eds.). (2018). *Hairy roots: an effective tool for plant biotechnology*. Springer. doi:10.1007/978-981-13-2562-5

Sujatha, G., Korac-Zdravkovic, S., Calic, D., Flamini, G., & Kumari, B. D. R. (2013). High-efficiency *Agrobacterum rhizogenes*-mediated genetic transformation in *Artemisia vulgaris*: Hairy root production and essential oil analysis. *Industrial Crops and Products*, *44*, 643–652. doi:10.1016/j.indcrop.2012.09.007

Thakore, D., Srivastava, A. K., & Sinha, A. K. (2017). Mass production of ajmalicine by bioreactor cultivation of hairy roots of *Catharanthus roseus. Biochemical Engineering Journal, 119*, 84–91. doi:10.1016/j.bej.2016.12.010

Thiruvengadam, M., Rekha, K., & Chung, I. M. (2016). Induction of hairy roots by *Agrobacterium rhizogenesis*-mediated transformation of spine gourd (*Momordica dioica* Roxb. Ex. Willd) for the assessment of phenolic compounds and biological activities. *Scientia Horticulturae, 198*, 132–141. doi:10.1016/j.scienta.2015.11.035 PMID:32287883

Thwe, A., Arasu, M. V., Li, X., Park, C. H., Kim, S. J., Al-Dhabi, N. A., & Park, S. U. (2016). Effect of different *Agrobacterium rhizogenes* strains on hairy root induction and phenylpropanoid biosynthesis in tartary buckwheat (*Fagopyrum tataricum* Gaertn). *Frontiers in Microbiology, 7*, 318–326. doi:10.3389/fmicb.2016.00318 PMID:27014239

Verma, P. C., Singh, H., Negi, A. S., Saxena, G., Rahman, L., & Banerjee, S. (2015). Yield enhancement strategies for the production of picroliv from hairy root culture of *Picrorhiza kurroa* Royle ex Benth. *Plant Signaling & Behavior, 10*(5), e1023976. doi:10.1080/15592324.2015.1023976 PMID:26039483

Vinterhalter, B., Milosevic, D. K., Jankovic, T., Plejevljakusic, D., Ninkovic, S., ... Vinterhalter, D. (2015). *Gentiana dinarica* Beck. hairy root cultures and evaluation of factors affecting growth and xanthone production. *Plant Cell, Tissue and Organ Culture, 121*(3), 667–679. doi:10.100711240-015-0737-z

Yan, H. J., He, M., Huang, W. J., Li, D. M., & Yu, X. E. (2016). Induction of hairy roots and plant regeneration from the medicinal plant *Pogostemon cablin. Pharmacognosy Journal, 8*(1), 50–55. doi:10.5530/pj.2016.1.11

Zlatic, N. M., & Stankovic, M. S. (2017). Variability of secondary metabolites of the species *Cichorium intybus* L. from different habitats. *Plants (Basel), 6*(4), 38–43. doi:10.3390/plants6030038 PMID:28891986

Zubricka, D., Mišianiková, A., Henzelyová, J., Valletta, A., De Angelis, G., D'Auria, F. D., Simonetti, G., Pasqua, G., & Cellarova, E. (2015). Xanthones from roots, hairy roots and cell suspension cultures of selected *Hypericum* species and their antifungal activity against *Candida albicans. Plant Cell Reports, 34*(11), 1953–1962. doi:10.100700299-015-1842-5 PMID:26194328

APPENDIX

1. What are the basic requirements for establishment of hairy root system?
2. What is reporter gene? Name few important reporter genes used in hairy root system.
3. Describe the mechanism of induction of hairy root system through *Agrobacterium rhizogenes.*
4. Describe the applications of hairy roots for the following: (i) functional analysis of genes, (ii) expression of foreign proteins, (iii) production of secondary metabolites, (iv) phytoremediation, and (v) germplasm conservation.
5. Describe how *Agrobacterium rhizogenes* has been used to transformed medicinal plants for secondary metabolite production.
6. What are the potential problems faced while using hairy root system in plant transformation?
7. Name few economically important plant species where hairy root system has been exploited commercially.

Chapter 9
Genome Editing

ABSTRACT

Targeted editing of the genomes of living organisms not only permits investigations into the understanding of the fundamental basis of biological systems but also allows to improve productively and quality of crops. This includes the creation of plants with valuable compositional properties and with traits that confer resistance to various biotic and abiotic stresses. Recently, several novel genome editing systems have been developed, which include zinc finger nucleases (ZFNs), transcription activator-like effector nucleases (TALNEs), and clustered regularly interspersed short palindromic repeats/ Cas9 (CRISPER/Cas9). These exciting new methods have proved themselves as effective and reliable tools for the genetic improvement of plants. The genome editing systems can also be used to exploit the genetic diversity present in the semi-domesticated and wild relatives of the cultivated crops by targeting homologous domesticated genes through allele-mining. In this chapter various tools available for gene editing, their merits, and demerits have been discussed.

INTRODUCTION

Genome editing may be defined as a set of advanced molecular biological techniques that facilitates precise, efficient, and targeted modifications at specific loci in a genome. Use of transcription activator such as effector nucleases (TALENs), and zinc-finger nucleases (ZNFs) has been in use for such purposes, but came to limelight recently due to discovery of CRISPER

DOI: 10.4018/978-1-7998-4312-2.ch009

(Clustered Regularly Interspersed Short Palindromic Repeats) system, which is comparatively simple and easy to use for gene editing. The gene editing technologies basically use sequence-specific nucleases (SSNs) that can induce to identify specific DNA sequences and create double-stranded breaks (DSBs). Organisms internal repair systems fixes the DSBs either by homologous recombination (HR) or non-homologous end joining (NHEJ) mechanisms. The HR can cause gene replacements or insertions, while NHEJ can cause insertions or deletions of nucleotides, which leads to gene knockouts. Since gene editing technologies alter target design genetically they are considered to be different from genetic engineering technologies used for development of genetically modified (GM) crops. After completion of editing the gene, it is not possible to distinguish between the edited gene and the naturally occurring mutants, in the segregating generations. Therefore, application of gene editing system should enhance the crop improvement process.

ZINC-FINGER NUCLEASES

Zinc finger nucleases (Z NFs) is considered to be the first generation genome editing tools that was developed through chemically engineered nucleases, having folded up to ββα configuration. The enzyme is derived from fusion of zinc-finger based DNA-recognition modules and the DNA-cleavage domain of the *Fok*I restriction endonucleases (Figure 1a). The α-helix of the protein gets inserted to the DNA after binding of the protein into the major groove of the double helix of DNA. Each zinc finger recognizes and binds to a triplet nucleotide sequence and assembles into a group at the specific binding site(s). The monomer of ZFN has two different functional domains: an artificial ZF Cys2-His2 domain (N-terminal end) and a non-specific *Fok*I DNA cleavage domain (C-terminal end). Dimerization of the *Fol*I domain is critical for enzymatic activity of ZEN. The individual zinc finger domains are interchangeable and according to the order of the domain, they can bind to specific sites sequences in the genome. Several zinc finger domains capable of recognizing large numbers of triplet nucleotides has been generated with the objectives to target large sequences of interest.

ZFNs have so far been used to modify rice, maize, soybean, rapeseed, tobacco, apple, fig petunia, and *Arabdiopsis*. Compared to other tools, ZFN has been found to been efficient, high specificity, and minimal non-target effects. Although ZNFs have been used successfully for development of herbicide tolerance and stacking useful traits in maize and identification of safe regions for gene integration in rice, the technique has remained complicated and technically challenging for crop improvement. Therefore, the current focus is on to improve the design and delivery system of ZNF technology for wider acceptability for crop improvement.

TRANSCRIPTION ACTIVATOR-LIKE EFFECTOR NUCLEASES

Transcription activator-like effector nucleases (TALENs) are a class of enzymes derived from the transcriptional activator-like effectors (TALE) repeats and the *Fok*I restriction enzyme (Figure 1a). The central domain of TALE protein is responsible for binding of DNA and nuclear location signal. It also serves as the activator of transcription of the target gene. In the TALE monomer, the DNA-binding domain is consist of a central repeat domain (CRD) which is responsible for binding of DNA and host specificity. The CRD is made up of 34 amino acid residues that are arranged in tandem repeats. Each of the 34 amino acid long repeats binds to one nucleotide in the nucleotide sequence of the target gene. In the 34 amino acid repeats, two amino acids (located in 12 and 13 positions) are highly variable (called, repeat variable diresidue, RVD) and are responsible for the recognition of specific nucleotide. The last tandem repeat which binds to the nucleotide at the 3'-end of the recognition site has only 20 amino acid residues, and thus it is called half repeat. Although, TALE protein can be designated to bind any DNA sequence of interest, it is interesting to note that 5'-most nucleotide base of the DNA sequence should be thymidine, where TALE protein binds. Any deviation from this requirement can affect the efficacy of TALE recombinase (TALE-R), TALE transcription factor (TALE-TF) and transcription activator-like effector nucleases (TALENs). Since each TALE repeats recognizes and targets a single nucleotide, thereby increasing the number of potential targets and allowing more flexible target design, compared to ZFNs (Zhang et al., 2018).

For creation of chimeric TALEN nuclease, plasmid vectors previously used for creation of ZEN was used. The sequence encoding the DNA-binding TALE domain was inserted into such plasmid vectors, thereby creating a synthetic chimeric sequence-specific nuclease genetic construct. Such genetic construct has the DNA-binding domain of TALEs and the catalytic domain of restriction endonuclease *Fok*I. Through this process artificial nucleases having

Figure 1. (a) Genome editing tools and DNA repair mechanisms. ZFNs and TALENs on the left panel use FokI endonuclease to cut DNA double strands. Since FokI functions as a dimer, when two ZFNs or TALENs bind their targets and bring the FokI monomers into close proximity, cleavage occurs. CRISPR/Cas9 system on the right panel employs sgRNA for DNA binding and Cas9 protein for DNA cleavage. While CRISPR/Cpf1 system uses crRNA for DNA binding and Cpf1 protein for DNA cleavage. On the middle panel, when DSB was produced by genome editing techniques, the plant's endogenous repair systems fix the DSB by NHEJ or HR. NHEJ introduces small indels into the DSB and results in frame-shift mutations or premature stop codons. HR can cause gene replacements and insertions in the presence of a homologous donor DNA spanning the DSB (b) Illustration of CRISPR/Cas9-mediated base editing. In the CBE system, nCas9 was fused to CD and UGI, and this complex could convert cytosine (C) in the targeting region to uracil (U), then U is changed to thymine (T) in DNA repair or replication processes, creating a C•G to T•A substitution. In the ABE system, nCas9 was fused to AD, and this system converts adenine (A) in the targeting region to inosine (I), which is treated as guanine (G) by polymerases, creating A•T to G•C substitutions. ABE adenine deaminases-mediated base editing, AD adenine deaminases, CBE cytidine deaminase-mediated base editing, CD cytidine deaminases, CRISPR clustered regularly interspaced short palindromic repeats, crRNA CRISPR RNA, DSB double-strand break, HR homologous recombination, nCas9 Cas9 nickase, NHEJ non-homologous end joining, sgRNA single-guide RNA, TALEN transcription activator-like effector nuclease, UGI uracil glycosylase inhibitor, ZFN zinc-finger nuclease

(Reproduced from Zhang et al. Genome Biology 2018, htttp;//creativecommons.org/ publication/ zero/1.0/).

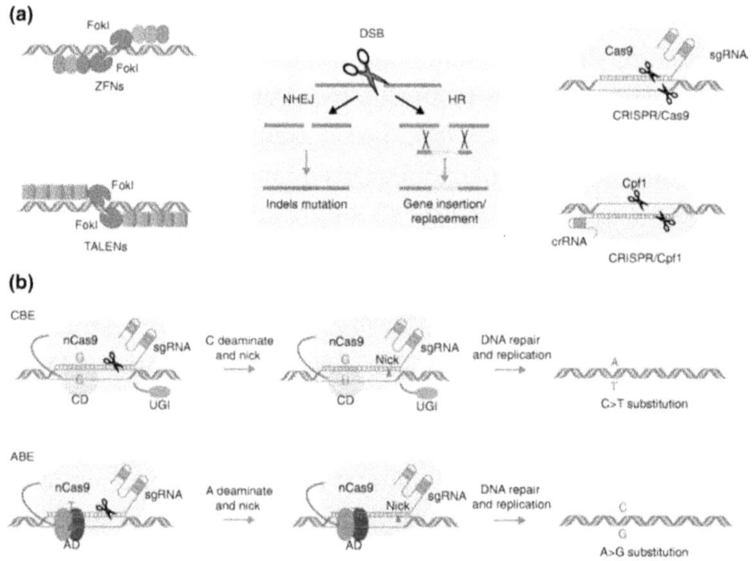

DNA-binding domain and different repeat variable diresidues (RVDs) can be created, which can target any nucleotide sequence that may be of interest.

The monomers with RVDs such as: Asparagine (Asn), and Isoleucine (Ile) (IN), Asparagine (Asn) and Glycine (Gly) (NG), Asparagine (Asn) and Asparagine (Asn) (NN), and Histidine (His) and Aspartic acid (Asp) (HD) bind to T, T, G and C bases, respectively. The first amino acid residue of all the RVDs, that is NI, NG, HD, and NN is involved in the stabilization of spatial conformation, although it does not directly bind to a nucleotide. The second amino acid residue binds to the nitrogenous bases of the nucleotide either through van der Waals force or through hydrogen bonding.

Through TALENs it is possible to introduce double stranded breaks in any location of the genome, provided that location carries the specific recognition sequence in the DNA-binding domain of TALEN. Another condition that has to be met is the presence of thymidine before 5'-end of the target sequence. However, it is possible to create mutants 5'-end thymidine constructs which can bind other nucleotides.

Genome of several plants species have been modified by using TALENs which include: rice, wheat, maize, barley, soybean, rapeseed, flex, sugarcane, tomato, tobacco, *Arabidopsis* and *Brachypodium.*

In 2012, first TALEN-mediated application foe genome editing was carried out in rice, where susceptible gene (*OsSWEET14*) for bacterial blight was edited to produce mutant rice cultivar having resistance to bacterial blight. Thereafter, TALEN was used in wheat to edit three *TaMLO* homologue to produce powdery mildew resistant cultivar, and in maize by editing *GL2* gene to produce glossy phenotype with reduced epicuticular wax in the leaves making the plant with higher efficiency of scarification and cell wall composition. TALEN has been used to modify nutritional profile in soybean, lower reducing sugar in potato, improve fragrant in rice, early flowering in *Brassica oleracea*, and high anthocyanin levels in tomato. Although TALEN has the potential for its use in crop improvement, construction of TALE repeats has remained a challenge. Further, the efficiency of using TALEN for gene targeting remained variable.

OLIGONUCLEOTIDE-DIRECTED MUTAGENESIS

In 2016, oligonucleotide-directed mutagenesis (ODM), a novel gene editing tool for the plants has become a reality. In ODM, a specific 20 to 100 base long oligonucleotide is used for targeted mutagenesis. The sequence of this oligonucleotide is identical to the target sequence in the genome, except a single base pair that is intended to e inserted in the genome to cause a mutation. Through this a site-directed editing of the gene (sequence) of interest is achieved. When the synthetic oligonucleotide having homology to a sequence of the targeted gene is exposed transiently in plant cells, it bind to the target sequence and activates the cells natural repair system. The repair system recognizes the mismatch (mutation) in the template DNA, and then copies the mismatch (mutation) into the target sequence. The desired targeted single nucleotide base editing is achieved through this process and plant genome becomes capable of conferring novel function or trait, while the synthetic oligonucleotide template is degraded. The plant cells with edited sequence are regenerated through tissue culture techniques, and varieties with desired traits are developed through traditional breeding methods.

CRISPER/Cas9 SYSTEM

CRISPER (Clustered regularly interspersed short palindromic repeats) were discovered in *E. coli* K12 in 1987. This unusual locus was found to include a set of short (29 nt) identical direct repeats that were randomly interspaced regularly by diverse short (32 nt) sequences. These loci were wide spread in prokaryotes and often found to be adjacent to a set of CRISPER-associated sequences (*cas* genes).

Elements of CRISPER/Cas System

Two major sets of sequences constitute the CRISPER/Cas system: the repeat-spacer sequences and the *cas* genes (Figure 1b). The main defining feature of CRISPER locus is the short CRISPER repeat sequences. The length and sequences within the CRISPER locus are highly conserved. The sequence is partially palindromic and thus can form secondary hairpin structure.

The CRISPER repeats are separated by spacer sequences, the length of which are highly conserved, but the sequences are highly diverse. In fact, these sequences (called pro-spacers) are derived from foreign genetic elements such as viruses and plasmids, which invade the bacteria over time. Pro-spacer sequences are identified from the nucleic acids of the invaders that carry CRISPER-targeting short sequence motif called "pro-spacer adjacent motif" (PAM). These sequences are of about 2-5 nt recognition motif, essential for CRISPER-mediated interference. Pro-spacers are spliced from the invading nucleic acids and are specifically integrated at the "leader" end of the CRISPER-repeat sequence. The leader sequence is generally A-T rich and located upstream of the first CRISPER repeat, which has a promoter binding site for transcriptional regulators and can promote transcription of the repeat-spacer sequences.

In majority of cases CRISPER loci are relatively short, having about 30 repeats and 1.6 kb size. However, in some extreme cases they may be quiet large having hundreds of repeats. Usually, in most organisms one or two CRISPER/Cas system exists. Although, in some exceptional cases more than two dozen of CRISPER/Cas system can exist.

In term of occurrence, sequence and number, *cas* genes are extremely diverse. Being a highly polymorphic protein family, they can carry out a wide variety of molecular functions. These include, recognition of RNA, binding with DNA, helicase and nuclease activities.

Diversity of CRISPER/Cas System

Although CRISPER/Cas systems are widespread in bacteria and archaea, there exist extensive differences in their content and occurrences. Based on the phylogeny and functional roles at various stages of CRISPER-mediated immunity, three main types and ten sub-types of CRISPER/Cas systems have been identified (Table 1). The Type I system is found most frequently in bacteria, the Type II exclusively in bacteria, and Type III most frequently in archaea.

There exist two universally occurring *cas* genes, *cas*1 and *cas*2, which are found in three CRASPER/Cas types. In addition, each CRISPER/Cas type is associated with a unique gene that selectively occurs in a given type such as: *cas*3 with Type I, *cas*9 with Type II, and *cas*10 with Type III systems.

MECHANISM of ACTION of CRISPER-MEDIATED INTERFERENCE

The CRISPER molecule is made up of short palindromic DNA sequences that are repeated along the molecule and are regularly spaced. Between these sequences are "spacers", foreign DNA sequences from organisms that have previously attacked the bacteria. The CRISPER molecule also includes

Table 1. Different types of CRISPER/Cas systems

Type	Gene	Sub-type	Host
I	Cas3	I-A	*Sulfolobus solfataricus*
		I-E	*Escherichia coli*
		I-F	*Pseudomonas aeruginosa*
II	Cas9	II-A	*Streptococcus thermophilies*
		II-B	*Streptococcus pyogenes*
III	Cas10	III-A	*Staphylococcus epidermidis*
		III-B	*Pyrococcus furiosus*

CRISPER-associated genes or *cas* genes. These encode proteins that unwind and cut DNA, and are called helicase and nucleases, respectively.

The CRISPER immune system protects the bacteria from repeated virus attacks through three steps as described in the following section (Figure 2).

1. **Adaptation**: when DNA from a virus invades the bacteria, the viral DNA is processed into short segments and integrated as a new spacer between the repeats. They serve as genetic memory of previous infection.
2. **Production of CRISPER RNA (crRNA)**: The CRISPER sequence undergoes transcription, including spacers and *cas* genes, creating a single-stranded RNA. The resulting single-stranded RNA is called CRISPER RNA (crRNA), which contains copies of the invading viral DNA sequence in its spacers.
3. **Targeting**: The CRISPER RNAs will identify viral DNA and guide the CRISPER-associated proteins (Cas nucleases) to them. The protein cleaves and destroys the targeted viral material.

It is possible to make use of the CRISPER/Cas9 systems recognition of specific DNA sequences in crop improvement programs. Instead of viral DNA as spacers, breeders can design their own sequences, based on their specific gene of interest. Once the sequence of the gene of interest is known, it can be easily used in CRISPER. It will then act just like a spacer for the system and guide the Cas9 protein (Cas nucleases) to a DNA matching sequence. The mechanism of gene-editing through CRISPER/Cas is described in Figure 3. Comparison of the parameters of different genome editing systems for plants is presented in Table 2.

ACTIVITIES CARRIED OUT THROUGH CRISPER/Cas9

Following activities can be carried out through CRISPER.Cas9 system.

Gene Knock-Out

Gene silencing using CRISPER started with the use of a single guide RNA (sgRNA) to the target genes and initiating a double stranded break using the Cas9 endonuclease. These breaks are then repaired by an innate DNA repair

Figure 2. The steps of CRISPR-mediated immunity. CRISPR loci contain clusters of repeats (black diamonds) and spacers (colored boxes) that are flanked by a "leader" sequence (L) and CRISPR-associated (cas) genes. During adaptation, new spacers derived from the genome of the invading virus are incorporated into the CRISPR array by an unknown mechanism. The synthesis of a new repeat is also required. During crRNA biogenesis a CRISPR precursor transcript is processed by Cas endoribonucleases within repeat sequences to generate small crRNAs. During targeting, the match between the crRNA spacer and target sequences (complementary protospacer) specifies the nucleolytic cleavage of the invading nucleic acid.

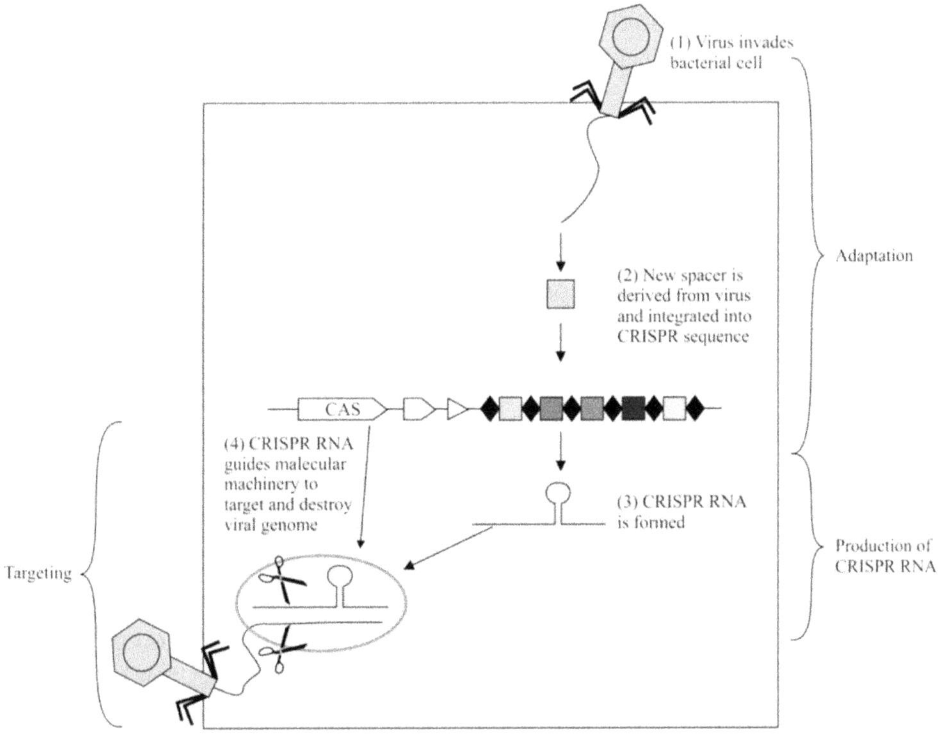

Figure 3. Mechanism of Gene Editing through CRISPR/Cas9 (For details see text)

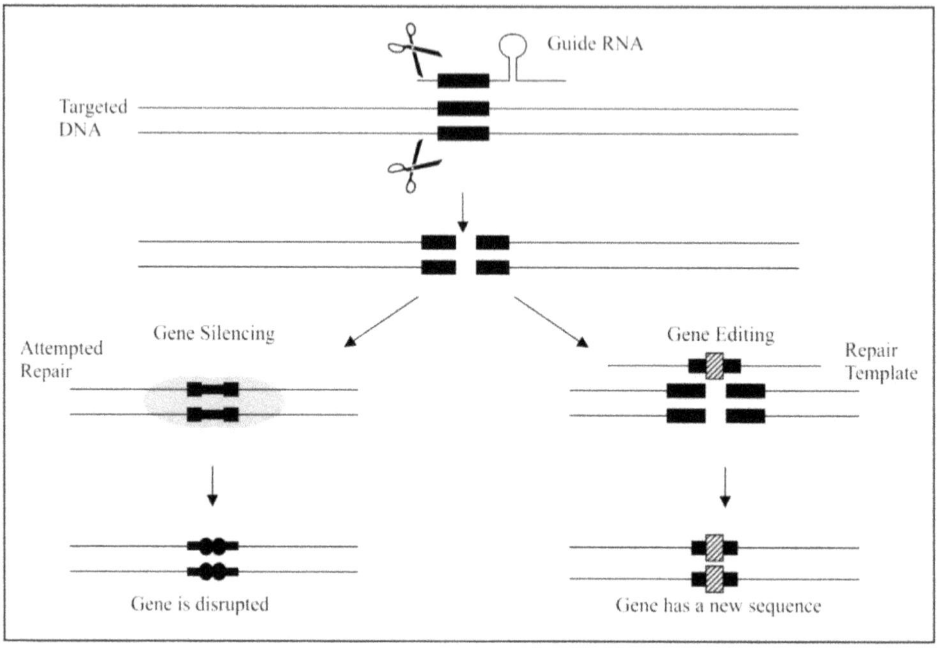

mechanisms, the non-homologous end-joining (NHEJ). However, NHEJ is error prone and results in genome deletions or insertions, which then translates into permanent silencing of the target gene.

DNA-Free Gene Editing

CRISPER can be used for DNA-free gene editing without the use of DNA vector, requiring only RNA or protein components. A DNA-free gene editing system can be a good choice to avoid the possibility of undesirable genetic alternations due to the plasmid DNA integrating at the cut site or random vector integrations.

Gene Insertion or "Knock-ins"

The CRISPER-induced double-stranded break can also be used to create a gene "knock-ins" by exploiting the cells' homologous-directed repair. The precise insertion of a donor template can alter the coding region of a gene. Several studies have demonstrated that single-stranded DNA can be used to create precise insertions using CRISPER/Cas9 system.

Table 2. Comparison of the parameters of different genome editing techniques for plants

Parameters	ZFNs	TELENs	ODM	CRISPER/Cas9
Components	Nonspecific Fok1 nuclease, Zn finger domains	Nonspecific Fok1 nuclease domain, TALE DNA-binding domains	Exocuclease polynucleotide	crRNA, Cas9 protein
Catalytic domain	Restriction endonuclease Fok1	Restriction endonuclease Fok1	No catalytic domain	HNH and RUVC
Structural protein	Dimeric protein	Dimeric protein	Nonprotein in nature	Monomeric protein
Length of target sequence (bp)	24-36	24-39	64-88	20-22
Cloning	Required	Required	Not required	Not required
Protein engineering	Required	Required	Not required	Required to test gRNA
gRNA production	Not required	Not required	Not required	Required
Mode of action	DS-breaks in target DNA	DS-breaks in target DNA	Info. Strand directs conversion(s) within target region	DS-breaks or SS-nicks in target DNA
Target recognition efficiency	High	High	High	High
Mutation rate	High	Medium	Medium	Low
Large scale library creation	Not possible	Difficult	Difficult	Possible
Multiplexing	Difficult	Difficult	Difficult	Possible

Transient Gene Silencing

By modifying the Cas9 protein so that it cannot cut DNA, transient gene silencing or transcriptional repression can also be done. The modified Cas9, led by a guide RNA, targets the promoter region of a gene and reduces transcriptional activity and gene expression. Transient activation or up-regulation of specific genes can be effectively carryout out.

TECHNOLOGICAL BREAK-THROUGHS FOR GENE EDITING

Genome editing technology has shown great potentiality for its use in crop improvement. However, the technology has several limitations such as low efficiency of HR, efforts on off-target, restrictive protospecer adjacent motif (PAM) sequences etc. To overcome these issues, several navel innovations have been introduced which are discussed in the following section.

Base Editing

In plants, efficiency of homologous recombination (HR) of double stranded breaks (DSBs) has been found to be much less template-free non-homologous end joining (NHEJ), thereby making it difficult to induce single nucleotide substitutions. On the other hand, studies on genome-wide association have shown that single-nucleotide changes are responsible for variations in elite traits in several crops. Therefore, development of efficient technique for precise point mutation is urgently required.

Recently the CRISPER/Cas9-mediated gene editing technology has been developed, which can precisely convert one base to another, without using DNA repair template. In this technology, the Cas9 nickase (nCas9) or dead Cas9 (dCas9) fused to an enzyme having base conversion activity is used. For example, cytosine (C) gets converted to uracil (U) through cytidine deaminases, which can pair with adenine (A) and subsequently adenine (A) can pair with thymidine (T). Thus G:C gets converted to A:T (Figure 1b). Similarly, adenine (A) can be converted to inosine (I) through adenine deaminases, which can pair with cytosine (C). subsequently, C and pair with G, thereby convert A:T to G:C. Base editing through cytidine-deaminase-mediation (CBE) has been adopted in *Arabidopsis*, rice, wheat, maize, watermelon and tomato to create among other things herbicide resistant plants. Comparative representation of the mechanism of CRISPR/Cas9 and base-editing system is presented in Figure 4 (Mishra et al., 2020).

Due to non-availability of naturally occurring enzyme which can catalyze adenine deamination in DNA, the adenine-deaminase-mediated base editing (ABE) is more complicated. In 2017, an efficient ABE technology has been developed by using several rounds of directed evolution and protein engineering. Thereafter, ABE has been used in *Arabidopsis*, rice and wheat to produce point mutations with altered desirable phenotypes.

Figure 4. Comparative representation of the mechanism of CRISPR/Cas9 and a base-editing system. (a) In a CRISPR/Cas9 system, the Cas9-sgRNA complex moves along the DNA strand and makes a double-stranded break (DSB) where the Cas9 encounters the appropriate protospacer adjacent motif (PAM) and the sgRNA matches the target DNA sequence. These DSBs are subsequently repaired either by nonhomologous end-joining (NHEJ) or by homology-directed repair pathway (HDR). (b) In a base-editing system, a catalytically dead Cas9 endonuclease (dCas9) fused to a catalytic cytidine deaminase domain is guided by a sgRNA molecule to make single-base substitutions without creating a double-stranded break in the DNA. *(Reproduced from Mishra et. al. Plant Biotechnology Journal, 18, 2020. doi:10.1111/pbi.13225).*

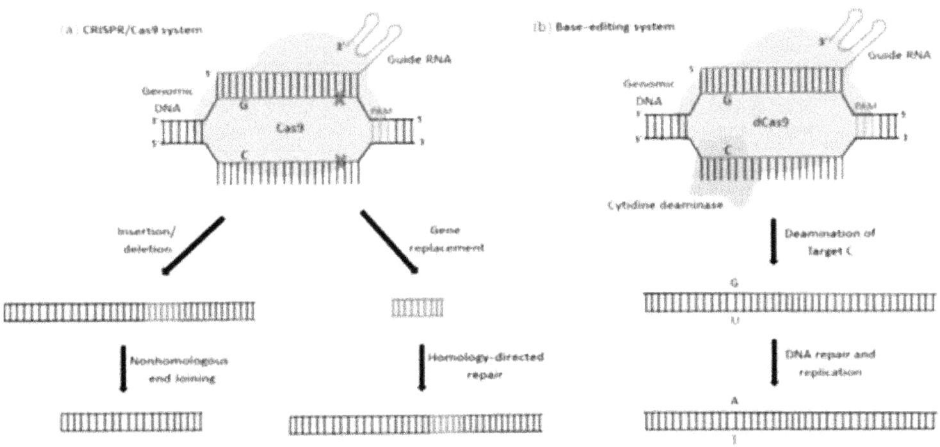

CBE is also used to generate nonsense mutations to disrupt or knockout expression of gene of interest. CBE has been found to be more specific than conventional SSN-mediated knockout method. Thus, CBE has provided new dimension for gene editing and broaden its potential applications for solving various challenges of crop improvement. The characteristics, catalytic window and functions of base editors have been presented in Table 3.

Cytosine Base Editors

Cytosine base editors are the vectors that catalyze the conversion of cytosine to thymine. The enzyme cytidine deaminase removes an amino group from cytosine thereby converting it to uracil, which results into a U-G mismatch. This mismatch gets resolved through DNA repair mechanism to form U-A base pairs. Thereafter, a A-T base pair is formed in the newly synthesized strand. This results into conversion of C-G to A-T in a programmable manner

Table 3. Characteristics, catalytic widow and functions of base editors

Base editors	Characteristics	Types of base substitutions	Catalytic window
	DNA base editor		
BE1	(APOEC1-XTEN-dCas9): Composed of a cytidine deaminase enzyme APOBEC1 (from rats) linked to a catalytically dead Cas9 (dCas9) by a amino acid XTEN linker	C to T	-17 to -13
BE2	(APOBEC-XTEN-dCas9-UGI): UGI is fused to the C terminus of BE1	C to T	-17 to -13
BE3	(APOEC-XTEN-Cas9-GI): Rapobec1 fused to the N-terminus of nickase Cas9 D10A through a 16-amino acid XTEN linker and a UGI fused to the C-terminus by a 4-amino acid linker	C to T	-16 to -12
BE4	Composed of rAPOEC1 fused to Cas9D10A through a 32-aa linker and two UGI molecule are linked to both C and N terminal of Cas9 nickase by a 9-aa linker	C to T	-17 to -13
ABE	TadA is fused to a catalytically impaired CRISPER/Cas9 mutant	A to G	-17 to -14
CRISPR-X	dCas9 is used to target a hyperactive AID variant to induce localized, diverse point mutations. The sgRNA backbone contains two MS2 RNA hairpins that each recruit two MS2 proteins fused to AID	C to T	-50 to +50
TAM	dCas9 is fused to human AID, co-expressed with UGI	C to T	-16 to -12
SaBE4-GAM	Gam protein fused to *Staphylococcus aurea* Cas9-derived BE4	C to T	-19 to -9
YEE-BE3	(W90Y+R126E+R132E): triple mutant	C to T	-15 to -13
	RNA base editor		
ADAR	Catalytically inactive Cas13 (dCas13) is fused to a naturally occurring ADAR (adenine deaminase acting on RNA)	A to I	-50 to +50

(Figure 5). The first-generation base editor (BE 1) was developed by David Liu and coworkers, in 2016. It was composed of an enzyme named cytidine deaminase APOBEC 1 (from rats) linked to a dCas9 by a 16 amino acid XTEN linker. The XTEN is a peptide which links the two proteins and maintains a balance between them. A group of naturally occurring cytidine deaminases such as, apolipoprotein B mRNA editing enzymes and catalytic polypeptide-like (APO-BEC) enzymes, found in vertebrates, acts to protect them from invading viruses. These enzymes act on single-stranded DNA/RNA as substrates (Mishra et al., 2020).

One of the limitations in BE1 was the frequent removal of uracil by uracil DNA glycosylase (UDG) which results in low editing efficiency. To overcome this limitation, a series of improved base editors were developed. The second generation base editor -BE2 (APOBEC-XTEN-dcas9-UG) was developed by adding a uracil DNA glycosylase inhibitor (UGI) to the C terminal end of the DNA targeting module. Addition of UGI leads to inhibition of the activity of UDG that catalyses the removal of U from DNA in cells and could initiate base excision repair (BER) pathway. UGI is an 83-residue protein derived from *Bacillus subtilis* bacteriophage PS1 that can block UDG activity in human cells. This inhibition of BER leads to threefold increase in the editing efficiency in human cells. Subsequently, BE3 base editor was developed, which was composed of rAPOBC1 fused to the N-terminus of nickase cas9 D10A through a 16-amino acid XTEN linker and a UGI fused to the C terminus by a 4-amino acid linker (Figure 6). The major improvement in BE3 was the replacement of dCas9 with Cas9 nickage (nCas9), which nicks the strand opposite to the deaminated cytidine (Mishra et al., 2020).

To increase the base-editing efficiency, fourth-generation base editors BE4 (derived from *S. pyogenes* Cas9p base editor) and SaBE4 (derived from *S. auteus* Cas9 base editor 4) were developed by linking rAPOBEC1 to Cas9D10A through a 32-aa linker and fusing two UGI molecules to both N

Figure 5. A comparison of three different approaches of base editing. (a) CBE – mediated base-editing strategy results in C-T conversions. (b) ABE –mediated base-editing strategy results in A-G conversions. (c) ADAR – mediated RNA base-editing results in A-1 conversion.
(Reproduced from Mishra et. al. Plant Biotechnology Journal, 18, 2020. doi:10.1111/pbi.13225)

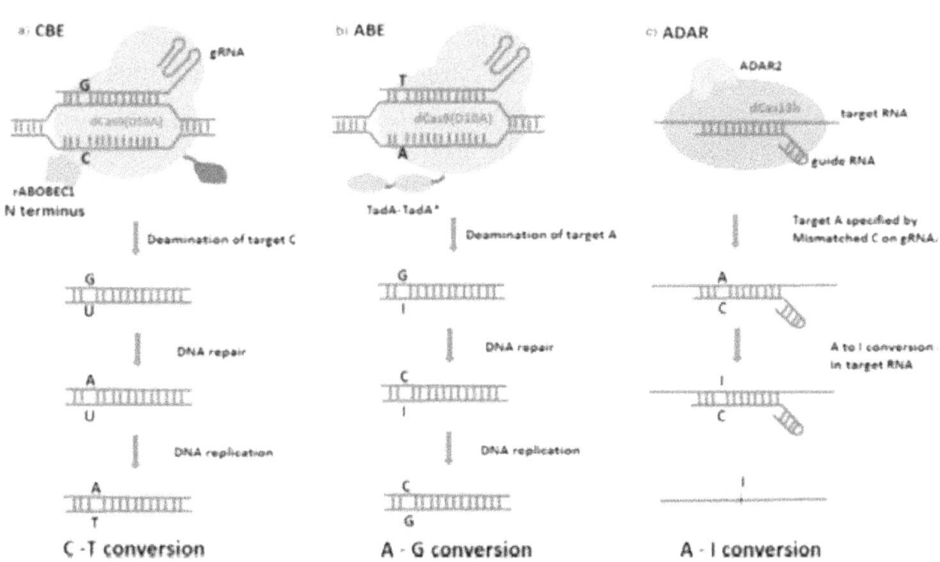

*Figure 6. Structural representation of base-editing platforms: (a) BE3 employs Cas9 nickase (nCas9D10A) along with a cytidine deaminase rAPOBEC1(orange) and an uracil DNA glycosylase inhibitor (UGI) (Green). (b) YEE-BE3 employs YEE-rAPOBEC1. (c) Target-AID employ PmCDA1 (d) SaBE4-gam employs SaCas9D10A, 2 9 UGI and has a Gam protein (red) fused to its terminus. (e) and (f) The TAM and CRISPR-X systems used dCas9 to recruit variants of the deaminase AID (AIDx or MS2-AID*D). (g) ABE is composed of ecTadA (WT)-ecTadA* (7.10) heterodimer fused to Cas9n. (h) Catalytically inactive Cas13 (dCas13) is fused to a naturally occurring ADAR2 (adenosine deaminase acting on RNA).*
(Reproduced from Mishra et. al. Plant Biotechnology Journal, 18, 2020. doi:10.1111/pbi.13225)

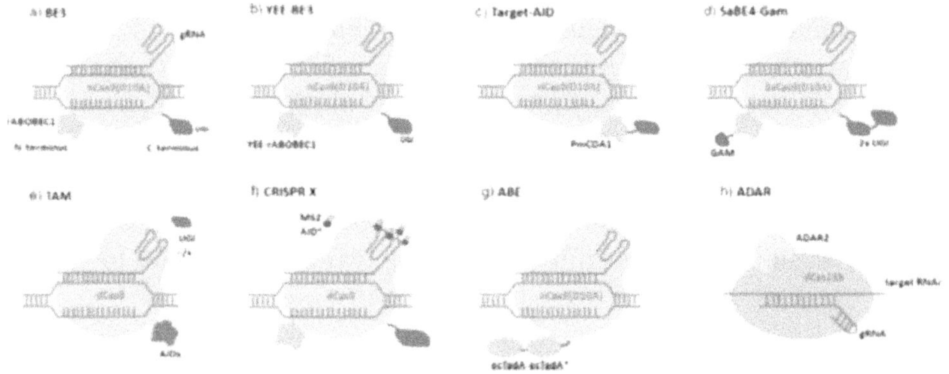

and C terminals of Cas9 nickase through a 9-aa linker. By using the UGI it is possible to block the access of UNG to the uracil intermediate, inhibiting BER, and thereby minimizing the formation of undesired by-products. In addition, a double-stranded DNA end-binding protein Gam (derived from bacteriophage Mu), was fused to the N terminal end of a Cas9 nickage that binds to the free ends of DSB and reduces indel formation during base-editing process, which improves purity of the product. Both E4-Gem and SaE4-Gam showed decrease in non-T product formation and increase C to T editing efficiency compared to BE4 and SaBE4.

Apart from introducing point mutation in a precise and programmable manner, deaminases are used to create a diverse library of point mutations localized to a targeted region of the genome (Mishra et al. 2020).

Adenine Base Editors

Capabilities of base editing were further expanded by the development of a class of AEEs that could modify adenine bases. Adenine DNA deaminases do not occur in nature. Li et al (2017) developed ABEs by using *E. coli* TadA through protein engineering and directed evolution technique (Figure 6). The *E. coli* TadA is a tRNA adenine deaminase that can convert adenine to inosine in the single-stranded anticodon loop of tRNA Arg. It also shares homology with APOBEC enzyme. Development of the first-generation ABEs was made by fusing a TadA with a catalytically impaired CRISPER/Cas9 mutant. Seventh-generation ABEs (ABE7.10) were developed for conversion of A-T to G-C in a wide range of targets with increased efficiency and product quality.

RNA Base Editors (ADAR)

Zhang et al. (2017) developed RNA base editors by using a catalytically inactive Cas13 (dCas13) and a naturally occurring adenosine deaminase acting on RNA (ADAR) to direct conversion of adenosine to inosine in mammalian cells. Cas13 is a type VI CRISPER-associated RNA-guided RNAs with abilities to bind RNA. The adenosine deaminase while acting on RNA (ADAR) family of enzymes, mediates endogenous editing of transcripts via hydrolytic deamination of adenosine to inosine. These enzymes can edit RNA precisely. The system used to edit RNA transcript is called RNA Editing for Programmable A to I Replacement (REPAIR). REPAIR is considered as a promising RNA editing platform having broad applicability for research, therapeutics and biotechnology (Figure 6).

Application of Base Editors in Crop Improvement

Cytosine and adenine base editors have been used successfully to edit specific genes in wade range of crop plants, including rice, maize, wheat, tomato, cotton, and watermelon (Table 4). In 2017, multiple herbicide resistance point mutations have been introduced through multiplex base editing, in rice. Genetic variations were also induced in rice by using a CRISPER/Cas9 toolkit comprised of rBE3 and rBE4 (rice base editors). Subsequently, rBE5 was used to target *Pi-d2* gene in rice, which harbor a point mutation modulating defense response to blast fungus. Recently, Zong and coworkers developed a new plant base editor, A3A-PBE by using human APOBEC3A, fused to Cas9 nickage to further enhance the base-editing efficiency in plants. The third-generation base editors, BE3, are used successfully to create C to A substitutions in various crop plants. However, the editing was possible for 5 nucleotide sequence only and thus editing activity was low. To overcome this, the previous base editor nCas9-PBE was modified to create A3A-PBE, in which the rat APOBEC1 was replaced with human APOBEC3A and the codons were optimized for cereals. The efficiency of A3A-PBE (in terms of C to T conversion) was found to be increased by more than 13 percent compared nCas-PBE, in rice and wheat genes. Further, the editing was possible for 17 nucleotides and which generate a low frequency of undesirable on-target indels.

Like CBEs, ABEs have also been successfully used for base editing in different crop plants. The ABEs developed for mammalian cells have been well adapted and optimized for editing plant cells for creating point mutations. A modified version of ABE7-10 (for mammalian cells), named ABE-P1 (ABE plant version 1) has been found to be highly efficient ABE for conversion of A-T to G-C in a programmable manner in rice. In 2018, Yan and coworkers developed a fluorescence-tracking ABE by using *E. coli* TadA variants and Cas9 variants for conversion of A to G. Application of this variant not only helped to induce A to G conversion but also made it convenient to select the base-edited plants through detection of fluorescence, in rice.

Li et al. (2019) synthesized a rice codon-optimized ABE-nCas9 tool to induce targeted A-T to G-C point mutation in rice genome. They cloned the rice-optimized ecTasA XTEB+N-TadA*7.10 into pHUN411 binary vector under the control of a maize ubiquitin promoter, and the rice amylose synthesis gene *Wx* was targeted by the vector. This system was found to very efficient in converting A-T to G-C without any off-target mutations.

DNA-Free Genome Editing Systems

In conventional genome editing system the DNA cassette having edited genetic components is delivered and integrated into the host genome. Since integration of the transgene occurs at random, it may generate undesirable genetic changes. In case the DNA cassette is degraded, the resulting fragments may get integrated into the host genome thereby produce undesirable effects. Expression of genome editing tools for long period results off-target effects in plants, due to the presence of abundant nucleases. Integration of foreign DNA into plant genome may also affect the regulatory mechanisms in GM plants. Therefore, DNA-free genome editing is a revolutionary technology, as the risk of producing undesirable off-target mutations can be reduced substantially.

Success in using DNA-free genome editing system has been accomplished through both protoplast-mediated transformation and particle bombardment.

Table 4. Genes targeted by adenine and cytidine base editors in different crops

Crops	Gene Targeted	Base Editors Used	Functions
Rice	*ALS*	CBE	Herbicide resistance
	C287	CBE	Herbicide resistance
	GL2IOsGRF4	ABE	Grain size and yield
	NRT1, 18, and SLR1	CBE	Enhance nitrogen use efficiency
	Pi-d2	CBE	Blast resistance
	OsACC-T1	ABE	Herbicide resistance
	OsCDC48	CBE	Regular Senescence and death
	OsMPK6	ABE	Pathogen response gene
	OsRLCK185, OsCERK1	CBE	Defense response
	OsPDS, OsSBEIIb	CBE	Nutritional improvement
	OsSPL14	CBE ABE	Herbicide resistance Plant architecture and grain yield
	SLR1	ABE	Dellla protein for plant height
	Wx	ABE	Amylose synthesis
Wheat	*TaLOX2*	CBE	Lipid metabolism
	TaDEP1, TaGW2	ABE	Panicle length and grain weight
Maize	*ZmCENH3*	CBE	Chromosomal segregation
Potato	*StALS, StGSS*	CBE	Herbicide resistance, Starch synthesis
Tomato	*SLALS1*	CBE	Herbicide resistance
Watermelon	*ALS*	CBE	Herbicide resistance

ABE= adenine base editor, CBE=cytidine base editor

In 2015, first DNA-free genome editing in plants was achieved by Woo and his colleagues. They transferred CRISPER/Cas9 ribonucleoproteins (RNPs) into protoplasts of *Arabidopsis*, lettuce, rice and tobacco. Thereafter, targeted mutations were produced in grape and apple by delivering purified CRISPER/Cas9 RNPs into protoplasts. Since efficient, regenerable protoplast systems are not available for many important crops, the technology is not getting the required boost.

Particle bombardment-mediated DNA-free genome editing has been accomplished in maize and wheat. Genome-edited plants have been produced by delivering CRISPER/Cas9 RNA and CRISPER/Cas9 RNPs into wheat embryos by particle bombardment. Similarly, by delivering CRISPER/Cas9 RNPs through particle bombardment, knockout mutants were produced in maize. Application of CRISPER/Cas9 RNPs have been found to cause few off-target effects in plants compared to CRISPER/Cas9.

Zong et al. (2018) have described a gene editing system by combing base editing and DNA-free genomic editing in wheat. They achieved base conversion frequency of 1.8% for C-to-T. Such developments generate lot of promise for crop improvement and commercialization of edited plants.

CRISPER/Cpf1 System

The type II CRISPER/SpCas9 system although efficient and simple, it can only recognize DNA sequence upstream of the 5'-NGG-3' PAMS. Therefore, it restricts it application to any potential target sites. The type V CRISPER/Cpf1 system has demonstrated its potential to overcome this problem. Unlike SpCas9 which create blunt ends, the Cpf1 recognizes T-rich PAMs and generate cohesive ends with 4 to 5 nucleotide overhangs (Figure 1a).

Recently, Cpf1 derived from *Francisella novicida* (FnCpf1) has been used successfully for targeted mutagenesis in rice and tobacco, and the Cpf1 ortholog from *Lachnospiraceae* bacterium (LbCpf1) induced targeted mutations in rice.

The FnCpf1 and LbCpf1 nucleases can generate precise gene insertions at the target site in rice at a much higher frequency than most other genome-editing nucleases. To expand the scope of CRISPER/Cpf1-mediated genome editing system, a variant (LbCPF1-RR) has been developed that enables the multiplex editing of the target genes having TYCV PAMs.

The CRISPER/Cpf1-mediated DNA-free genome editing system can be combined with base editing and DNA-free genome editing systems to achieve specific objectives in crop improvement programs. Recently, CRISPER/Cpf1-mediation and DNA-free genome editing system has been combined to achieve specific objectives in rice.

A variant of Cpf1, named as Cas12a, has been developed which differs from classical CRISPER/Cas system in several aspects: (1) the nucleases are smaller ranging from 135 to 158 kDa, (2) a naturally occurring single guide RNA is present in the system, (3) cutting of Cas12a results into staggered cuts Cas9 cutting with blunt ends, (4) for Cas12a system the protospacer adjacent motifs has to be rich in thymidine, whereas it should be rich in guanine for

Cas9, and (5) from the recognition site of Cas12a, the DNA is cut distal site, whereas DNA is cut at proximal site by Cas9.

Prime Editing Technology

Anzalone et al. (2019) from Brood Institute of MIT and Harvard has developed a new CRISPER genome editing system by combining CRISPER/Cas9 and a reverse transcriptase into a single system. They called this system as "prime editing", which can edit human genome very efficiently with precision, in highly versatile fashion. The technique is so efficient that it can rectify up to 89 percent of known disease causing genetic variations.

Prime editing differs from earlier gene editing systems in that it uses RNA to direct the insertion of new DNA sequences into human genome. This new RNA is called prime editing guide (pegRNA), which directs a catalytically impaired Cas9 protein, fused to reverse transcriptase to the target site. Unlike double-stranded breaks created by CRISPER/Cas9 system, the modified Cas9 cuts the DNA only in one strand. The pegRNA contains additional RNA nucleotides which encodes the new edited sequence. The reverse transcriptase reads the RNA extensions and then incorporates the corresponding DNA nucleotides into the target site.

The "prime editing" has shown lot of promise and the technology is available freely to all the researchers and non-profit organizations. Thus it will be interesting to see how the will take advantage of this technology for crop improvement.

CRISPER/Cas9 for CROP IMPROVEMENT

CRISPER/Cas9 has played a huge part in the increase in genome editing studies in recent years. The system has broad applications in plant and animal improvement, as well as in the medical field. As a relatively young technique, various discoveries and innovations for its efficient use in wider applications are in the offing. Researchers have found that the CRISPER/Cas9 system can be applied to nearly every organism.

The CRISPER, derived from *Streptococcus pyogenes* have been developed as a versatile genome editing technology for wide range of applications. Compared to ZENs and TALENs, CRISPER system has been found to be simple, efficient, low cost, and capable of targeting multiple genes. Therefore,

CRISPER has become the choice of breeders for crop improvement. Accordingly, CRISPER has been used for address variety of issues in breeding several crop plants including rice, wheat, maize, barley, sorghum, soybean, potato, tomato, rapeseed, flex, cotton, cucumber, lettuce, grapes, grapefruit, apple, orange, watermelon and *Camelina*. Application of CRISPER has been mostly to induce null alleles by knocking the gene. This can be carried out by introducing small indels into the target gene to create frame-shift mutation or by introducing stop codons to interrupt mRNA prematurely (Figure 9.1a). Both gene knockout mutants and insertion and replacement mutants have been produced through genome-editing technologies in different plant species. Mutants thus produced have been shown to be useful for improvement of several crop plants (Table 5). CRISPER/Cas9 system has been adopted for gene editing for improvement of several traits, such as yield, plant architecture, plant aesthetics, and disease resistance in important crop plants.

Usually, cells grown in culture are used for genetic modification through CRISPER system. The viral vector or plasmid having high and stable synthesis of CRISPER/Cas9 system elements are introduced into the plant cells. Alternatively, cultured protoplasts and a plasmid containing CRISPER/Cas9 elements are used to obtain genetically modified plants. The third approach could be use of *Agrobacterium* sp. having a plasmid that contains CRISPER/Cas9 system.

CRISPER has been used successively in rice for increasing grain number, dense erect panicle, large grain size, overall yield, resistance against blast and bacterial blight, resistance to herbicide. Ying Wang and coworkers have designed several CRISPER sgRNAs and successfully deleted fragments of the dense and erect panicle1 (DEP1) gene in the *Indica* rice line IR58025B. Improvements in yield-related traits, such as dense and erect panicles and reduced plant height, were observed in the mutant plant produced.

CRISPER/Cas9-mediated gene editing can used to disrupt disease-causing genes, known as "S-gene" for development of disease resistant plants. in rice, ethylene-responsive gene *OsERF922* were generated through CRISPER/Cas9 tool, which showed reduced blast lesions and increased resistance against blast disease. Similarly, bacterial blight-resistant plants were produced by targeted mutagenesis of *SWEET13* gene. Peng and coworkers have the edited effector binding elements (EBEs) by CRISPR/Cas9 system in the *CsLOB1* gene promoter region, and thereby increased disease resistance in *Citrus sinensis* against *Xanthomonas citri*.

CRISPER technology was also used successively to develop resistance to powdery mildew in wheat and tomato, improved self-life and heat stability

Table 5. Specific traits of crop plants improved by genome-editing technique

Plants	Target gene	GE Technique	Type of DNA repair	Trait modified
Brassica oleracea	*FRIGIDA*	TALENs	NHEJ	Early flowering
Camelina sativa	*FAD2*	CRISPER/Cas9	NHEJ	Decreased poly-unsaturated fatty acids
Cassava	*EPSPS*	CRISPER/Cas9	HR	Herbicide resistance
Citrus paradise	*CsLOB1*	CRISPR/Cas9	NHEJ	Citrus canker resistance
Cucumber	*eiF4E*	CRISPER/Cas9	NHEJ	Virus resistance
Cucumis sativus	*EIF4E*	CRISPR/Cas9	NHEJ	Broad virus resistance
Grape	*VvWRKY52*	CRISPER/Cas9	NHEJ	*Botrytis cinerea* resistance
Grapefruit	*CsLOB1 promoter*	CRISPER/Cas9	NHEJ	Alleviated citrus canker
Gossypium hirsutum	*Gh14-3-3d*	CRISPR/Cas9	NHEJ	Verticillum wil resistance
Grapefruit	*CsLOB1*	CRISPER/Cas9	NHEJ	Citrus canker resistance
Maize	*ZmTLP*	ZFNs and CRISPER/Cas9	HR	Herbicide tolerance
Maize	*ZmTLP*	ZFNs	HR	Trait stacking
Maize	*ZmGL2*	TALENs	NHEJ	Reduced epicuticular wax in leaves
Maize	*ZmMTL*	TALENS	NHEJ	Induction of haploid plant
Maize	*Wx1*	CRISPER/Cas9	NHEJ	High amylopectin content
Maize	*TMSS*	CRISPER/Cas9	NHEJ	Thermosensitive male-sterile
Maize	*ARGOS8*	CRISPER/Cas9	HR	Draught stress tolerance
Maize	*ALS*	CRISPER/Cas9	HR	Herbicide resistance
Maize	*ZmHKT1*	CRISPR/Cas9	NHEJ	Salinity tolerance
Maize	*PPR, RPL*	CRISPR/Cas9	NHEJ	Reduced zein protein
Mushroom	*PRO*	CRISPER/Cas9	NHEJ	Anti-browning
Orange	*CsLOB1*	CRISPER/Cas9	NHEJ	Citrus canker resistance
Potato	*Vinv*	TALENs	NHEJ	Minimizing reducing sugar
Potato	*Wx1*	CRISPER/Cas9	NHEJ	High amylopectin content
Potato	*ALS*	CRISPER/Cas9	HR	Herbicide resistance
Potato	*CP and Rep sequences*	CRISPR/Cas9	NHEJ	Yellow leaf curl resistance
Potato	*SIJAZ2*	CRISPR/Cas9	NHEJ	Bacterial speck resistance
Potato	*SIMIo1*	CRISPR/Cas9	NHEJ	Powdery mildew resistance
Potato	*SINPR1*	CRISPR/Cas9	NHEJ	Drought tolerance
Potato	*GBSS*	CRISPR/Cas9	NHEJ	Increased amylopectin/amylose
Rice	*OsQQR*	ZFNs	HR	Trait stacking
Rice	*OsSWEET14*	TALENs	NHEJ	Bacterial blight resistance
Rice	*OsBADH2*	TALENs	NHEJ	Fragrant rice
Rice	*LAZY1*	CRISPER/CAS9	NHEJ	Tiller spreading
Rice	*Gn1a, GS3,DEP1*	CRISPER/Cas9	NHEJ	Enhance grain number. Large grain size, dense erect panicle
Rice	*SEIIb*	CRISPER/Cas9	NHEJ	High amylose content
Rice	*OsERF922*	CRISPER/Cas9	NHEJ	Rice blast resistance
Rice	*OsSWEET13*	CRISPER/Cas9	NHEJ	Bacterial blight resistance
Rice	*OsMATL*	CRISPER/Cas9	NHEJ	Induction of haploid plants
Rice	*ALS*	CRISPER/Cas9	HR	Herbicide resistance
Rice	*EPSPS*	CRISPER/Cas9	NHEJ	Herbicide resistance
Rice	*eIF4G*	CRISPR/Cas9	NHEJ	Tungro virus resistance
Rice	*OsNAC041*	CRISPR/Cas9	NHEJ	Salinity tolerance
Rice	*OsOTS1*	CRISPR/Cas9	NHEJ	Salinity tolerance
Rice	*OsOTS2*	CRISPR/Cas9	NHEJ	Salinity tolerance
Rice	*OsAnn3*	CRISPR/Cas9	NHEJ	Cold tolerance
Rice	*SAPK2*	CRISPR/Cas9	HDR	Drought and salinity tolerance
Rice	*OsMPK2* *OsPDS* *OsBADH2*	CRISPR/Cas9	HDR	Multiple stress tolerance
Rice	*OsAAP3*	CRISPR/Cas9	NHEJ	Grain yield
Rice	*OsCCD7*	CRISPR/Cas9	NHEJ	High tillering

continued on following page

Table 5. Continued

Plants	Target gene	GE Technique	Type of DNA repair	Trait modified
Rice	*GW5*	CRISPR/Cas9	NHEJ	Grain weight
Rice	*Hd2, Hd4, Hd5*	CRISPR.Cas9	NHEJ	Early heading
Rice	*OsSWEET11*	CRISPR/Cas9	NHEJ	Grain weight
Rice	*OsGRF4*	CRISPR/Cas9	NHEJ	Grain size
Rice	*IPA, GS3, DEP1, Gn1a*	CRISPR/Cas9	NHEJ	Improved yield
Rice	*GS3, GW2, GW5, TGW6*	CRISPR.Cas9	NHEJ	Grain weight
Rice	*SBEIIb*	CRISPR/Cas9	NHEJ	Amylose, starch resistance
Rice	*OsNAC14*	CRISPR/Cas9	NHEJ	Draught tolerance
Rice	*SAPK1, SAPK2*	CRISPR/Cas9	NHEJ	Salinity tolerance
Rice	*Waxy*	CRISPR/Cas9	NHEJ	Enhanced glutinosity
Sugarcane	*COMT*	TALENs	NHEJ	Improved cell wall composition
Sugarcane	*COMT*	TALENs	NHEJ	Improved saccharification efficiency
Soybean	*FAD2-1A, FAD2-1B*	TALENs	NHEJ	High oleic content
Soybean	*FAD2-1A, FAD2-1B, FAD3A*	TALENs	NHEJ	High oleic, low linoleic content
Soybean	*ALS*	CRISPER/Cas9	HR	Herbicide resistance
Soybean	*Drb2a, Drb2*	CRISPR/Cas9	NHEJ	Draught and salt tolerance
Soybean	*GmFT2*	CRISPR/Cas9	NHEJ	Delayed flowering
Soybean	*FAD2-1A, FAD2-1B*	CRISPR/Cas9	NHEJ	Improved oil quality
Tomato	*ANT1*	TALENs	HR	High anthocyanin
Tomato	*SlMLO1*	CRISPER/Cas9	NHEJ	Powdery mildew resistance
Tomato	*SlJAZ2*	CRISPER/Cas9	NHEJ	Bacterial speck resistance
Tomato	*SP5G*	CRISPER/Cas9	NHEJ	Early harvest time
Tomato	*SlAGL6*	CRISPER/Cas9	NHEJ	Parthenocarpy
Tomato	*lncRNA1459*	CRISPR/Cas9	NHEJ	Long self-life
Tomato	*SO, SP5G, CLV3, WUS, GGP1*	CRISPER/Cas9	NHEJ	Tomato domestication
Tomato	*SlMAPK3*	CRISPR/Cas9	NHEJ	Drought tolerance
Tomato	*SlCBF1*	CRISPR/Cas9	NHEJ	Cold tolerance
Tomato	*SGR1, LCY-E, Blc, LCY-B1*	CRISPR/Cas9	NHEJ	Increased lycopene
Tomato	*SlGAD2, SlGAD3*	CRISPR/Cas9	NHEJ	Enhanced γ-aminobutyric acid
Wheat	*EDR1*	CRISPER/Cas9	NHEJ	Powdery mildew resistance
Wheat	*GW2*	CRISPER/Cas9	NHEJ	Increased grain weight and protein content
Wheat	*TaDREB2, TaDREB3*	CRISPR/Cas9	NHEJ	Draught tolerance
Wheat	*TaGW2*	CRISPR/Cas9	HR	Grain weight
Wheat	*α-gliadin*	CRISPR.Cas9	HR	Low gluten

GE Technique= Genome editing technique, HR= Homologous recombination, NHEJ= Non-homologous end joining, TALEN= Transcription activator-like effector nuclease, ZFN= Zinc-finger nuclease, CRISPER= Clustered regularly interspersed short palindromic repeats.

in soybean, high amylopectin in maize, resistance to ipomovirus (cucumber vein yellowing virus) and papaya ring spot mosaic virus-W in cucumber, draught resistance in maize, and herbicide resistance in flex.

Shah and coworkers edited the *TaMLO* gene through CRISPER/Cas technique and produced wheat lines resistant to powdery mildew disease caused by *Blumeria graminis.* Similarly, in wheat and tomato lines having resistance against powdery mildew were developed through CRISPER/Cas9-mediated multiplex gene editing of *EDR1* gene and *MLO* gene, respectively.

In hexaploid wheat, geminiviral-based DNA replicons was utilized for transient expression of the CRISPR/Cas9 system against wheat dwarf virus (WDV), and 12 fold up-regulation was observed in ubiquitin gene expression. Stable over-expression of sgRNA and Cas9 that particularly target Geminivirus genome to prevent its growth has been adopted to develop virus resistant plants. On the other hand, the CRISPR/Cas9 system has been used to mutate viral genomes to reduce infection caused by them. Recently, a new ortholog of Cas9 has been discovered in *Francisella novicida* (FnCas9) to edit RNA virus genome. FnCas9 can inhibit the replication of the tobacco mosaic virus and cucumber mosaic virus and provide immunity against them.

Cai et al. (2018) have used the CRISPER/Cas9 system to induce mutations on GmFT2a, an integrator in the photoperiod flowering pathway of soybean. The developed soybean plants showed late flowering, resulting in increased vegetative size. The mutation was also found to be stably inherited in the following generations.

Tain et al (2018) used the CRISPER/Cas9 system to target CIPDS, the phytoene desaturase in watermelon, to achieve the albino phenotype. All genome-edited watermelons harbor mutations in CIPDS and showed full or mosaic albino phenotype. This study served as a proof of concepts of using the CRISPER/Cas9 system in watermelon breeding.

Citrus plants resistant to citrus canker caused by *Xanthomonas citri* sub sps. Citri (Xcc), a serious disease of citrus, has been developed through CRISPER/Cas9. The promoter of the CsLOB1 gene, which promotes canker development, was targeted for the development of canker resistant citrus plants.

CRISPER/Cas9 was also used to generate mutants in the flowering suppressor SELF-PRUNING5G (*SP5G*) gene in tomato to manipulate photoperiod response. The mutations brought about by CRISPER/Cas9 caused rapid flowering and enhanced the compact growth habit of field tomatoes, resulting in a quick burst of flower production and early yield.

Crop yield is a complex, multi-genic, and quantitative character that is influences by several features. It has been demonstrated that the CRISPER/Cas9 technology can be effectively utilize to enhance crop yield.

The CRISPER/Cas9 technique has been used to knockout the genes that are known to negatively regulate yield-related traits such as, tiller number (*OsAAP3*), panicle size (*TaDEP1*, *OsDEP1*), grain weight (*TaGASR7*, *TaGW2*), and grain number (*OsGn1a*). In rice, by employing a multiplexing gene editing strategy, three genes including GS2, GS3, GA5, and TGWA6 were mutated simultaneously and headed towards trait pyramiding and enhancing grain size and weight. Similarly, Li and coworkers applied the CRISPER/Cas9 system to knockout three yield-related genes (*Hd2*, *Hd4*, and *Hd5*), which resulted in early heading in rice. In wheat, knockout of the gene *GASR7* via CRISPER/Cas9 technique increased kernel weight.

The CRISPER/Cas9 technology has also been used for quality improvement in crops such as storage quality, nutritional value, starch content and fragrance. In rice, the cooking and eating quality has been improved by mutating the *Waxy* gene using CRISPER/Cas9 techniques. The nutritional value (high amylose content) of rice has also been improved by knocking out the *SBEllb* gene. In 2018, Sanchez and coworkers produced low-gluten wheat through CRISPER/Cas-mediated gene editing of the glutin –encoded gene family *α-gliadin*. High yielding soybean plants with improved levels of oleic acid was developed by disrupting the *FAD2-1B* and *FAD2-1A* genes employing CRISPER/Cas9 system. Sorghum nourishment quality has been improved by employing CRISPR/Cas9 technique, targeting *k1C* genes which were responsible for poor digestibility and hindered production of improved amino acids.

CRISPER/Cas9-MEDIATED GENE EDITING IN PLANTS

Several efforts have been successfully employed for target gene editing in many crop plants via CRISPER.Cas9-based genome editing tools. Factors which have been reported to affect the editing ability of the CRISPER/Cas9 system include targeted DNA, GC contents, Cas9 codons, sgRNA structure, and expression of Cas9 and sgRNA. Therefore, all these factors should be highly optimized to develop increased efficiency of CRISPER/Cas9 system.

Designing CRISPER/Cas9 Delivery System

Initial attempts to use CRISPER/Cas9 in plants were not encouraging due to low efficiency. With the improvement of technology, highly efficient vector delivery systems for CRISPER/Cas9 have been designed for genome editing in plants. Some of the important development includes, viral infection suppression, disruption of gene of cis-elements, genomic deletion, gene knockout, and multiplex genome editing. Discovery of new Cas9 variants, efficient screening methods for knockout mutants, vector selection, and construction and employment of the most appropriate delivery system for Cas9 expression cassette has led to more precise, accurate, and targeted delivery of the Cas9 system in plants. Construction, screening and delivery of the CRISPER/Cas system into plants are described in the following section

Cargo-Vector for CRISPER/Cas9 System

Both single-vector and binary-vector systems are used in CRISPER/Cas9-mediated gene editing. For plant transformation, any specific vector constructed with several gRNAs and Cas9 protein expression cassette, can be applied. Various structural constructs of gRNAs can be utilized for various Cas9 proteins to design a unique gRNA - Cas9 nuclease, for accuracy and easiness in experimental design. In single vector system, RNA polymerase III-driven promoters (U6/U3) are designed for gRNA expression, whereas in the case of Cas9 gene expression ubiquitin and CaMV355 promoters based on RNA polymerase II are used. Recently, certain new adjustments are made in the single-vector system, which include polymerase II and dual polymerase II promoters. Single polymerase II vectors are used to control the expression of gRNA and the Cas9 gene at the same time, while the dual polymerase II vectors utilizes two different promoters to control expression of gRNA and Cas9 gene. With the addition of these new technologies has helped to decrease the vector length and thereby improve the efficiency of the system.

Bioinformatics Tools for Designing CRISPER/Cas9 Construct

One of the most critical steps in developing highly precise genome editing system is the designing of a sgRNA construct for CRISPER/Cas9. Several online tools with plant databases are available, which permit the designing of sgRNA for identification of new target sites (Table 6). For example, Xie et al. (2014) developed a web tool named CRISPER-PLANT to design efficient sgRNA constructs for CRISPER/Cas9-based gene editing. Similarly, a novel web tool was developed by Michno et al (2015) for rapid detection of target loci in soybean for CRISPER/Cas9-mediated gene editing, and the CRISPER-P was developed by Lei et al. (2014) for designing of sgRNA for every plant having an available sequenced genome and to help evaluate off-targets.

Table 6. List of various sgRNA (single guide RNA) designing bioinformatics tools for the CRISPER/Cas system

Tool Name	Year	Description and Function	Web Link
CRISPER Design	2013	Precise sgRNA construction for target sites, assess off-target sites	http://www.genome-engineering.org
sgRNAcas9	2014	Rapid design of sgRNA with less off-target	https://www.biootools.com/col.jsp?id=103/
CHOPCHOP	2014	Detect optimal target sites for sgRNA, produce potential scores for target sites	https://chopchop.cbu.uib.no/
CRISPR-P	2014	Generate synthetic sgRNA, predict potential sites for enzyme cut	https://www.cbi.hzau.cn/crisper
GPP Web Portal	2014	Produce potential sgRNA score	https://www.roadinstitute.org/rnai/public/analysis-tools/sgrnadesign
SSFinder	2014	High-throughput detection of target sites	https://code.google.com/p/ssinder
E-CRISP	2014	Potential target site evaluation	https://www.-crisp.org/E-CRISP/designcrispr.html
Cas-OFFinder	2014	Based on RNA-guided endonucleases, robust for detecting off-target sites	http://www.rgenome.net/cas.oinder/
CRISPERseek	2014	Screen sgRNA for targeted sequences, produce cleavage scores for predicted off- targets	https://www.bioconductor.org/packages/release/bioc/html/CRISPERseek.html
CRISPER-PLANT	2014	Construct specific sgRNAs for particular plant species	https://www.genome.arizona.edu/crisper
CRISPERdirect	2014	Design sgRNA with minimum off-target	https://crisper.dbcls.jp
Azimuth	2015	Design sgRNA for both on-target and off-target models	https://research.microsoft.com/en-us/projects/azimuth/
CCTop	2015	Predict target sgRNA sequence based on possible off-target	https://crisper.cos.uni-heidelerg.de
Cas-Designer	2015	RNA-guided endo- nucleases, provides all information about off-targets and out-of frame scores	http://rgenome.net/csas-designer/
CRISPy	2016	Target prediction for sgRNA, geographical representation of results	http://crispy.secondarymetaolites.org
phytoCRISP-Ex	2016	UNIX-based standalone, Cas9 target prediction	http://www.phytocrispex.biologie.ens.fr/CRISP-Ex/
CRISPR-DO	2016	Specific for both coding and non-coding targets, provides information regarding off-targeted sites and its functional conservation	http://cistrome.org/crisper/
CRISPRpred	2017	Efficient designing of sgRNA based on target in silico prediction	https://githubb.com/khaled-buet/ CRISPRpred
CRISPR-P 2.0	2017	Predict on-target scores, analyze and detect guide sequence	http://cbi.hzau.edu.cn/CRISPR2
sgRNA Score 2.0	2017	Design sgRNA for several PAM sites	https://crispr.med.harvard.edu/ sgRNAScoreV2
CRISPER-Local	2018	Design sgRNA for non-reference cultivars, predict sgRNA that can target multiple genes	http://crisper.hzau.edu.cn/CRISPR-Local/
CRISPRInc	2019	Design sgRNA for lncRNAs, works for all species	http://www.crisprlnc.org

Construction of sgRNA Expression Cassette

The most important step in CRISPER/Cas9-mediated gene editing system is construction of a unique sgRNA cassette. This works as a guide for the Cas9/

sgRNA complex which comprises of 98 nucleotides with 20 nucleotide target sequence. In plants, RNA polymerase III is used to transcribe sgRNA with the help of U3 or U6 promoters. Since the expression cassettes of sgRNA-U3/U6 promoters are of very small length (about 300-600 bp), adopter ligation or overlapping PCR can be applied to construct these expression cassettes. In 2015, a robust cloning-free approach for sgRNA expression cassette based on PCR technique was developed by Ma and coworkers. Direct cloning of sgRNA expression cassette into binary vectors for the CRISPER/Cas9 system developed using Gibson assembly or Golden Gate cloning strategy. Gao and Zhao (2014) utilized a ribozyme mechanism to generate sgRNA by transcription of pre-RNA through RNA polymerase II, through which it is possible to ligate inducible or constitutive promoters to obtain the desired function of sgRNA.

Construction of Cas9 Expression Cassette

In the case of Cas9 nuclear localization in eukaryotes, the coding sequence of Cas9 (4107 bp) must be fused with the nuclear localization signal. In plants, user-biased codons were used to design highly efficient and optimized Cas9 expression cassette for improved gene editing. For example, in rice utilization of codon-optimized Cas9p has been improved by enhancing GC content. Usually, constitutive promoters such as 35S *Cauliflower mosaic virus* (CaMV) and ubiquitin from *Arabidopsis thaliana*, maize, and rice can be used to control the expression of Cas9 in monocot and dicot plants, for highly targeted gene editing, through callus-based transformation approaches.

Transformation Approaches for CRISPER/Cas9-Based Vector Delivery into Plants

In genome editing through CRISPER/Cas9-mediated system, the cargo-vector containing the expression cassette of both sgRNA and the *Cas9* gene must be directed to target sites in plant cells. For transformation of the cargo-vector, floral dip and biolistic approaches are generally adopted. Currently, advanced strategies such as ribonucleoprotein complex, virus-mediated delivery, and plasmid delivery systems are used for transformation of plants. PEG-mediated and biolistic transformation techniques are also being used for direct delivery of *Cas9* gene expression cassettes. However, in certain cases regeneration from protoplast may be difficult due to heritable targeted mutations. *Agrobacterium*-mediated transformation has been found to be the most efficient approach for stable transformation of the CRISPER/Cas9 system in plants.

Strategies for Mutant Screening

Numerous mutant libraries have been created by CRISPER/Cas9 system in several important crop plants such as rice and tomato. Scientists are required to screen huge number of mutants, using a strategy that includes the detection of off-target and on-target edits. To assist in screening and identify of the desired mutants, different techniques have been developed, which include high resolution melting analysis (HRMA), annealing at critical temperature polymerase chain reaction (ACT-PCR), polyacrylamide gel electrophoresis (PAGE)-mediated genotyping, restriction enzyme site loss technique, and T7 endonucleases I (T7E1) approach. However, there exist certain advantages and limitations for each technique.

Rapid detection of the mutant is possible only when it manifest a clear phenotype. for example, in rice and tomato a visible albino phenotype was observed when a gene phytoen desaturate mutated via the CRISPER/Cas9 system, which was applied as a phenotype marker to detect CRISPER/Cas9 edited plants. In another approach, high-throughput sequencing can be used to screen all the mutants generated by the CRISPER/Cas9 system, with precision and accuracy. Whole genome sequencing is quite beneficial and helpful for detection of DNA-free plants edited by CRISPER/Cas9 system.

REPAIR OF CLEAVED GENOME SITES

In any gene editing system, the repair of the DNA breaks created by the nucleases is an important step. Usually, the DNA breaks are repaired by the endogenous repair mechanism of the cells such as: homologue-dependent repair (HDR) or non-homologous end-joining (NHEJ) mechanisms. In HDR mechanism, a sequence having homology to target is used as a template to repair DNA breaks or lesions. The template that contains the desired sequence is flanked by sequences having homology to both sides of the break point, thereby forcing the insertion of the desired sequence into the target site. Thus in HDR, homologous recombination enables insertion or gene recovery. In NHEL mechanism, the ends of the cleaved DNA are joined together, which results into insertion or deletion of nucleotides, resulting shifting of the gene reading frame. This process may lead to gene knockout. Since the HDR mechanism is comparatively simple and efficient and it is preferred by molecular plant breeders.

SOFTWARE TOOLKIT for CRISPER-BASED GENOME EDITING

Many steps are involved in in executing CRISPER/Cas9 and Cpf1 systems in plants. These include, (1) selection of specific target sites(s) having no highly homologous sequences which may act as the off-target site(s) in the genome; (2) designing and synthesis of oligonucleotides similar to the target sequence, (3) preparation and expression of cassette(s) for sgRNA (single guide RNA) for generating target specific sequence specificity, (4) construction of plant-transformation/expression vector(s), and (5) evaluation of the outcome of gene (genome) editing approach. Several software such as CRISPER-P, E-CRISPER, Breaking-Cas are available to carry-out some of the activities mentioned above. However, most of the toolkits cannot carry-out all the steps simultaneously. Xie et al. (2017) developed a web-based software package named CRISPER-GE (Genome Editing) which is capable to expedite all experimental designs and analyze the mutants for CRISPER/Cas9/Cpf1-based genome editing in plants. Details about the tool are discussed in Chapter 10.

SAFETY ASPECTS of GENOME EDITING SYSTEM

Genome editing systems is expected to preserve the native genomic structure. Therefore, they are considered to be safe technologies for crop improvement. However, there exist certain concerns related to the biosafety issues of the crops developed through such techniques. Some of the important aspects of genome editing system are discussed in the following section.

Non-Target Effects

Careful selection of sites for the creating double stranded breaks in the genomic DNA is essential to minimize the non-target effects of the genome editing systems. This can be achieved through prior bioinformatics analysis.

It is important to avoid sites with repeated sequences and sites having high homology with other regions of the genome. Several software are now available to assist selection of target sites and nuclease design and their validation.

Regulations on Plants Created by Genome Editing

Through genome editing technology it is possible introduce stably inherited point mutations (modifications) into the plant genome and thereafter remove the transgenic region easily. This allows creation of non-transgenic plants with improved traits. The technology is much faster than the conventional breeding methods. The plants developed through genome editing technology are near identical to the plants developed through classical breeding methods. However, their safety must be ascertained based on the developed product rather than the process involved. It is important to note that ODM-derived products plants are indistinguishable from the plants derived through conventional breeding methods. Therefore, it is apparent that plants derived through ODM should not be brought under any regulations. Through CRISPER/Cas9 technology it is possible to obtain marker-free (for example, antibiotic resistance) modified plants. Accordingly, the existing regulations on transgenic plants should also be not applicable to plants developed through CRISPER/Cas9 technology. Therefore, plants derived through genome editing systems should not be considered as genetically modified (GMO) plants.

FUTURE PROSPECTS

Following areas have great potential for the application of gene editing systems for crop improvement.

Multiplexing and Trait Stacking in Crop Breeding

In plants, manipulation of agronomic traits requires precise engineering of complex metabolic pathways, controlled by complex genetic network. Therefore, it is important to develop molecular tools with the capability to manipulate multiple genes simultaneously. CRISPER and its variations have the capability to manipulate multiple genes simultaneously and therefore gaining popularity among plant breeders.

Wang and his coworkers have exploited the multiplexing capability of CRISPER to create a system enabling the clonal reproduction from F1 hybrids in rice. This helped to preserve the favorable high degree of heterozygosity. This was achieved by altering three meiotic genes to function as mitosis like cell division, thereby producing diploid gametes and tetraploid seeds. Increase in additional ploidy level was prevented by targeting of a gene involved in fertilization.

Generation of Genomic Diversity for Crop Breeding

CRISPER/Cas has been demonstrated to have the potential to generate broad range of allelic diversity at specific loci. Shen and coworkers could simultaneously edit eight yield and quality traits genes in rice. They further could isolate homozygous mutated alleles of all the eight targeted genes. The mutants isolated represent octuple, septuple, and sixtuple mutants as well as heterozygous mutants for all the targeted genes. Thus they could generate ample genetic diversity for selection within one generation. It has also been shown that editing the same QTls can deliver different results on the basis of the genetic background.

High-Throughput Mutant Libraries

Complete genome sequencing for many crops have now been completed, but the functions of many genes sequences are unknown, which may control important agronomic characters. Therefore, the next step should be to systematically analyze the functions of the sequenced genes. Gene-knockout is an effective strategy for determining gene functions. Hence, construction of large-scale mutant libraries of the whole genomes of different crops shall provide the opportunity for the study of functional genomics of crop plants.

Gene Regulation

Gene editing systems can also be used to study the mechanism of regulation of gene expression in plants. Repression and activation of genes can be achieved by fusing transcriptional repressors or activators to the DNA-binding domains of genome-editing constructs (such as ZFP, TALE or dCas9), which target the regulatory regions of endogenous genes. Recently, CRISPER/Cas9 technology was used to alter the *cis*-regulatory control of quantitative trait loci in tomato.

Genome editing could also be used to reveal the function of many non-canonical RNAs that are linked to crop improvement. Since most no-coding transcripts are of nuclear origin, they lack open reading frames. The genome editing tools can modulate transcription directly and thus optimally suitable to interrogate the function of each RNAs.

Another application of CRISPER/Cas9 system could be in the formation of conditional alleles, to provide spatial and temporal control of gene expression, to understand the function of lethal genes. Application of inducible or tissue specific promoters for Cas9/sgRNA expression can be used to understand regulation of gene expression in a specific tissue, in development stages and under varying environmental conditions.

It is now possible to label endogenous genes with fluorescent proteins and visualize their *in vivo* expression. Through fluorescent labeled dCas9, it is possible to study the changes of genome dynamics during developmental stages of plants under different environmental conditions. Application of dCas9 can facilitate to understand the selection process of activation/repression of effector domains to specific genomic loci for regulation of endogenous gene expression in plants.

Genome editing technologies can be successfully used for epigenetic editing through selection of proteins responsible for modification of histones and methylation of DNA. These technologies have emerged as a novel way to understand regulating cellular functions in plants. CRISPER/Cas9 system can be used to understand epigenetic regulation and for identification of proteins attached to enriched chromatin carried out through enrichment of chromatin target sites. It is also possible to use CRISPER/Cas9 to identify regulatory proteins which binds to specific DNA sequences to control expression of genes.

Utilization of Genetic Diversity from Wild and Uncultured Species

Out of over 300,000 plant species, only about 200 are commercially used. Three major crops rice, wheat and maize provide energy for human consumption. Improvement in these three crops may not be possible for ever to meet the increasing demand. Therefore, it is important to explore the possibilities of utilizing vast genetic diversity in the wild and uncultured plant species. Genome editing systems have the potential to resolve the issue. One of the approaches could be to target the so-called domestication genes in the wild and uncultured plant species and accelerate the domestication process through CRISPER. During the domestication process of crop plants, certain favorable traits such as favorable plant architecture, simultaneous flowering, easy in harvesting, larger fruit size, high yield etc. were given importance during the selection process. With increase in our understanding of the genetic basis of these domestication traits, it has become much easier to identify several such domestication genes.

Recently, Zsogon et al. (2018) have demonstrated de-novo domestication of *Solanum pimpinellifolium*, an ancient stress tolerant tomato relative, through application of CRISPER/Cas9 system. Similarly, Li et al. (2018) have demonstrated de-novo domestication of four wild tomato accessions having resistance against specific stress conditions through CRISPER. More recently, Lemmon and coworkers have achieved de-novo domestication in *Physalis pruinosa,* an orphan crop of Solanaceae family, after identification of domestication genes and their alteration through CRISPER technology.

Plant Breeding Innovations Through CRISPR/Cas9

CRISPR-Cas9 has received a lot of attention in recent years due to its range of applications, including biological research, breeding and development of agricultural crops and animals, and human health applications. These include gene silencing, DNA-free CRISPR/Cas9 gene editing, homology-directed repair (HDR), and transient gene silencing or transcriptional repression (CRISPRi).

Traditional plant breeding methodologies have contributed towards advancement in agriculture during the past several decades. Traditional methods basically depend on the existence of genetic variability in the natural population and induced mutations thereof. Mutations are usually rare and occur at random. Moreover, desirable mutations may not occur in the elite varieties. The process of creating variability and their utilization is often time consuming and laborious process. In contrast, genome editing easier, faster and can be used precisely to modify the targeted gene(s) in any crop for its improvement. A variety of genome editing techniques is now available for crop improvement (Figure 7). Thus it is important for the plant breeders to select the optimum system for a given plant species, based on the objective of the breeding program. After selecting the appropriate genome-editing tools, the target sequences are designed and introduced in to the vector along with

Figure 7. (a) The network of genome editing methods and the corresponding genome editing tools. (b) Flow chart illustrating the successive steps in plant genome editing, and the estimated time needed for each step. RNP ribonucleoprotein, TALEN transcription activator-like effector nuclease, ZFN zinc-finger nuclease.
(Reproduced from Zhang et al. Genome Biology 2018, http;//creativecommons.org/ publication/ zero/1.0/).

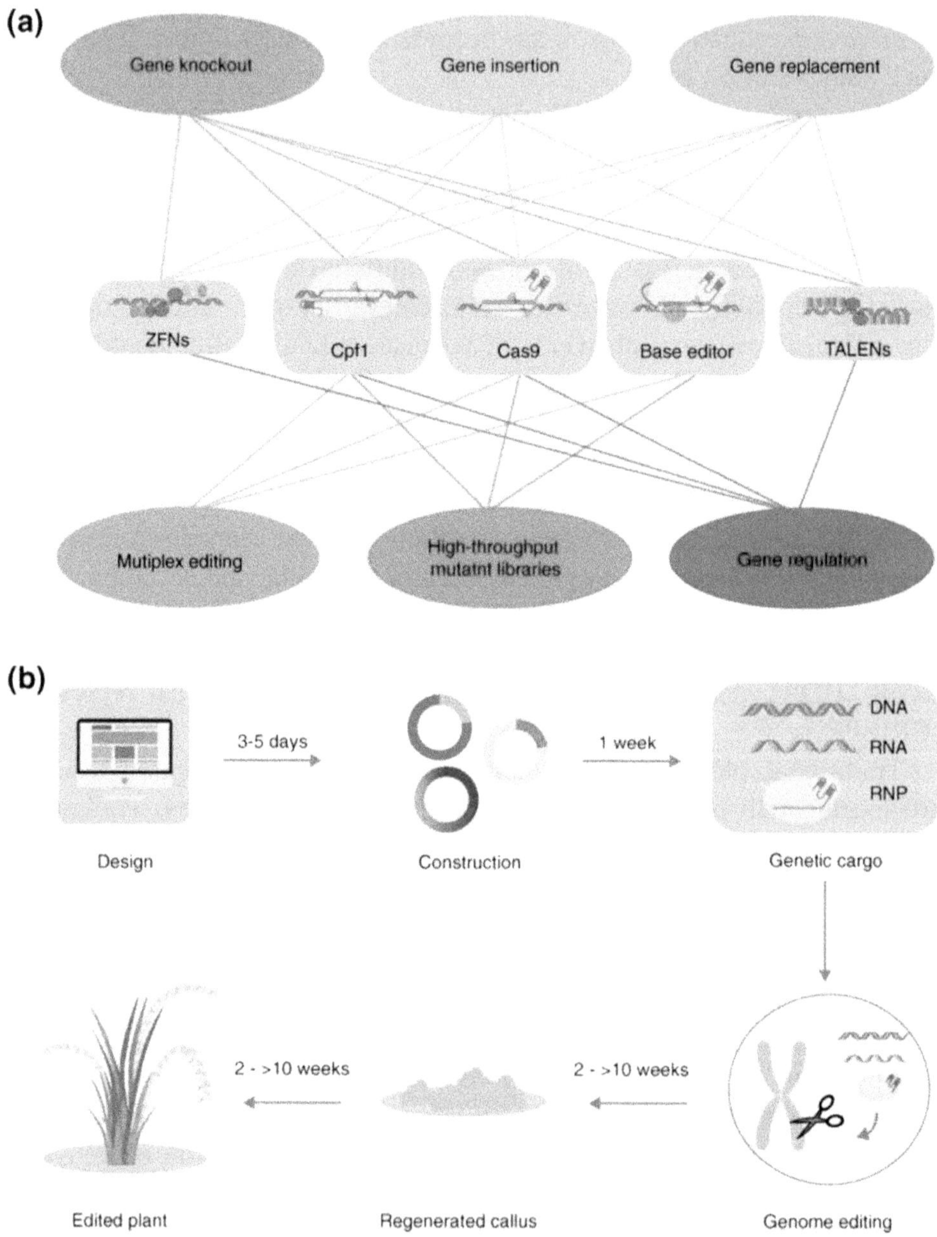

the appropriate genetic cargo (DNA, RNA, or RNPs). After introduction of the genetic cargo in to the plant cells, the target sequence shall get modified (Zhang et al. 2018). The calli developed from the edited cells should then be regenerated to produce the edited plants (Figure 7).

It is well known that protoplast-based regeneration system is not available for many important crop plants. Therefore, alternative method should be developed for gene-editing technology. Use of pollen or immature embryos that can coax to germinate *in vitro* could be a successful method for the application of gene-editing in plants. Since regeneration process is not required, such approach could provide better results. With the progress and development of genome-editing tools, it is expected to speed up the crop breeding process to meet the ever-increasing demand for food.

Application of Cas-CLOVER Technology

CRISPR is considered to be the most precise and efficient tool for creating genetic modifications in animals, plants and microorganisms. However, it may also create some off-target mutations, because it uses a single RNA guide.

Figure 8. Comparative assessment of the CRISPR/Cas9, TALEN, and Cas-COVER systems of gene editing. For explanation see text.
(Reproduced from: Li et. al. 2018. Precision CRISPR Congress Poster Presentation, Boston, MA)

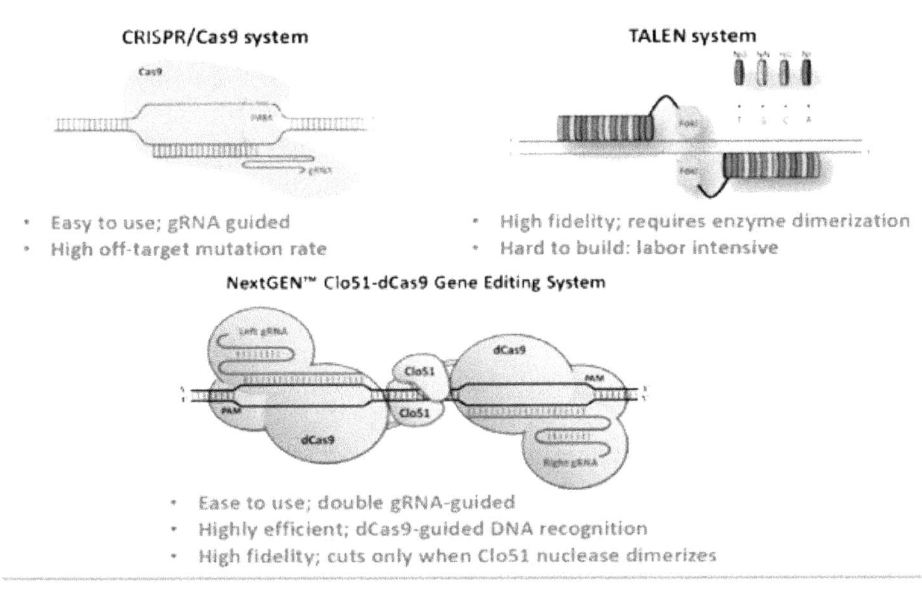

To overcome this problem, a new unique gene-editing tool called Cas-CLOVER was developed by Hera Biolabs in 2018. Technically it is fully dimeric Clo51-dCas9 genome editing technology, which is practically similar to CRISPR-Cas9. But it utilizes a different nuclease protein called Clo51 endonuclease. Cas-CLOVER is a combination of proteins that involves a nuclease-inactivated Cas9 protein melded to the Clo51 endonuclease. Cas-CLOVER tool requires two RNA guides and their activity dependents on specific spacer lengths. Unlike CRISPR, it is considered to have robust editing efficiency, with high fidelity. In addition, it creates no off-target mutation as demonstrated by Next Generation Sequencing (NGS). Basic features of Cas-CLOVER system are as follows (Li et al., 2018):

1. Exclusively licensed technology and suv0licensable by Hera
2. Functionally similar other CRISPR/Cas9 technoligies, but uses a different nuclease protein called C1o51, which is covered under a set of patents distinct from other CRISPR/Cas9 technologies
3. Cas-CLOVR is a fusion protein that comprises a nuclease-inactivation Cas9 protein fused to the C1o51 endonuclease
4. Cas-CLOVER achieves greater specificity through utilization of two guide RNAs as well as a nuclease activity that requires dimerization of subunits associated with each guide RNA

Thus, Cas-CLOVER genome editing tool is considered not only to be novel but also as an advanced tool. It has been shown to be high specificity and efficiency in both proliferating and resting T cells. It has the capacity to overcome the various limitations of CRISPR and TALENS. The capability to provide higher fidelity genome editing proved to be safer in clinical applications. Cas-CLOVER is expected to open new doors of research in genome editing science, in various organisms including plants. Comparative assessments of CRISPR/Cas9, TALENs and Cas-CLOVER technologies is presented in Figure 8.

CONCLUSION

The new CRISPR-based gene editing tools have become increasingly efficient and flexible in plant cells. Apart from targeted gene mutagenesis, other types of genetic modifications such as base substitution, gene knocking and replacement have now become available in several plant species. It is now theoretically possible to obtain precise gene editing at any locus without requiring any selection marker.

CRISPR-based gene editing technology can be classified into two categories; gene-mutagenesis and gene-correction tools. Gene mutagenesis is used to introduce full or partial loss-of-function mutations into the target gene (e.g. canonical CRISPER/Cas9 system and base editing system). On the other hand, for gene-correcting purpose, the modifications on the target gene have to be very precise. The base editing systems and fragment deletion, insertion and replacement tools can be used for this purpose. Although much progress has been made, the low homology repair (HR) frequency in plant cells is still a problem for precise gene editing.

Base editing has become efficient means of precisely converting one base to another. Diverse base editing systems have been established successfully to alter/modify genes in plants for both biological functional analysis and crop improvement. Several variants of Cas9 and mutations of cytidine deaminase and adenine deaminase have been developed and incorporated into CBE and ABE base editors thereby expanding their editing scope and improve their specificity and editing efficiency in plants. With the introduction of new base editing technology and improvement in their specificity, base editing will have unparalleled potential crop improvement.

One of the big challenges in crop breeding is to achieve efficient delivery of CRISPR components into the reproductive cells to generate heritable gene modifications. The second big challenge is to decide which gene(s) to edit in order to improve a particular character. It is expected that with the aid of CRISPR gene editing systems, the underlying genes for complex traits shall be identified and subsequently edited for crop improvement.

REFERENCES

Anzalone, A., Randolph, P., Davis, J., Sousa, A. A., Koblan, L. W., Levy, J. M., … Li, D. R. (2019). Search-and replace genome editing without double-stranded breaks or donor DNA. *Nature*. doi:1038/s41586-019-1711-4

Cai, Y., Chen, L., Shi Sun, S., Wu, C., Weiwei Yao, W., Jiang, B., Tianfu Han, T., & Hou, W. (2018). CRISPR/Cas9-Mediated Deletion of Large Genomic Fragments in Soybean. *International Journal of Molecular Sciences*, *19*(12), 3835–3846. doi:10.3390/ijms19123835 PMID:30513774

Gao, Y., & Zhao, Y. (2014). Self-processing of ribozyme-flanked RNAs into guide RNAs in vitro for CRISPR-mediated genome editing. *Journal of Integrated Plant Breeding*, *56*(4), 343–349. doi:10.1111/jipb.12152 PMID:24373158

Lei, Y., Lu, L., Lin, H. Y., Li, S., Xing, F., & Chen, L. L. (2014). CRISPER-P: A web tool for synthetic single-guide RNA design of CRISPER-system in plants. *Molecular Plant*, *7*(9), 1494–1496. doi:10.1093/mpsu044 PMID:24719468

Li, J., Sun, Y., Du, J., Zhao, Y., & Xia, L. (2017). Generation of targeted point mutations in rice by a modified CRISPR/Cas9 system. *Molecular Plant*, *10*(3), 526–529. doi:10.1016/j.molp.2016.12.001 PMID:27940306

Li, R., Liu, C., Zhao, R., Wang, L., Chen, L., Yu, W., Zhang, S., Sheng, J., & Shen, L. (2019). CRISPER/Cas9-mediated SINPR1 mutagenesis reduces tomato plant drought tolerance. *BMC Plant Biology*, *19*(1), 38–46. doi:10.118612870-018-1627-4 PMID:30669982

Li, T., Yang, X., Yu, Y., Si, X., Zhai, X., Zhang, H., ... Xu, C. (2018). Domesti. cation of wild tomato is accelerated by genome editing N*ature. Biotechnology (Faisalabad)*, *36*, 1160–1163.

Li, X., Wang, X., Tong, M., Tan, Y., Down, J. D., Shedlock, D. J., & Ostertag, E. M. (2018). Cas-CLOVER™: A high-fidelity genome editing system for safe and efficient modification of cells for immunotherapy. *Precision CRISPR Congress Poster Presentation.*

Michno, J. M., Wang, X., Liu, J., Curtin, S. J., Kono, T. J., & Stupar, R. M. (2015). CRISPR/Cas mutagenesis of soybean and *Medicago truncatula* using a new web-tool and modified Cas9 enzyme. *GM Crops and Food: Biotechnology in Agriculture and the Food Chain*, *6*(4), 243–252. doi:10.1080/21645698.2015.1106063 PMID:26479970

Mishra, R., Joshi, R. K., & Zhao, K. (2020). Base editing in crops: Current advances, limitations and future implications. *Plant Biotechnology Journal*, *18*(1), 20–31. doi:10.1111/pbi.13225 PMID:31365173

Tian, S., Jiang, L., Cui, X., Zhang, J., Guo, S., Li, M., ... Zong, M. (2018). Engineering herbicide-resistant watermelon variety through CRISPER/Cas9-mediated base-editing. *Plant Cell Reports*, *67*, 1068–1079. PMID:29797048

Xie, K., Zhang, J., & Yang, Y. (2014). Genome-wide prediction of highly specific guide RNA spacers for CRISPER-Cas-mediated genome editing in model plants and major crops. *Molecular Plant*, *7*(5), 923–926. doi:10.1093/mpsu009 PMID:24482433

Xie, X., Ma, X., Zhu, Q., Zeng, D., Li, G., & Liu, Y.-G. (2017). CRISPER-GE: A convenient software toolkit for CRISPER-based genome editing. *Molecular Plant*, *10*(9), 1246–1249. doi:10.1016/j.molp.2017.06.004 PMID:28624544

Zhang, K., Raboanatahiry, N., Zhu, B., & Li, M. (2017). Progress in genome editing technology and its application in plants. *Front. Plant Sci.*, *8*, 177–184. doi:10.3389/fpls.2017.00177 PMID:28261237

Zhang, Y., Massel, K., Godwin, I. D., & Gao, C. (2018). Applications and potential of genome editing in crop improvement. *Genome Biology*, *19*, 201–2011. doi:10.118613059-018-1585-z PMID:30501614

Zong, Y., Song, Q., Li, C., Jin, S., Zhang, D., Wang, Y., Qiu, J.-L., & Gao, C. (2018). Efficient C-to-T base editing in plants using a fusion of ncas9 and human apobec3a. *Nature Biotechnology*, *36*(10), 950–963. doi:10.1038/nbt.4261 PMID:30272679

Zsogon, A., Cermak, T., Naves, E. R., Notini, M. M., Edel, S. W., Weinl, S., ... Peres, L. F. O. (2018). D*e* novo. domestication of wild tomato using genome editing Nature. *Biotechnology*, *36*, 1211–1216.

ADDITIONAL READING

Ali, Z., Eid, A., Ali, S., & Mahfouz, M. M. (2018). Pea early-browning virus-mediated genome editing via the CRISPER/Cas9 system in *Nicotiana benthamiana* and *Arabidopsis*. *Virus Research*, *244*, 333–337. doi:10.1016/j.virusres.2017.10.009 PMID:29051052

Anderson, M., Turesson, H., Olsson, N., Falt, A. S., Ohlsson, P., Gonzalez, M. N., ... Hofvander, P. (2018). Genome editing in potato via CRISPER-Cas9 ribonucleoprotein delivery. *Physiologia Plantarum*, *164*(4), 378–384. doi:10.1111/ppl.12731 PMID:29572864

Barrangou, R., & Horvath, P. (2014). Functions and applications of RNA-guided immune system. In R. A. Mayers (Ed.), *Encyclopedia of molecular cell biology and molecular medicine: RNA biology* (pp. 1–24). Wiley-VCH Verlag GmBh & Co.

Bharat, S.S., Li, S., Li, J. Yan, L., & Xia, L. (2019). Base editing in plants: current status and challenges, *The Crop Journal*, Retrieved from: . 10.002. doi:10.1016/j.cj.2019

Blin, K., Pedersen, L. E., Webbber, T., & Lee, S. Y. (2016). CRISPy-web: An online resource to design sgRNA for CRISPER applications. *Synthetic and Systems Biotechnology*, *1*(2), 118–121. doi:10.1016/j.synbio.2016.01.003 PMID:29062934

Bonawitz, N. D., Ainley, W. M., Itaya, A., Chennareddy, S. R., Cicak, T., Effinger, K., & Pareddy, D. R. (2018). Zinc finger nuclease-mediated targeting of multiple transgenes to an endogenous soybean genomic locus via non-homologous end joining. *Plant Biotechnology Journal*, *17*(4), 750–761. doi:10.1111/pbi.13012 PMID:30220095

Braatz, J., Harloff, Mascher, M., Stein, N., Himmelbach, A., & Jung, C. (2017). CRISPER-Cas9 targeted mutagenesis leads to simultaneous modification of different homogenous gene copies in polyploidy oilseed rape (*Brassica napus*). *Plant Physiology*, *174*(2), 935–942. doi:10.1104/pp.17.00426 PMID:28584067

Butt, H., Jamil, M., Wang, J. Y., Al-Babili, S., & Mahfouz, M. (2018). Engineering plant architecture via CRISPER/Cas9-mediated alternation of strigolactone biosynthesis. *BMC Plant Biology*, *18*(1), 174–186. doi:10.118612870-018-1387-1 PMID:30157762

Chari, R., Yeo, N. C., Chavez, A., & Church, G. (2017). Sigma score 2.0- a species independent model to predict CRISPER/Cas9 activity. *ACS Synthetic Biology*, *6*(5), 902–904. doi:10.1021/acssynbio.6b00343 PMID:28146356

Chen, L., Li, W., Katin-Grazzini, L., Ding, J., Gu, X., Li, Y., Gu, T., Wang, R., Lin, X., Deng, Z., McAvoy, R. J., Gmitter, F. G. Jr, Deng, Z., Zhao, Y., & Li, Y. (2018). A method for the production and expedient screening of CRISPER. Cas9-mediated non-transgenic mutant plants. *Horticulture Research*, *5*(1), 13–18. doi:10.103841438-018-0023-4 PMID:29531752

Chen, W., Zhang, G., Li, J., Zhang, X., Huang, S., Xiang, S., Hu, X., & Liu, C. (2019). CRISPERInc: A manually curated database of validated sgRNAs for IncRNAs. *Nucleic Acids Research*, *47*(D1), D63–D68. doi:10.1093/nar/gky904 PMID:30285246

Curtin, S. J., Xiong, Y., Michoo, J. M., Campbell, B. W., Stec, A. O., Cermak, T., ... Stuper, R. M. (2018). CRISPER/Cas9 and TALENs generate heritable mutations for genes involved in small RNA processing of *Glycine max* and *Medicago truncatula. Plant Biotechnology Journal*, *16*(6), 1125–1137. doi:10.1111/pbi.12857 PMID:29087011

Du, H., Zeng, X., Zhao, M., Cui, X., Wang, Q., Yang, H., Cheng, H., & Yu, D. (2016). Efficient targeted mutagenesis in soybean by TALENs and CRISPER/Cas9. *Journal of Biotechnology*, *217*, 90–97. doi:10.1016/j.jbiotec.2015.11.005 PMID:26603121

Feng, Z., Zhang, B., Ding, W., Lin, X., Yang, D. L., Wei, P., ... Zhu, K. J. K. (2013). Efficient genome editing in plants using a CRISPER.Cas system. *Cell Research*, *23*(10), 1229–1232. doi:10.1038/cr.2013.114 PMID:23958582

Gehrke, J. K., Cervantes, O., Clement, M. K., Wu, Y., Zeng, J., Bauer, D. E., Pinello, L., & Joung, J. K. (2019). An APOBEC3A-Cas9 base editor with minimized bystander and off-target activities. *Nature Biotechnology*, *36*(10), 977–982. doi:10.1038/nbt.4199 PMID:30059493

Grechko, V. V. (2002). Molecular DNA markers in phylogeny and systematics. *Russian Journal of Genetics*, *38*(8), 851–868. doi:10.1023/A:1016890509443 PMID:12244688

Grohmann, L., Keilwagen, J., Duensing, N., Dogand, E., Harting, F., Withelm, R., ... Sprink, T. (2019). Detection and identification of genome editing in plants: Challenges and opportunities. *Frontiers in Plant Science*, *10*, 1–8. doi:10.3389/fpls.2019.00236 PMID:30930911

He, J., Zhao, X., Larocu, A., Laroche, A., Lu, Z. X., Lin, H. K., & Li, Z. (2014). Genotyping-by-sequencing (GBS), an ultimate marker-assisted selection (MAS) tool to accelerate plant breeding. *Frontiers of Plant Breeding*, *5*, 1–6. doi:10.3389/fpls.2014.00484 PMID:25324846

He, Y., Zhu, M., Wang, L., Wu, J., Wang, Q., Wang, R., & Zhao, Y. (2018). Programmed self-elimination of the CRISPER/Cas9 construct greatly accelerates the isolation of edited and transgene-free rice plants. *Molecular Plant*, *11*(9), 1210–1213. doi:10.1016/j.molp.2018.05.005 PMID:29857174

Hess, G. T., Tycko, J., Yao, D., & Bassik, M. C. (2017). Methods and applications of CRISPER-mediated base editing in eukaryotic genomes. *Molecular Cell*, *68*(1), 26–43. doi:10.1016/j.molcel.2017.09.029 PMID:28985508

Hsu, P. D., Lander, E., & Zhang, F. (2014). Development and applications of CRISPR-Cas9 for Genome Engineering. *Cell*, *157*(6), 1262–1278. doi:10.1016/j.cell.2014.05.010 PMID:24906146

Hsu, P. D., Scott, D. A., Weinstein, J. A., Ran, T. A., Konermann, S., Agarwal, V., ... Zhang, F. (2013). DNA targeting specificity of RNA-guided Cas9 nucleases. *Nature Biotechnology*, *31*(9), 827–832. doi:10.1038/nbt.2647 PMID:23873081

Hu, X., Meng, X., Liu, Q., Li, J., & Wang, K. (2018). Increasing the efficiency of CRISPER-Cas9-VQR precise genome editing in rice. *Plant Biotechnology Journal*, *16*(1), 292–297. doi:10.1111/pbi.12771 PMID:28605576

Hua, K., Tao, X., Yuan, F., Wang, D., & Zhu, J. K. (2018). Precise A-T to G-C base editing in the rice genome. *Molecular Plant*, *11*(4), 627–630. doi:10.1016/j.molp.2018.02.007 PMID:29476916

Hua, Y., Wang, C., Huang, J., & Wang, K. (2017). A simple and efficient method for CRISPER/Cas9-induced mutant screening. *Journal of Genetics and Genomics*, *44*(4), 207–213. doi:10.1016/j.jgg.2017.03.005 PMID:28416245

Huang, J., Li, J., Zhou, J., Wang, L., Yang, S., Hurst, L. D., Li, W.-H., & Tian, D. (2018). Identifying a large number of high yielding genes in rice by pedigree analysis, whole-genome sequencing, and CRISPER-Cas9 gene knockout. *Proceedings of the National Academy of Sciences of the United States of America*, *115*(32), E7559–E7567. doi:10.1073/pnas.1806110115 PMID:30037991

Jacob, T. B., Zhang, N., Patel, D., & Martin, G. B. (2017). Generation of a collection of mutant tomato lines using pooled CRISPER libraries. *Plant Physiology*, *174*(4), 2023–2037. doi:10.1104/pp.17.00489 PMID:28646085

Ji, X., Si, X., Zhang, Y., Zhang, H., Zhang, F., & Gao, C. (2018). Conferring DNA virus resistance with high specificity in plants using virus-inducible genome-editing system. *Genome Biology*, *19*(1), 197–202. doi:10.118613059-018-1580-4 PMID:30442181

Jiang, G. L. (2013). Molecular marker and marker-assisted breeding in plants. Retrieved from: . doi:10.5772/52583

Jung, J. H., & Altpeter, F. (2016). TALEN mediated targeted mutagenesis of the caffeic acid O-methyltransferase in highly polyploidy sugarcane improves cell wall composition for production of bioethanol. *Plant Molecular Biology*, *92*(1-2), 131–142. doi:10.100711103-016-0499-y PMID:27306903

Kamburova, V. S., Nikitina, E. V., Shermatov, S. E., Buriev, Z. T., Kumpatla, S. P., Emani, C., & Abdurakhmonov, Y. (2017). Genome editing in plants: an overview of tools and applications. *International Journal of Agronomy*. https://doi:org./10.11 5 5/2017/7315351.

Kannan, B., Jung, J. H., Moxley, G. W., Lee, S., & Altpeter, F. (2018). TALEN-mediated targeted mutagenesis of more than 100 COMT copies/alleles in highly polyploidy sugarcane improves saccharification efficiency without compromising biomass yield. *Plant Biotechnology Journal*, *16*(4), 856–866. doi:10.1111/pbi.12833 PMID:28905511

Kim, H., Kim, S. T., Ryu, J., Kang, B.-C., Kim, J.-S., & Kim, S.-G. (2017). CRISPER/CPF1-mediated DNA-free plant genome editing. *Nature Communications*, *8*(1), 14406–14411. doi:10.1038/ncomms14406 PMID:28205546

Li, R., Zhang, L., Wang, L., Chen, L., Zhao, R., Sheng, J., & Shen, L. (2018). Reduction of tomato-plant chilling tolerance by CRISPER-CAS9-mediated SICBF1 mutagenesis. *Journal of Agricultural and Food Chemistry*, *66*(34), 9042–9051. doi:10.1021/acs.jafc.8b02177 PMID:30096237

Li, X., Wang, Y., Chen, S., Tian, H., Fu, D., Zhu, B., Luo, Y., & Zhu, H. (2018). Lycopene is enriched in tomato fruit by CRISPER/Cas9-mediated multiplex genome editing. *Frontiers in Plant Sciences*, *9*, 559–564. doi:10.3389/fpls.2018.00559 PMID:29755497

Liang, Z., Chen, K., Zhang, Y., Lin, J., Yen, K., Qin, J. L., & Geo, C. (2018). Genome editing of bread wheat using biolistic delivery of CRISPER/Cas9 in vitro transcripts or ribonucleoproteins. *Nature Protocols*, *13*(3), 413–430. doi:10.1038/nprot.2017.145 PMID:29388938

Lin, C. S., Hsu, C. T., Yang, L. H., Lee, L. Y., Fu, J. Y., Chang, Q. W., ... Zhang, R. (2018). Application of protoplast technology to CRISPER/Cas9 mutagenesis: From single-cell mutation detection to mutant plant regeneration. *Plant Biotechnology (Sheffield, England)*, *16*(7), 1295–1310. doi:10.1111/pbi.12870 PMID:29230929

Liu, H., Ding, Y., Zhou, Y., Jin, W., Xie, K., & Chen, L. L. (2017). CRISPER-P 2.0: An improved CRISPER/Cas9 tool for genome editing in plants. *Molecular Plant*, *10*(3), 530–532. doi:10.1016/j.molp.2017.01.003 PMID:28089950

Lu, H. P., Lin, S. M., Xu, S. L., Chen, W. Y., Zhou, X., Tan, Y. Y., ... Shu, Q. Y. (2017). CRISPER-S: An active interference element for a rapid and inexpensive selection of genome-edited, transgene-free rice plants. *Plant Biotechnology Journal*, *15*(11), 1371–1373. doi:10.1111/pbi.12788 PMID:28688132

Lu, Y., Ye, X., Gao, R., Huang, J., Wang, W., Tang, J., ... Qian, Y. (2017). Genome-wide targeted mutagenesis in rice using the CRISPER/Cas9 system. *Molecular Plant*, *10*(9), 1242–1245. doi:10.1016/j.molp.2017.06.007 PMID:28645638

Ma, J., Koster, J., Qin, Q., Hu, S., Li, W., Chen, C., Cao, Q., Wang, J., Mei, S., Liu, Q., Xu, H., & Liu, X. S. (2016). CRISPER-DO for genome-wide CRISPER design and optimization. *Bioinformatics (Oxford, England)*, *32*(21), 3336–3338. doi:10.1093/bioinformatics/btw476 PMID:27402906

Ma, X., Zhang, Q., Zhu, Q., Liu, W., Chen, Y., Qin, R., ... Liu, Y. G. (2015). A robust CRISPER/Cas9 system for convenient, high-efficient multiplex genome editing in monocot and dicot plants. *Millennium Plant*, *8*, 1274–1284. PMID:25917172

Makarova, K. S., Haft, D. H., Barrangou, R., Brouns, S. J., Carpentier, E., Horvath, P., ... Koonin, E. V. (2011). Evolution and classification of the CRISPER-Cas system. *Nature Reviews. Microbiology*, *9*(6), 467–477. doi:10.1038/nrmicro2577 PMID:21552286

Mao, Y., Botella, J. R., Liu, Y., & Zhu, J. K. (2019). Gene editing in plants: Progress and challenges. *National Science Review*, *6*(3), 421–437. doi:10.1093/nsr/nwz005

Mao, Y., Botella, J. R., & Zhu, J. K. (2017). Heritability of targeted gene modifications induced by plant-optimised CRISPER systems. *Cellular and Molecular Life Sciences*, *74*(6), 1075–1093. doi:10.100700018-016-2380-1 PMID:27677493

Meng, X., Yu, H., Zhang, Y., Zhuang, F., Song, X., Gao, S., Gao, C., & Li, J. (2017). Construction of a genome-wide mutant library in rice using CRISPER/Cas9. *Molecular Plant*, *10*(9), 1238–1241. doi:10.1016/j.molp.2017.06.006 PMID:28645639

Metje-Sprik, J., Menz, J., Modrzejewski, D., & Sprink, T. (2019). DNA-free genome editing: Past, present and future. *Frontiers in Plant Science*, *9*, 1–9. doi:10.3389/fpls.2018.01957 PMID:30693009

Molla, K. A., & Yang, Y. (2019). CRISPR/Cas-mediated base editing: Technical considerations and practical applications. *Trends in Biotechnology*, *37*(10), 1121–1142. doi:10.1016/j.tibtech.2019.03.008 PMID:30995964

Moose, S. P., & Mumm, R. H. (2008). Molecular plant breeding as a foundation for 21st century crop improvement. *Plant Physiology*, *147*(3), 969–977. doi:10.1104/pp.108.118232 PMID:18612074

Nadeem, M. A., Nawaz, M. A., Shahid, M. Q., Dogan, Y., Comertpay, G., Yidiz, M., ... Chung, G. (2018). DNA molecular markers in plant breeding: Current status and recent advances in genomic selection and genome editing. *Biotechnology, Biotechnological Equipment*, *32*(2), 261–285. doi:10.1080/13102818.2017.1400401

Okuzaki, A., Ogawa, T., Koizuka, C., Kaneko, K., Inabas, M., Imamura, J., & Koizuka, N. (2018). CRISPER/Cas9-mediated genome editing of the fatty acid desaturase 2 gene in *Brassica napus*. *Plant Physiology and Biochemistry*, *131*, 63–69. doi:10.1016/j.plaphy.2018.04.025 PMID:29753601

Patwardhan, A., Ray, S., & Roy, A. (2014). Molecular marker in phylogenetic studies-a review. *Journal of Phylogenetics & Evolutionary Biology*, *2*, 2–9.

Pawluk, A., Davidson, A. R., & Maxwelll, K. L. (2018). Anti-CRISPER: Discovery, mechanism and function. *Nature Reviews. Microbiology*, *16*(1), 12–17. doi:10.1038/nrmicro.2017.120 PMID:29062071

Perez-de-Castro, A. M., Vilanova, S., Canizares, J., Pascual, J. M., Blanca, M. J., Diez, J., Prohens, J., & Pico, B. (2012). Application of genomic tools in plant breeding. *Current Genomics*, *13*, 179–195. doi:10.2174/138920212800543084 PMID:23115520

Qi, L. S., Larson, M. H., Gilbert, L. A., Doudna, J. A., Weissman, J. S., Arkin, A. P., & Lim, W. A. (2013). Repurposing CRISPER as an RNA-guided platform for sequence-specific control of gene expression. *Cell*, *152*(5), 1173–1183. doi:10.1016/j.cell.2013.02.022 PMID:23452860

Qi, W., Zhu, T., Tian, Z., Li, C., Zhang, W., & Song, R. (2016). High-efficiency CRISPER/Cas9 multiplex gene editing using the glycine tRNA-processing system-based strategy in maize. *BMC Biotechnology*, *16*(1), 58–73. doi:10.118612896-016-0289-2 PMID:27515683

Qi, Y. (Ed.). (2019). *Plant Genome Editing with CRISPR Systems, Methods and Protocols*. Springer. doi:10.1007/978-1-4939-8991-1

Rahman, M. K., & Rahman, M. S. (2017). CRISPERpred: A flexible and efficient tool for sgRNAs on-target activity prediction in CRISPER/Cas9 systems. Public Library of Science (*PLoS*). *ONE*, *12*, e0181943. doi:10.1371/journal.pone.0181943 PMID:28767689

Ran, Y., Patron, N., Kay, P., Wong, D., Buchanan, M., Cao, Y. Y., ... Webb, S. R. (2018). Zinc fi.nger nuclease-mediated precision genome editing of an endangered gene in hexaploid bread wheat (*Triticum aestivum*) using a DNA repair template. *Chih Wu Sheng Li Hsueh T'ung Hsun*, *16*, 2088–2101.

Rastogi, A., Muruk, O., Bowler, C., & Tirichine, L. (2016). A web-based and stand-alone application to find specific target sequences for CRISPER/Cas editing. *BMC Bioinformatics*, *17*(1), 261–278. doi:10.118612859-016-1143-1 PMID:27363443

Razzaq, A., Saleem, F., Kanwal, M., Mustafa, G., Yousaf, H. M. I., Hameed, M. K., ... Joyia, F. A. (2019, August 19). (201.9). Modern. trends in plant genome editing: An inclusive review of the CRISPER/Cas9 toolbox. *International Journal of Molecular Sciences*, *20*(16), 4045–4089. doi:10.3390/ijms20164045

Sauer, N. J., Mozoruk, J., Miller, R. B., Warburg, Z. J., Walker, K., Beetham, P. R., Schöpke, C. R., & Gocal, G. F. (2016). Oligonucleotide-directed mutagenesis for precision gene editing. *Plant Biotechnology Journal*, *14*(2), 496–502. doi:10.1111/pbi.12496 PMID:26503400

Schiml, S., Fauser, F., & Puchta, H. (2014). The CRISPR/Cas system can be used as nuclease for *in planta* gene targeting and as paired nickases for directed mutagenesis in *Arabidopsis* resulting in heritable progeny. *The Plant Journal*, *80*(6), 1139–1150. doi:10.1111/tpj.12704 PMID:25327456

Shah, Q., Wang, Y., Li, J., Zhang, Y., Chen, K., Liang, Z., ... Gao, C. (2013). Targeted genome modification of crop plants using a CRISPER-Cas system. *Nature Biotechnology*, *31*(8), 686–688. doi:10.1038/nbt.2650 PMID:23929338

Shen, L., Wang, C., Fu, Y., Wang, J., Lin, Q., Zhang, X., ... Wang, K. (2018). QTL editing confers opposing yield performance in different rice varieties. *Journal of Integrative Plant Biology*, *60*(2), 89–93. doi:10.1111/jipb.12501 PMID:27628577

Shimatani, Z., Ariizumi, T., Fujikura, U., Konodo, A., Ezura, H., & Nishida, K. (2019). Targeted base editing with CRISPER-deaminase in tomato. *Methods in Molecular Biology (Clifton, N.J.)*, *1917*, 297–307. doi:10.1007/978-1-4939-8991-1_22 PMID:30610645

Stemmer, M., Thumberger, T., Dei Soi Keyer, M., Wittbrodt, J., & Mateo, J. L. (2015). CCTop: An intuitive, flexible and reliable CRISPER/Cas9 target prediction tool. Public Library of Science *(PLoS). ONE*, *10*, e0124633. doi:10.1371/journal.pone.0124633 PMID:25909470

Sun, J., Liu, H., Liu, J., Cheng, S., Peng, Y., Zhang, Q., Yan, J., Liu, H.-J., & Chen, L. L. (2019). CRISPER-local: A local single-guide RNA (sgRNA) design tool for non-reference plant genome. *Bioinformatics (Oxford, England)*, *35*(14), 2501–2503. doi:10.1093/bioinformatics/bty970 PMID:30500879

Varshney, R., Nayak, S. N., Hoisington, D., & Graner, A. (2009). Molecular plant breeding: methodology and achievements. In D. J. Somers (Ed.), *Methods in molecular biology* (pp. 283–304). Humana Press.

Wolter, F., & Puchta, H. (2017). Knocking out consumer concerns and regulator's rule: Efficient use of CRISPER/Cas ribonucleoprotein complexes for genome editing in cereals. *Genome Biology*, *18*(1), 682–698. doi:10.118613059-017-1179-1 PMID:28245842

Wolter, F., Schindele, P., & Puchta, H. (2019). Plant breeding at the speed of light: The power of CRISPER/Cas to generate directed genetic diversity at multiple sites. *BMC Plant Biology*, *19*(1), 176–183. doi:10.118612870-019-1775-1 PMID:31046670

Yan, F., Kuang, Y., Ren, B., Wang, J., Zhang, D., Lin, H., Yang, B., Zhou, X., & Zhou, H. (2018). Highly efficient A-T to G-C base editing by Cas9n-guided tRNA adenosine deaminase in rice. *Molecular Plant*, *11*(4), 631–634. doi:10.1016/j.molp.2018.02.008 PMID:29476918

Zhang, A., Liu, Y., Wang, F., Li, T., Chen, Z., Kong, D., Bi, J., Zhang, F., Luo, X., Wang, J., Tang, J., Yu, X., Liu, G., & Luo, L. (2019). Enhanced rice salinity tolerance via CRISPER/Cas9-targeted mutagenesis of the OsRR22 gene. *Molecular Breeding*, *39*(3), 47–56. doi:10.100711032-019-0954-y

Zhang, S., Zhang, R., Song, G., Gao, J., Li, W., Han, X., Chen, M., Li, Y., & Li, G. (2018). Targeted mutagenesis using the *Agrobacterium tumefaciens*-mediated CRISPER-Cas system in common wheat. *BMC Plant Biology*, *18*(1), 302–312. doi:10.118612870-018-1496-x PMID:30477421

Zhang, Z., Hua, L., Gupta, A., Tricoli, D., Edwards, K. J., Yang, B., & Li, W. (2019). Development of an Agrobacterium-delivered CRISPER/Cas9 system for wheat genome editing. *Plant Biotechnology Journal*, *17*(8), 1623–1635. doi:10.1111/pbi.13088 PMID:30706614

Zong, Y., Wang, Y., Li, C., Zhang, R., Chen, K., Ran, Y., Qiu, J.-L., Wang, D., & Gao, C. (2017). Precise base editing in rice, wheat and maize with a Cas9-cytidine deaminase fusion. *Nature Biotechnology*, *35*(5), 438–440. doi:10.1038/nbt.3811 PMID:28244994

APPENDIX

1. Explain how zinc-finger-nucleases (ZFNs) can be used for genome editing?
2. Explain how transcription activator-like effector nucleases (TALE) can be used for genome editing?
3. Explain the mechanism by which CRISPER/Cas system operates in the bacteria as an effective immune system in bacteria.
4. What are the activities that can be carried out through CRISPER/Cas system?
5. What is Base Editing system? Explain its usefulness.
6. What is DNA-free genome editing system? Explain how this system is useful.
7. Describe the CRISPER/Cpf1 system of genome editing.
8. Explain how multiplexing and trait stacking can help in crop improvement?
9. Explain how different genome editing systems can effectively be used for crop improvement?
10. Describe the main types and sub-types of CRISPER/Cas model systems along with the bacterial hosts.

Chapter 10
Software Tools to Assist Breeding Decisions

ABSTRACT

Plant breeders are usually faced with the problem of predicting the performance of new individuals with untested gene combinations. Therefore, it is important to follow an integrated breeding approach by combining molecular tools, molecular mapping, and MAS. It is also required to develop tools for modeling and simulation analysis by utilizing all pre-existing and newly generated data. Several software tools have been developed that integrates breeding simulations and phenotype prediction models using genomic information. Reliable phenotype prediction models for the simulation were constructed from actual genotype and phenotype data. Such simulation-based genome-assisted approach to breeding will help optimize plant breeding in all important agricultural crops. Software tools have also been developed for designing target sites or evaluating the outcome of genome/gene editing system. This chapter provides an overview of the key software support tools that will assist the plant breeders in decision making during the process of conducting various breeding program.

INTRODUCTION

Important steps involved in molecular breeding include: identification and selection of beneficial genetic variations and utilization of such variations more effectively and efficiently for crop improvement. Marker-assisted selection

DOI: 10.4018/978-1-7998-4312-2.ch010

(MAS) has been recognized as a reliable identification tool and thus has lot of potential for its utilization in crop breeding. To achieve the desired goal, MAS has to be combined with other advanced techniques such as genome scans, advanced biometrical analysis and quantitative genetics modeling that will require complex software. Effective molecular breeding procedure will depend on the following parameters (Singh et al. 2010).

1. Identification of new sources of variation and development of strong marker-trait associations.
2. Management of large numbers of genotypes and manipulation of genotype, phenotype and pedigree data.
3. Selection of desirable recombinants by combining genotypic and phenotypic information.
4. Development of breeding systems to minimize population size, number of generations and overall cost, but to maximize genetic gain for traditional and novel traits.

Supporting tools are required to manage and optimize various components of plant molecular breeding procedures. Many of these tools come in the form of software. Some of the important software having applications in molecular plant breeding can be obtained from the sites: http://bioreseach.ac.uk/browse/mesh/D012984.html, http://linkage.rockefeller.edu/soft/list.html.

MANAGEMENT AND EVALUATION OF GERMPLASM

With the increase in the size of germplasm collections, genebank curators find it difficult to manage and provide required information to the breeders and researchers. MAGE (marker-assisted germplasm evaluation) has played an important role in the process of acquisition, maintenance, distribution, and use of germplasm in the gene banks. For efficient management of the various activities of the gene bank through MAGE it is important to have the following resources in place (Xu 2014).

1. Characterization of suitable genetic markers for large number alleles, polymorphic information content (PIC value), size and range of the alleles, signal strength, working conditions and information for multiplexing,

2. Generate high-density molecular maps for selection of markers evenly spread over the entire genome or densely spread over the specific region of interest,
3. Establishment of association of the marker and the trait of agronomical importance,
4. Establishment of high-throughput genotyping system, and
5. Establishment of an efficient data management and analysis system.

Several open-source and commercial software packages such as GeneFlow SAS, STATISTICA, NTSYS, JMP, STRUCTURE, PowerMaker etc. are available for evaluation and analysis of germplasm. STRUCTURE can be used to investigate population structure using multi-locus genotype data. Which in turn provide information about presence of distinct populations, study hybrid zones, identify migrants and admixed individuals, and estimate frequency of the alleles in specific population. With PowerMaker it is possible to perform statistical analysis of marker data derived from germplasm accessions, and deliver data-driven integrated analysis environment (IAE) for marker data. It also can handle a variety of data from SSRs, SNPs and RFLPs. A host of other statistical analysis, having relevance to management and utilization of germplasm, can be carried out through PowerMaker. GGT (Geographical Genotype) software allows transforming molecular marker data into simple chromosomal drawings. Through POPDIST it is possible to calculate the number of different genetic identities, reconstruction of phylogeny measures and distance measures. ADEGENET is used to handle molecular markers data for multivariate analysis.

MANAGEMENT OF BREEDING POPULATION

Computational tools have been progressively used to assist plant breeders in decision making during various breeding programs. These include choice of parental lines, types of crosses to be undertaken, nature of breeding system to be followed, establishment and maintenance of heterozygotic groups, selection of lines for development of synthetic cultivars, prediction of performance of progeny and hybrid etc. (Damme et al. 2011, Lorenz 2013, Fu et al. 2017)

Establishing Heterotic Patterns

Development of highly heterotic hybrid depends largely on having large genetic diversity in the pool of germplasm of potential parents. However, in many crops it is not possible to predict the level of hybrid vigor, based on the analysis of the parental lines. For example, crosses between inbred lines derived from complementary heterotic groups are performed to generate commercial maize hybrids. Therefore, one of the primary strategies in maize hybrid breeding is to construct heterotic groups. But, which genotype from the heterotic group will provide the maximum heterosis is not possible to predict in many instances.

Availability of gene-based markers may provide an opportunity to establish parent-hybrid performance relationships at molecular level. Establishment of genome-wide heterozygosity and specific allelic (linkats) combinations may be utilized to determine maximum heterosis and hybrid vigor in some crops. Determination of heterotic patterns is a continuous process and each cycle is composed of following steps: identification of broad heterotic group through cluster analysis, determination of heterotic pattern through combining ability and heterosis analysis, and updating and maintenance of heterotic groups (Kizilkaya et al. 2010).

Predicting Hybrid Performance

Breeding for hybrid cultivars involves two major steps: identification of parental lines and selection of the best combinations of these parental lines for hybrid development. These procedures demand large amount of work for field evaluation, test crossing and progeny tests. In the breeding process testcrossing has to be carried out at many stages starting from the very first generation. Since hybrid performance is highly unpredictable in most of the crops, breeders have to spend lot of time in testcrossing, as currently there is no alternative to evaluate the hybrid performance. Therefore, development of alternative method for predicting hybrid performance is one of the challenges of the hybrid plant breeders.

Development of a reliable hybrid performance method without performing hundreds or thousands of single crossing has been the objective of many studies using marker data and combination of phenotypic and marker data in rice and maize. Hybrid performance is believed to be governed by many genes. Therefore, genome-wide genotyping of the parental lines is should provide an opportunity to establish parent-hybrid performance relationships at the molecular level. Analysis of genome-wide heterozygosity and allelic combinations may provide clues for breeding more heterotic and vigorous hybrid crop plants. Thus, utilization of parental genotyping should substantially reduce the quantum of testcross-based phenotyping analysis (Huang et al. 2015, Kadam et al. 2016, Lin et al. 2016).

Animal breeders have successfully used Best Linear Unbiased Prediction (BULP) procedure for evaluating genetic merits of dairy cattle. Traditionally, intrapopulation additive genetic models have been used with BLUB in animal breeding. In 1994, BLUP was used in maize breeding with interpopulation genetic models involving general and specific combining ability tests, and found to be useful for predicting single-cross performance. The predicted single cross performance can be used to predict the performance of F_2 x tester combinations, double crosses and three-way crosses as well (Andorf et al. 2019).

In certain situations, tools are required for selective genotyping and pooled DNA analysis. The software package GenePool (http://genepool.tgen.org/) provides analytical tools which can be used for detection of shifts in relative allele frequency between pooled genomic DNA of specific cases and their controls through SNP-based genotyping microarrays. The Pooled DNA Analyser (PDA) is another package which can be used for analysis of pooled DNA data (http://www.ibms.sinica.edu.tw/~csjfann/first%20flow/programlist.htm).

CONSTRUCTION OF GENETIC MAPS THROUGH MOLECULAR MARKERS

Prerequisite steps for application of MAS are, construction of genetic maps through molecular markers and utilization of these maps in marker-trait association analysis. Some of the tools used for the above mentioned purpose are discussed in the following section.

Tools for Genetic Map Construction

Conventionally genetic maps are constructed using information obtained from the segregating populations of specific crosses. The MAPMAKER/EXP was the first and most widely used software for construction of genetic maps, which was developed in 1987 by the Whitehead Institute. Almost all genetic maps constructed using first generation molecular markers, RFLP, have utilized this software. In 1993, alternative software MAP MANAGER CLASSIC was developed which has interactive program to map Mandelian loci using intercross with codominant markers, backcross or recombinant inbred lines (RILs) (http://www.mapmanager.org/mapmgr.html).

For construction of genetic maps from distorted segregation information obtained from backcrossing, double haploid and RIL population, specialized software such as MAPDISTO (we/ftp:http://mapdisto.free.fr/) is used. This software can analyze of marker data showing segregation distortion due to differential viability of gametes or zygotes, and compute and draw genetic maps using graphical interface.

It is also possible to combine maps or data from multiple populations, derived from different crosses, into a single map through the software JOINMAP. The mapping population may consist of BC_1, F_2, RIL, F_1 and F_2 derived double haploid (DH) and out-breeder full sib family. JOINMAP can also integrate maps derived from several other functions such as automatic phase determination for out-breeder full-sib family, linkage group determination etc. The CMap is another software package with comparative functions, developed as a web-based tool. It provides information on comparison of genetic and physical maps.

QTL Mapping Based on Genetic Linkage

Association between target traits and molecular markers are based on genetic linkage. There exist several software packages for studying association between marker genotype and the phenotypic traits. Some of the commonly used software is: QTL, MAPQTL, Cartographer, Qgene and PLAQTL. While MCQTL performs QTL mapping in multi-multi-dimensional allelic situations, all the others are capable of handling bi-allelic populations. The MAPMAKER/QTL is the most frequently used software, which is based on maximum likelihood estimation of linkage between marker and phenotype using interval mapping. Another popular software MAPL can be used to get information on segregation ratio, linkage tests, value of recombination, marker groups, and marker orders by metric-dimensional scaling through analysis of variance (ANOVA) and interval mapping.

The QTL Cartographer (http://statgen.ncsu.edu/qtlcart/cartographrer.html) is most widely used software which performs several statistical methods using multiple markers simultaneously. The software package OneMap (http://www.ciagre.usp.br/~aafgarci/OneMap/) can be used for construction of linkage maps in out-crossing plant species.

To meet the requirements of special situations of QTL, several mapping software were developed. These include MCQTL, MapPop, and QTLNetwork. The MCQTL allows the analysis of populations derived from inbred lines and can link the families on the assumption that QTL locations are the same in all families. MapPop is used for selective mapping and bin mapping by selecting good samples from the mapping populations. QTLNetwork can be used for mapping and visualizing the genetic architecture of complex traits on populations derived from crossing inbred lines.

A web-based tool WEBQTL was developed for exploring the genetic modulation of several thousand phenotypes collected over 30 years period by several hundred investigators, for mice. Similar web-bases tools should be developed for plants as well.

eQTL Mapping

Availability of whole genome sequences in many crop plants has opened additional scope for revealing links between phenotypes and genotypes (genes) through tools such as linkage analysis, positional cloning and microarray. It has become essential to develop a convenient bioinformatics tool to display the relationships between eTraits, markers and genes. It has also become essential to integrate current results with the information generated from earlier studies on the organism. To address these issues, Muller et al. (2006) developed the eQTL Explorer. This provides scope to store expression profiles, linkage data and information derived from external sources from a relational database. This helps in simultaneous visualization and interpretation of the combined data through a Java geographical interface. In 2007, a web-based eQTL Viewer was developed which plots results of eQTL mapping. With this thousands of eTraits can be viewed in a single display, with readily identifiable *cis*- and *trans*- regulation patters. It is also empowered to present annotations, organize eTrait in biological groups such as biochemical pathways. All these features have made eQTL Viewer applicable to understand genome-wide transcriptional regulation patters.

The web-based software PhenoGen can be used to identify candidate genes that control complex characters, based on co-occurrence of differentially expressed genes in microarray experiments and phenotypic QTL expression.

The PGMapper can automatically match phenotypes to genes from a defined genome region or a group of given genes derived from known databases such as OMIM and PubMed.

Linked-Disequilibrium Based QTL Mapping

Linked-disequilibrium (LD) or association mapping has become increasingly popular during recent years. The unstructured populations having unrelated individuals or randomly selected individuals can be used for such mapping. Before initiating LD mapping, the genotyped units are subjected to statistical analysis to remove the population structure, as it can cause false positive associations due to virtual correlations rather actual linkage. The software STRUCTURE can be used for this purpose.

Genome-Wide Association Mapping

To determine the linkage between genetic variations and agronomic traits, genome-wide association (GWA) studies are widely undertaken. A powerful GWA study shall include the measurement of hundreds of thousands of SNPs in several thousand individuals. Highly sophisticated analyzing tools are required for high volume data. GOLDSURFACE2 (GS2) can be used for analysis and visualization of GWA studies. The program is developed in Java and can be used in all platforms. Other tools available for GWA studies are GENOMIZER (htto://www.ikmb.uni-kiel.de/genomizer), PLANK (http://pngu.mgh.harvard.edu/purcell/plink), MAPBUILDER (http://bios.ugr.es/MBapBuilder), CATS (Calculator for Association with Two Stage design, http://www.sph.umich.edu/csg/abecasis/CaTS) (Massman et al. 2013, Perez and Compos 2014).

Integrated Haplotype and LD Analysis

Usually it becomes difficult to handle large amounts of SNP data for analysis of haplotypes and their association with the traits of interest. To overcome such problems, HaploBuild (http://snp.bumc.u.edu/modules.php?name=HaploBuild) was developed which can be used for construction and testing of haplotypes for SNPs found in close proximity but may not necessarily be contiguous. The software, HAPLOVIEW (http://www.broad.mit.edu/personal/jcbarret/haploview) can simplify and expedite the process of haplotype analysis. HAPSTAT (http://www.bios.unc.edu/~lin/hapstat/) can be used for statistical analysis of haplotype-disease association. Otter related software are: DPPH (Direct method for Preferred Phylogeny Haplotyping, http://www.c-sif.cs.ucdavis.edu/~gusfield/dpph.html), EHAP (detecting association between haplotypes and phenotypes, http://www.compgen.pitt.edu), HAPLOBLOCK (http://bioinfo.cs.technion.ac.il/haploblock), HAPLOT, HAPLOREC, and HAP (http://research.calit2.net/hap).

SUPPORT TOOLS FOR MARKER-ASSISTED SELECTION

MAS being a major procedure of molecular breeding need various decision making tools at various stages. Stages at which these tools shall be useful include foreground and background selection, identification of recombinants having favorable allelic combinations etc. In large scale MAS huge data are generated, which need to be analyzed and integrated into other type of data, within short time. Although software has been developed to assist in decision making in such situations, they do not fulfill the breeder's requirements, as they can resolve only some procedures of MAS. Thus development of support tools for MAS is one of the challenges for efficient and quick adopting MAS.

Support tools for MAS are required for the following purposes (Perez-de-Castro et al. 2012).

1. To determine minimum sample size for foreground and background selection,
2. To estimate genetic gains,
3. To construct selection indices for multiple traits and whole genome selection,
4. To estimate the content of the recipient genome of selected individuals at each generation of integration,
5. To identify desirable plants based on genotype and phenotype,
6. To estimate cost-benefit,
7. To make simulation studies.

PLABSIM a simulation tool used for MAS programs. This can be used to investigate the effect of varying population size, determine the position and density of markers, selection strategies to be adopted on the genetic composition of the breeding products and on number of marker data point required. PLABISM has the following features: simulate of any diploid genome with an arbitrary number of loci at arbitrary positions on arbitrary chromosome number, combine an arbitrary number of selection steps with a selection strategy, reproduce schemes for all common breeding methods, select genotypes at defined loci, calculate selection indices from allelic frequencies at several loci, and analyze simulated data for a broad range of genetic parameters including population size, position and density of markers, number of marker data points required and selection strategies for the genetic composition of the breeding product.

iMAS (www.generationcp.org) can assist in developing and applying MAS. iMAS integrates freely available software for identification and application of trait-linked markers. This also helps the user to operate the software and interpret the results correctly. Other software applicable to MAS include: POPMIN a program that allows numerical optimization of population sizes in marker assisted backcross breeding program, BCSIM, a backcross simulation software for evaluation of marker-assisted backcross programs, the GGT (Graphical Genotypes software) which allows to transform molecular marker data to simple chromosomes (He et al. 2014).

BREEDING BY DESIGN

The concept of 'breeding by design' offers the possibility to predict the outcome of a set of crosses on the basis of relevant information on molecular markers. The process involves three steps: mapping all the loci involved in determining all the relevant phenotypic traits, determination of the allelic variations at those loci, and making crosses as per the design. The first step can be completed either by using mapping populations segregating for the phenotypic trait of interest or candidate gene approach (in which information from model plant species is exploited) or linkage disequilibrium (LD) mapping. However, it has been found that application of GWA (Genome Wide Association) studies allow more efficient way to accomplish the first step. The second step cannot be completed by on the basis of bi-parental populations, as only two alleles per locus can segregate. Therefore, the analysis should include plant materials representing the variability of the species.

In 'breeding by design', after the loci of interest have been mapped, and contribution of each allelic variant determined, crosses can be designed by combining favorable alleles to obtain superior genotypes (Yamamoto et al. 2015). According to the breeding objective, this breeding strategy can be defined and executed. So far this procedure has been used for seed length in soybean, heading date in rice, improvement of quality in maize etc.

Genome-Wide Selection

MAS require the identification of markers associated with the trait of interest. This is considered to be one of the weaknesses of MAS. The approach through this this step can be eluded, is called genome-wide selection (Figure 1). In genome-wide selection, estimation of effects of all loci on phenotypes, haplotypes and available markers are considered simultaneously. It is required to have both phenotypic and genotypic data for the reference population. The parameters for the simulation model can be determined on the basis of available data set. After establishing the model, it becomes possible to determine the genomic value of each individual, representing the breeding populations. Availability of good number of molecular markers shall ensure the accurate prediction.

It has been possible to compare the predictions of genotypes values obtained using genome-wide selection with true genotypic value due to availability of large phenotypic databases for different crops. In several studies conducted on *Arabidopsis*, maize and barley, it was observed that the results obtained through genome-wide study was more accurate compared to results obtained from previous studies conducted on the basis of selection of markers with effects on the phenotypes.

Models used in genome-wide selection can predict breeding values, and in some instances, can detect the regions associated to the trait. Usually, from breeders' perspective, application of MAS with the available molecular markers is sufficient. However, with the availability of new high throughput–omics technologies, breeders have can adopt new strategies to search for candidate genes, based on microarray for differential expression of genes. Exploitation of these strategies could facilitate in candidate gene identification involved in the trait of interest and make MAS even more efficient (Massman et al. 2013, Desta and Ortiz 2014) .

SIMULATION AND MODELING

Plant breeding through simulation modeling has become necessary due to generation of vast and diverse information. Simulation modeling can assist

Figure 1. Genome selection scheme. The model predicts the phenotype of plants in a breeding population on the basis of the genotyping results: this is the genomic estimated breeding value (GEBV), used to select the desired phenotype. (Reproduced from: Perez-de-Castro et. al. Current Genomics, 2012).

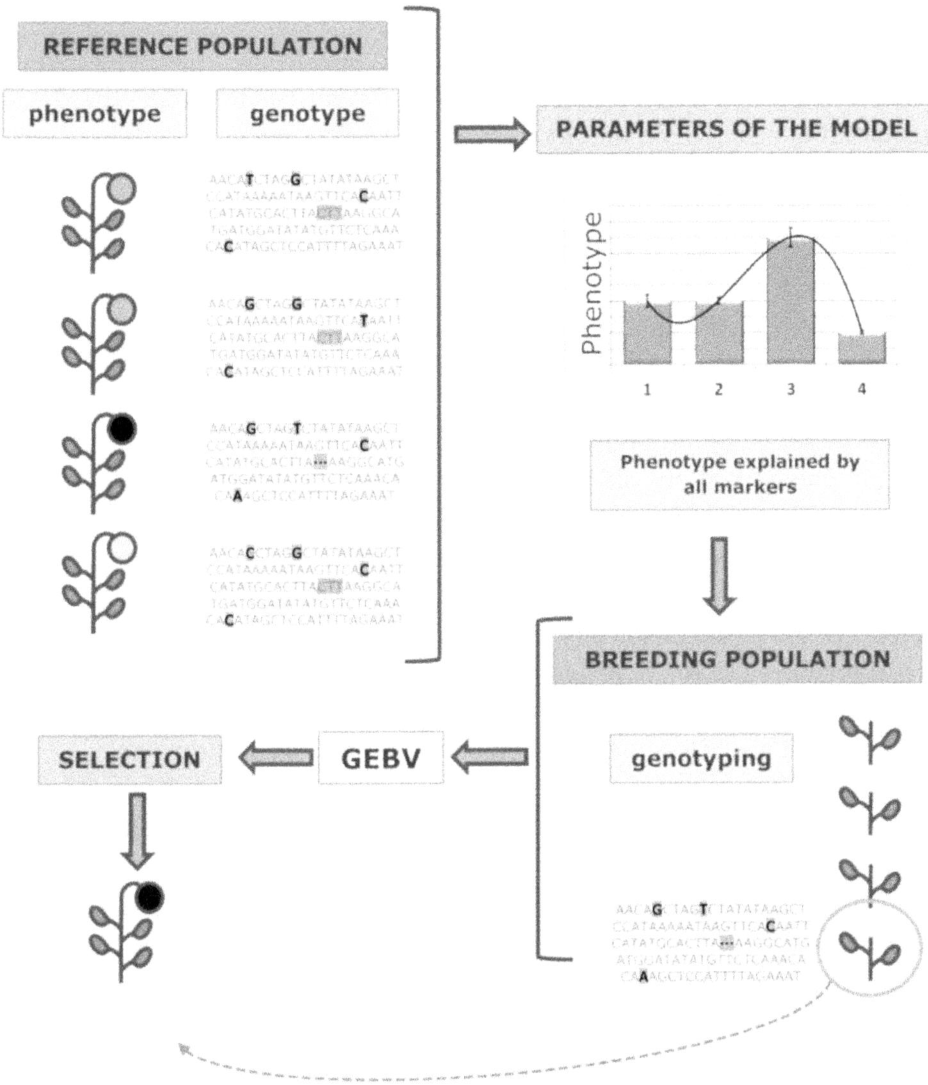

in decision making at different stages of breeding processes. In this section present status on the application of simulation modeling in plant breeding has been discussed.

Importance of Simulation Modeling

Over time, the number of genes and QTLs for various traits identified has increased substantially. Therefore, there is a need for software tools to handle and analyze multitude of information for crop improvement. Computer simulation can help the breeders to obtain information on probable crossings with selected of genotypes, which will drastically reduce the time and resource requirements of field experiments. Usually, in quantitative genetics various assumptions (on: linkage, multiple alleles, epistasis, pleiotropy, experimental design) are made to test the validity of mathematical and statistical theories. Computer simulation can relax use of some of the assumptions and their effects on breeding programs (Sun et al. 2011).

The whole plant physiology modeling can be used to partition complex traits into simple components and to know how these components interact with each other and contribute towards overall expression of the trait in different environmental conditions. The areas in which crop modeling can help to improve the efficiency of plant breeding program are: characterize environment to define the target population for each component of the environment, assess the value of specific putative traits in improved plant type, and enhance integration of molecular genetic methodologies.

Simulation Through Genetic Models

Simulation, using simple genetic models has been used to study many special plant breeding issues. Genetic details identified for simulation to elucidate differences in growth and development among cultivars is as follows.

1. Genetic model without any reference to species
2. Genetic model having specific reference to species but without reference to genotypes
3. Genetic model representing genetic differences by cultivar-specific parameters
4. Genetic model representing genetic differences by specific alleles, having linear effects of the genes on the parameters of the model

5. Genetic model representing genetic differences by genotypes with simulated gene action based on knowledge about gene regulation and effect of gene products, and
6. Genetic model representing genetic differences by genotypes with simulated gene action at the level of interactions of regulators, products of the gene and other metabolites

The level i and ii are used only when genetic representation of species are required. Level iii and iv (GeneGro Version 1) are used for the current crop models. Level v is used on the basis of knowledge on gene action (represented in GeneGro Version 2). Level vi is used for unicellular organisms.

It is possible to create a QTL-based crop physiology model, by combining eco-physiological modeling with genetic mapping, which could be used for resolving genetic basis of complex environment dependent yield related traits. Using this tool it has been possible to predict specific leaf area in barley, leaf growth response to varying temperature and water deficit in maize, stay-green response to nitrogen in sorghum, and duration of pre-flowering in barley. By excluding environment and gene dependence parameters, it was possible to develop breeding strategies to enhance rate of yield over several cycles of selection. By combining CROPGRO-Soybean with a linear model, which can predict cultivar-specific parameters as a function of E-loci, it has been possible to predict cultivar performance and refine crop breeding systems.

It is possible to predict how different genotypes interact with environments to produce different phenotypes through the genotype-to-genotype (GP) model. Based on the information on genes, characteristics of germplasm, parents, breeding objectives, target environments etc., the breeding procedures and selection methods can be simulated and optimized, and predict development of new cultivars.

The simulation platform QUGENE has been developed for quantitative analysis of genetic models, through its two stage architecture. The first stage deals with define the genotype-environment (GE) system and generate the starting population of individuals. In the second stage, application modules are included, for investigation and analysis of the starting population of individuals within the GE system. Several breeding modules were developed using QUGENE software. These modules can simulate performance of breeding lines in a given environment and also predict the effect of long-term selection over many breeding cycles and seasons, to optimize and improve efficiency breeding methodology.

QULINE can be used for simulation for mass selection, pedigree selection, bulk selection, backcross breeding, top cross breeding, DH breeding and MAS. To simulation in QULINE, method of propagation applicable for the crop must be defined. The propagation methods with decreasing genetic diversity are: Non-self-random mating, Random mating, Double cross, Top cross, Backcross, Single cross, Self-pollinated, Double haploid (DH), and clonal (asexual) reproduction. For each cross, QULINE randomly determine the male and female parents, from a defined initial population or may select few preferred parents from a crossing block. Identification of the preferred parents from the crossing blocks can be made based on the defined selection criteria.

APPLICATION OF GENOTYPING-BY-SEQUENCING IN PLANT BREEDING

Genotyping-by-sequencing (GBS) is one of the most powerful platforms for undertaking plant breeding programs, from single gene markers to whole genome profiling. GBS is a faster and low cost tool, which allows plant breeders to implement GWAS, genetic linkage analysis, genome diversity study, molecular marker discovery and genome selection (GS). GBS is a robust method across a range of species, as prior knowledge about the genome of the species is not required, and SNP discovery and genotyping can be completed together (He et al. 2014).

Since GWAS require hundreds of thousands of markers to generate enough information and coverage, the emergence of NGS technologies has greatly improved such marker resolution. Progressively more and more crops (rice, maize, wheat, barley, potato, and cassava) are being optimized by GBS for the low-cost, efficient, and large scale of genome sequencing.

GBS has also been shown to be an efficient and valid tool for genomic diversity studies. Among other applications of GBS in plant breeding, identification of high density SNP markers to construct genetic linkage maps has been found to be very useful. The GBS technology has been found to be useful for genetic analysis and marker development in maize, soybean, lupine, lettuce, and rapeseed. GBS can be effectively applied even in the absence of reference genome sequences or without having previous DNA polymorphism data.

Application of GS through GBS has become a major supplement to traditional breeding approaches, and is an important feature to convert genome-assisted breeding into commercial crops having large and complex genomes. Application GBS approach in barley and wheat has resulted development of high density markers in species without having a reference genome.

Although GBS has been found to have novel qualities for enhancing the efficiency of plant breeding approaches, some potential drawbacks have been identified. First, it is difficult to align true alleles of each single locus in the case of large, complex and polyploidy genomes. However, among all the tools, GBS offers highest potential to resolve such issues. Second, in the case of mutation at the restriction site, the genomic DNA of that site cannot be PCR amplified. Consequently, the SNPs of this region will not be available. In such a scenario, a heterozygote may appear as homozygote. Although, the drawback stated above is common to all the other methods applicable for such analysis (He et al. 2014).

SOFTWARE TOOLKIT CRISPER-GE

CRISPER-GE is a set of powerful tools and following functions can be carried-out through this: (1) design target single guide RNAs (sgRNAs) (targetDesign), (2) predict off-target sites (offTarget), (3) design primers for construction of the sgRNA expression cassettes (primerDesign), (4) amplify target site-containing genomic fragments, (5) determine mutant sequences from the sequencing chromatograms of genome PCR amplicons containing target site(s) (DSDecodeM), and (6) download genomic sequences of certain regions from reference genomes (seqDownload). By using CRISPER/Cas9/Cpf1 vector system, it is possible to achieve complete solution for genome editing in plants through CRISPER-GE (Xie et al. 2017).

The tool targetDesign of the CRISPER-GE software assist in choosing appropriate target site(s) for the Cas9/Cpf1 nucleases. The associated program offTarget helps in the selection of specific target site(s) and prevents cleavage at non-target sites (sequences) of the genome. The targetDesign tool assist to find all the possible target sites in a given sequence and thereby predict the off-target sequences (sites). With the help of the offTarget algorithm it is possible to estimate scores for predicting all the off-target sites in the assigned genome.

There exist several other genome editing design tools such as E-CRISPER, CRISPER-P, and Breaking-Cas (Heigwar et al. 2014, Oliveros et al. 2016).

However, CRISPER-GE has the following advantages over the other software: (1) through targetDesign and offTarget programs it is possible to design target sites and predict potential off-target sites, (2) in case reference genome sequence or the genome sequence of any close related species is not available, it is possible to design target site(s) through targetDesign. However, in such cases prediction of potential off-target sites may not be possible in the target genome (Xie et al. 2017).

CONCLUSION

Plant breeders or agro-technicians are using plant breeding software for their daily activities in breeding, testing, and analytical. They use digital tools to collect data, validate the information, take decisions and monitor their activities. The plant breeding software is not only their daily work instrument, but also provides the dashboards to organize and monitor the breeding programs.

During breeding programs several operations have to be followed: crosses and hybridization, inbred line production, test crosses and trials, and analysis. The plant breeding software supports all such activities at every key stage. The plasticity of the software enables to configure it according to the specific requirements of the breeders. For example, when breeders choose parental lines for your next crossings, they would like to do it their way: selection of specific males and females parents which fulfill their criteria, provide attention to specific traits and combining abilities to make their selection etc. With plant breeding software, these activities can be done quickly, without hazardous data manipulation, on a powerful crossing matrix. Sometimes even new useful features are discovered through such systems.

To successfully implement genomics-assisted breeding (GAB) in crop improvement programs, it is important to have efficient and effective analytical and decision support tools (ADSTs), to evaluate and select plants for developing next-generation crops. Although phenotyping remains expensive and time consuming procedure, availability of software for prediction of allelic effects on phenotypes opens new doors to enhance genetic gain across crop cycles, building on reliable phenotyping approaches, and building pedigree information.

REFERENCES

Andorf, C., Beavis, W., & Hufford, D. (2019). Technological advances in maize breeding: Past, present and future. *Theoretical and Applied Genetics*. Advance online publication. doi:10.100700122-019-03306-3 PMID:30798332

Desta, Z. A., & Ortiz, R. (2014). Genomic selection: Genome-wide prediction in plant improvement. *Trends in Plant Science*, *19*(9), 592–601. doi:10.1016/j.tplants.2014.05.006 PMID:24970707

Fu, Y. B., Yang, M. H., Zeng, F., & Biligetu, B. (2017). Searching for an accurate marker-based prediction of an individual quantitative trait in molecular plant breeding. *Frontiers in Plant Breeding*, *8*, 1–12. doi:10.3389/fpls.2017.01182 PMID:28729875

He, J., Zhao, X., Laroche, A., Lu, Z. Y., Liu, H., & Li, Z. (2014). Genotyping by sequencing (GBS), an ultimate marker assisted selection (MAS) tool to accelerate plant breeding. *Frontiers in Plant Science*, *5*, 1–9. doi:10.3389/fpls.2014.00484 PMID:25324846

Heigwar, F., Kerr, G., & Boutros, M. (2014). E-CRISPER: Fast CRISPER target site identification. *Nature Methods*, *11*(2), 122–123. doi:10.1038/nmeth.2812 PMID:24481216

Kadam, D. C., Potts, S. M., Bohn, M. O., Lipka, A. E., & Lorenz, A. J. (2016). Genomic prediction of single crosses in the early stages of a maize hybrid breeding pipeline. *G3: Genes. Genomes and Genetics*, *6*, 3442–3453.

Kizilkaya, K., Fernando, R. L., & Garrick, D. J. (2010). Genomic prediction of simulated multibreed and purebred performance using observed fifty thousand single nucleotide polymorphic genotype. *Journal of Animal Science*, *88*(2), 544–551. doi:10.2527/jas.2009-2064 PMID:19820059

Lin, G., Zhao, Y., Gowda, M., Friedrich, C., Longin, H., Reif, J. C., & Mette, M. F. (2016). Predicting hybrid performance for quality traits through genomic-assisted approaches in central European wheat. *Public Library of Science (PLoS) ONE*, *1*, e0158635.

Lorenz, A. J. (2013). Resource allocation for maximizing prediction accuracy and genetic gain of genomic selection in plant breeding: a simulation experiment. *G3: Genes. Genomes and Genetics*, *3*(3), 481–491. doi:10.1534/g3.112.004911 PMID:23450123

Massman, J. M., Gordillo, A., Lorenzana, R. E., & Bernardo, R. (2013). Genomewide predictions from maize single-cross data. *Theoretical and Applied Genetics*, *126*(1), 13–22. doi:10.100700122-012-1955-y PMID:22886355

Muller, M., Goal, A., Thimma, M., Dickens, N. J., Hitman, T. J., & Mangion, J. (2006). eQTL explorer: Integrated mining of combined genetic linkage and expression experiments. *Bioinformatics (Oxford, England)*, *2*(4), 309–311. doi:10.1093/bioinformatics/btk007

Oliveros, J. C., Franch, M., Tabas-madrid, D., San-Leon, D., Montoliu, L., Cubas, P., & Pazos, F. (2016). Breaking-Cas-interactive design of guide RNAs for CRISPER-Cas experiments for ENSEMBL genomes. *Nucleic Acids Research*, *44*(W1), W267–W271. doi:10.1093/nar/gkw407 PMID:27166368

Pérez, P., & Campos, G. (2014). Genome-wide regression & prediction with the BGLR statistical package. *Genetics*, *198*(2), 483–495. doi:10.1534/genetics.114.164442 PMID:25009151

Perez-de-Castro, A. M., Vilanova, S., Canizares, J., Pascual, J. M., Blanca, M. J., Prohens, J., & Pico, B. (2012). Application of genomic tools in plant breeding. *Current Genomics*, *13*, 179–195. doi:10.2174/138920212800543084 PMID:23115520

Singh, R. K., Singh, R., Ye, G. U., & Selvi, A. (2010). *Molecular plant breeding, methods and application*. Studium Press LLC.

Sun, X., Peng, T., & Mumm, R. H. (2011). The role and basics of computer simulation in support of critical decisions in plant breeding. *Molecular Breeding*, *28*(4), 421–436. doi:10.100711032-011-9630-6

Xie, X., Ma, X., Zhu, Q., Zeng, D., Li, G., & Liu, Y. G. (2017). CRISPER-GE: A convenient software toolkit for CRISPER-based genome editing. *Molecular Plant*, *10*(9), 1246–1249. doi:10.1016/j.molp.2017.06.004 PMID:28624544

Xu, Y. (2014). *Molecular plant breeding*. CAB International.

Yamamoto, E., Matsunaga, H., Onogi, A., Kajiya-Kanegae, H., Minamikawa, M., Suzuki, A., ... Fukuoka, H. (2015). A simulation-based breeding design that uses whole-genome prediction in tomato. *Scientific Reports*, *6*, 1–11. PMID:26787426

ADDITIONAL READING

Barrangou, R., & Horvath, P. (2014). Functions and applications of RNA-guided CRISPER-Cas immune system. In R. A. Meyers (Ed.), *Encyclopedia of molecular cell biology and molecular medicine: RNA biology* (pp. 34–56). Wiley-VCH Verlag GmbH & Co.

Damme, V. V., Gomez-Paniagua, H., & Vicente, M. C. (2011). The GCP molecular marker toolkit, an instrument for use in breeding food security crops. *Molecular Breeding*, *28*(4), 597–610. doi:10.100711032-010-9512-3 PMID:22162942

Fritsche-Neto, R., Akdemir, D., & Jannink, J. L. (2018). Correction to: Accuracy of genomic selection to predict maize single-crosses obtained through different mating designs. *Theoretical and Applied Genetics*, *131*(7), 1603–1612. doi:10.100700122-018-3118-2 PMID:29796770

Fritsche-Neto, R., Akdemir, D., & Jannink, J. L. (2018a). Accuracy of genomic selection to predict maize single-crosses obtained through different mating designs. *Theoretical and Applied Genetics*, *131*(5), 1153–1162. doi:10.100700122-018-3068-8 PMID:29445844

Garber, M., Grabherr, M. G., Guttman, M., & Trapnell, C. (2011). Computational methods for transcriptome annotation quantification using RNA-Seq. *Nature Methods*, *8*(6), 469–477. doi:10.1038/nmeth.1613 PMID:21623353

Gorjanc, G., Battagin, M., Dumasy, J. F., Antolin, R., Gaynor, R. C., & Hickey, J. M. (2017). Prospects for cost-effective genomic selection via accurate within-family imputation. *Crop Science*, *57*(1), 216–228. doi:10.2135/cropsci2016.06.0526

Hickey, J. M., Dreisigacker, S., Crossa, J., Hearne, S., Babu, R., Prasanna, B. M., Grondona, M., Zambelli, A., Windhausen, V. S., Mathews, K., & Gorjanc, G. (2014). Evaluation of genomic selection training population designs and genotyping strategies in plant breeding programs using simulation. *Crop Science*, *54*(4), 1476–1488. doi:10.2135/cropsci2013.03.0195

Huang, X., Yang, S., Gong, J., Zhao, Y., Feng, Q., Gong, H., Li, W., Zhan, Q., Cheng, B., Xia, J., Chen, N., Hao, Z., Liu, K., Zhu, C., Huang, T., Zhao, Q., Zhang, L., Fan, D., Zhou, C., ... Han, B. (2015). Genomic analysis of hybrid rice varieties reveals numerous superior alleles that contribute to heterosis. *Nature Communications*, *6*(1), 6258. doi:10.1038/ncomms7258 PMID:25651972

Jan, H. U., Abbadi, A., Lucke, S., Richard, A. N., & Showdon, R. J. (2016). Genomic prediction of testcross performance in canola (*Brassica napus*). *Public Library of Science (PLoS). ONE*, *11*(1), e0147769. doi:10.1371/journal.pone.0147769 PMID:26824924

Larièpe, A., Moreau, L., Laborde, J., Bauland, C., Mezmouk, S., & Bauland, C. (2017). General and specific combining abilities in a maize (*Zea mays* L.) test-cross hybrid panel: Relative importance of population structure and genetic divergence between parents. *Theoretical and Applied Genetics*, *130*(2), 403–417. doi:10.100700122-016-2822-z PMID:27913832

Liu, G., Zhao, Y., Gowda, M., Longin, F. H., Reif, J. C., & Mette, M. F. (2016). Predicting hybrid performance for quality traits through genomic-assisted approaches in central European wheat. *Public Library of Science (PLoS). ONE*, *11*, e0158635. doi:10.1371/journal.pone.0158635 PMID:27383841

Longin, C. F. H., Mi, X., & Würschum, T. (2015). Genomic selection in wheat: Optimum allocation of test resources and comparison of breeding strategies for line and hybrid breeding. *Theoretical and Applied Genetics*, *128*(7), 1297–1306. doi:10.100700122-015-2505-1 PMID:25877519

Massman, J. M., Jung, H. J. G., & Bernardo, R. (2013). Genomewide selection versus marker-assisted recurrent selection to improve grain yield and stover quality traits for cellulosic ethanol in maize. *Crop Science*, *53*(1), 58–66. doi:10.2135/cropsci2012.02.0112

Metje-Sprink, J., Menz, J., Modrzejewski, D., & Sprink, T. (2019). DNA-free genome editing: Past, present and future. *Frontiers in Plant Science*, *9*, 1–9. doi:10.3389/fpls.2018.01957 PMID:30693009

Meuwissen, T., Hays, B., & Goddrd, M. (2001). Prediction of total genetic value using genome-wide dense marker maps. *Genetics*, *157*, 1819–1829. PMID:11290733

Muleta, K. T., Pressoir, G., & Morris, G. P. (2019). Optimizing genomic selection for a sorghum breeding program in Haiti: a simulation study. *G3: Genes. Genomes and Genetics*, *9*, 391–401. PMID:30530641

Rabier, C. E., Barre, P., Asp, T., Charnet, G., & Mangin, B. (2016). On the accuracy of genome selection. *Public Library of Science (PLoS). ONE*, *11*, e0156086. doi:10.1371/journal.pone.0156086

Schoop, P., Riedelsheimer, C., Utz, H. F., Schon, C. C., & Melcchiner, A. E. (2015). Forecasting the accuracy of genome prediction with different selection targets in the training and prediction set as well as truncation selection. *Theoretical and Applied Genetics*, *128*(11), 2189–2201. doi:10.100700122-015-2577-y PMID:26231985

Singh, B., Bohra, A., Mishra, S., Joshi, R., & Pandey, S. (2015). Embracing new-generation 'omics' tools to improve draught tolerance in cereal and food-legume crops. *Biologia Plantarum*, *59*(3), 413–428. doi:10.100710535-015-0515-0

Spindel, J., Begum, H., Akdemir, D., Virk, P., Collard, B., Redona, E., ... McCouch, S. R. (2015). Genome selection and association mapping in rice (*Oryza sativa*): Effect of trait genetic architecture, training population composition, marker number and statistical model on accuracy of rice genome selection in elite, tropical rice breeding lines. *Public Library of Science (PLoS). Genet*, *11*, e1004982. PMID:25689273

Spindel, J. E., Begum, H., Akdemir, D., Coolard, B., Redona, E., Jannink, J. L., & McCouch, S. (2016). Genome-wide prediction models that incorporate de novo GWAS are a powerful new tool for tropical rice improvement. *Heridity*, *116*(4), 395–408. doi:10.1038/hdy.2015.113 PMID:26860200

Technow, F. (2019). Use of F2 bulks in training sets for genomic prediction of combining ability and hybrid performance, *G3: Genes, Genomes. Genetics*, *9*, 1557–1569. PMID:30862623

Technow, F., Riedelsheimer, C., Schrag, T. A., & Melchinger, A. E. (2012). Genomic prediction of hybrid performance in maize with models incorporating dominance and population specific marker effects. *Theoretical and Applied Genetics*, *125*(6), 1181–1194. doi:10.100700122-012-1905-8 PMID:22733443

Technow, F., Schrag, T. A., Schipprack, W., Bauer, E., Simianer, H., & Melchinger, A. E. (2014). Genome properties and prospects of genomic prediction of hybrid performance in a breeding program of maize. *Genetics*, *197*(4), 1343–1355. doi:10.1534/genetics.114.165860 PMID:24850820

Thavamanikumar, S., Dolferus, R., & Thumma, B. R. (2015). Comparison of genomic selection models to predict flowering time and spike grain number in two hexaploid wheat double haploid popunations. *G3, Gene. Genomes and Genetics*, *5*, 1991–1998. doi:10.1534/g3.115.019745 PMID:26206349

Varshney, R. K., Singh, V. K., Hickey, J. M., Xun, X., Marshall, D. F., Wang, J., Edwards, D., & Ribaut, J. M. (2016). Analytical and decision support tools for genomics-assisted breeding. *Trends in Plant Science*, *21*(4), 354–360. doi:10.1016/j.tplants.2015.10.018 PMID:26651919

Varshney, R. K., Terauchi, R., & McCouch, S. R. (2014). Harvesting the promising fruits of genomics: Applying genome sequencing technologies to crop breeding. *Public Library of Science (PLoS). ONE*, *12*, e1001883. PMID:24914810

Wang, M., Mao, Y., Lu, Y., Tao, X., & Zhu, J. K. (2017). Multiplex gene editing in rice using the CRISPER-Cpf1 system. *Molecular Plant*, *10*(7), 1217–1222. doi:10.1016/j.molp.2017.03.001

Windhausen, V. S., Atlin, G. N., Hicky, J. M., Crossa, J., Jannink, J. L., Sorrells, M. E., ... Melchinger, A. E. (2012). Effectiveness of genome prediction of maize hybrid performance in different breeding populations and environments. *G3: Gene. Genomes and Genetics*, *2*, 1427–1436. doi:10.1534/g3.112.003699 PMID:23173094

APPENDIX

1. What resources are required for effective management of various activities of gene bank through MAGE?
2. Describe the areas of plant breeding programs to which computational tools have been used?
3. Explain how computational tools can be used for predicting hybrid performance.
4. What computations tools are available for construction of genetic maps through molecular markers?
5. What is eQTL mapping? Explain the application of eQTL mapping in plant breeding.
6. What is genome-wide association mapping? Explain its applications in plant breeding.
7. Explain for what purposes computerized support tools are required for MAS.
8. Explain for what purpose “breeding by design” is used?
9. What is genome-wide selection? Describe the applications of genome-wide selection.
10. What is the importance of simulation modeling in plant breeding?
11. Describe the areas identified for genetic simulation modeling to elucidate differences in growth and development among cultivars?
12. Describe the areas in which genotyping-by-sequencing can be applied in plant breeding.

Chapter 11
Genomics, Proteomics, and Metabolomics

ABSTRACT

Genomics could be viewed as the study of the randomness of DNA sequences. It may be possible to predict the structure of a gene product from the nucleotide sequences and thereby predict its function. The terms "structural genomics" and "functional genomics" were coined to denote the assignment of structure and function to a gene product, respectively. Proteomics focuses on the products of gene, which are basically proteins. Proteins are responsible for the development of phenotype, and proteomics is the bridge between genotype and phenotype. The transcribed mRNAs and their abundance are called transcriptome. Proteomics also deals with the interaction between proteins called intractomics. Metabolomics is concerned with identification, abundance, and localization of all the molecules excluding lipids and polysaccharides in the cell. In this chapter, the basic concepts and analysis of the genomic, proteomic, and metabolomics data for their practical utilization are discussed.

INTRODUCTION

Two important aspects of gene mapping are to identify mutants and establish linkage through appropriate crossing. In situations where the above mentioned methodology is difficult to apply, somatic hybridization and recombinant DNA technology was used to map DNA sequences to specific chromosomes. Initially, most of these sequences were not actually full-length genes but

DOI: 10.4018/978-1-7998-4312-2.ch011

marker sequences such as restriction fragment length polymorphism (RFLPs), single nucleotide polymorphisms (SNPs) and other molecular markers. Once assigned to chromosomes, these markers were used in pedigree analysis to establish linkage between the markers and disease phenotypes for genetic disorders in human. The existing available techniques would be laborious, time consuming, and an insurmountable task.

The study of genomes by using a newly developed method called DNA sequencing has revolutionized the gene mapping in all organisms. The first sequencing of the 5400 nucleotide was made of the virus ΦX174. Sequencing of several other viruses was completed thereafter. But the technology was slow and labor-intensive, limiting its uses to small genomes. It took about two decades to develop the computer-based automated DNA sequencing method, which was amenable for sequencing of large and complex genomes of the eukaryotes.

Advances in recombinant DNA technologies coupled with development of computer aided automated DNA sequencing methods has created a new area of research called genomics. Genomics is basically concerned with the analysis of nucleotide sequences of genes. This involves comparison of nucleotide sequences of the genes, and analysis of the succession of symbols in sequences. Initially attempts were made to elucidate the function of sequences whose functions is unknown, by comparing with the sequences of known function. It is based on the principle that similar sequences encode similar protein structures, and thus they perform similar functions. However, this principle may not be true universally. The second way is to compare sequences known to code for same protein in different organisms, in order to deduce phylogenic relationships. A third approach is to compare the sequences of healthy and diseased organisms, in an attempt to assign genetic causes to specific disease.

In its purest form, genomics could be viewed as the study of the randomness of DNA sequences. This endeavor is still inchoate, since the irregularities and their relation to function are not understood. However, it may be possible to predict the structure of a gene product from the nucleotide sequences. This can then be used to predict its function. Thus the term "structural genomics" and "functional genomics" were coined to denote the assignment of structure and function to a gene product, respectively.

Proteomics focuses on the products of gene, which are basically proteins. Usually, only 10 percent of the genes are actually translated into protein, in any given cell, under a given set of conditions. On the other hand, a given gene sequence can give rise to tens of different proteins, by varying the arrangements of the exons and by post-translational modifications. Thus, proteins are responsible for the development of phenotype, and proteomics is the bridge between genotype and phenotype.

At a particular epoch, the transcribed mRNAs and their abundance are called transcriptome, and all the translated proteins and their abundance or net rates of synthesis are called proteome. Usually there exist huge differences between the transcriptome and proteome. Separating and identifying the proteins from one another through different technique is an important step for generating primary data, which can be used for their analysis. For example, comparison of proteomes of diseased and healthy organisms may lead to identification of the molecular basis of a disease.

Proteomics also deals with the interaction between proteins called intractomics. Information about the affinity of each protein with every other protein in the cell, and non-protein material such as lipid bilayers, polysaccharides, RNA and DNA, constitute the primary data of intractomics. The investigation of protein glycosylation is called glycomics. Investigation of protein products is called metabolomics. Metabolomics is concerned with identification, abundance and localization of all the molecules excluding lipids and polysaccharides, in the cell. In this chapter the basic concepts and analysis of the genomic, proteomic and metabolomics data for their utilization in plant breeding have been discussed.

GENOMICS

The term genomics was first coined by Tom Rodrick in 1986 (Kusha 1998). Genomics involves sequencing and analysis of an organism's genome. The genome is the entire DNA content present within one cell of an organism. Genomics also involves the study of intra- and inter-allelic interactions such as epistasis, heterosis, and pleiotropy within a genome. Various aspects involved in genomics are presented in the following section.

Genome Sequencing and Elimination of Errors

After sequencing the genome of an organism it is usually a practice to announce a draft genomic sequence several years before a final sequence is announced. Most of the errors in the draft sequence are eliminated in the final sequence, to the acceptable level by the competent body. Typically to ensure high level of accuracy, chromosome segments are sequenced more than once. Multiple sequencing data are then compiled to find the final sequence. Often complimentary DNA strands are sequenced separately to eliminate error. Since huge number of nucleotides has to be analyzed, it has been often realized that utmost care has to be taken to come to the final sequence. It may involve repetition of the sequencing for several times, even after final compilation of results. Once satisfied, a genome is analyzed to identify functional genes, regulatory elements, and other features having important information.

Genome Databases

Even before whole-genome sequencing, information on DNA sequence of specific regions of the chromosomes of various organisms was accumulating through gene cloning and recombinant DNA technology. However, there was no systematic effort to make the sequence data available to large number of scientists working to understand functional aspects of DNA and its protein products. Demand for information has generated development of software which can analyze large data. In the meantime, both private and public databases started developing. Once the genomics emerged as a new approach for analyzing DNA, bioinformatics took over, and developed into a full-fledged area of research.

Today bioinformatics has several applications. Important among them are: to compare DNA sequences and establish sequence homology, to identify gene(s) in a genomic DNA sequence, to identify regulatory regions like promoters and enhancers, to identify structural sequences (e.g. telomeric sequence), to predict amino-acid sequence of a putative polypeptide encoded by a cloned gene sequences, to analyze protein structure and predict protein functions, to deduce evolutionary relationships between genes and organisms etc.

As genome sequence data accumulated, several DNA-sequence databases were developed and made available online freely. This has helped to generate and share information globally. National Centre for Biotechnology Information (NCBI) based at Washington D.C., USA developed and maintained the database called GenBank, one of the largest publicly available database of DNA sequence. It contains more than 400 billion bases of sequence data from about 2,80,000 eukaryotes, prokaryotes, plasmids and organelles. After identifying a gene sequence, they are named and deposited into the GenBank.

The GenBank provides an accession number which helps in access and retrieve the sequence for analysis.

The other two sequence databases are: the Nucleotide Sequence Database (EMBL) and DNA Databank of Japan (DDBJ). These are repository for raw sequence data, but each entry is extensively annotated and has features that highlight important properties of each sequence.

Similarly, SWISS-PROT and TrEMBL are major primary databases for the storage of protein sequences. Secondary protein databases are generated from the primary databases and are deposited in the databases like PROSITE, PRINTS and BLOCKS. Some other important databases on genomics are presented in Table 1.

Advances in DNA sequencing technique have revolutionized its use and today genomes of about 50,000 species have been sequenced. Information thus obtained can be utilized to (Barth et al. 2015): (1) define and understand genetic diversity and molecular basis of disease in humans, animals and plants, (2) manipulate the genome to deliver enhanced yields of crop and biopharmaceuticals, and (3) identify novel personalized medicines targeted to the patient populations with a genetic profile that predicts positive clinical outcome.

Identification of Genes

The ultimate aim is to identify all biochemically active genes of a genome by algorithmically processing the sequence, and predict the reactions and reaction products of those portions coding for proteins. In eukaryotes, presence of exon-intron structure makes it difficult to predict the course of the key operations of transcription, splicing and translation from sequence alone. Gene prediction can be divided into two methods; intrinsic (template) and extrinsic (lookup).

The principle of the extrinsic method is to identify a gene by finding a sufficiently similar known sequence in the existing databases. In other words, a gene of unknown function is compared with the database of sequences with known function. This approach reflects a widely used, but not necessarily correct, assumption that similar sequences have similar function. A major limitation of this approach is that about a third of the newly sequenced data from different organisms do not find sufficiently similar known sequences with the existing data available in the databanks (Barth et al. 2015).

Table 1. List of genomics and proteomics databases

Name of the database	Uniform Resource Locator (URL)	Description of the databases
ACeDB	http://www.acedb.org	Sequence of *C. elegans*, *S. pombe*, human and genome information
COGs (Cluster of Orthologous Group of proteins)	htto://www.ncbi.nlm.nih.gov/COG	COGs consist of individual protein or paralogues from at least three lineages
Entrez Gene	https://www.ncbi.nlm.nih.gov/entrez/ query.fcgi?db=gene	Gene specific information on completely sequenced genomes
Entrez Genome	https://www.ncbi.nlm.nih.gov/sites/ entrez?db=genome	Viral, pro- and eukaryotic genomes
ERGO	http://www.ergo-light.com	Provide links to functional role of enzymes, protein alignments, phylogenetic trees, gene clusters, potential operons and functional domains
FlyBase	http://flybase.org	Integrated resource for genetic, molecular and descriptive data of Drosophilidae
Genome Project Database	https://www.ncbi.nlm.nih.gov/entrez/ query. fcgi?CMD=search&DB =genomepri	Collection of complete and incomplete large scale sequencing, assembly, annotation, and mapping data for cellular organisms
GO	http://www.geneontology.org	Gene Ontology Consortium database
KEGG (Kyoto Encyclopedia of Genes and Genomes)	http://www.genome.ad.jp/kegg	To understand higher order functional meanings and utilities of the cell or organism from its genome information
TIGR Gene Indices	http://compbio.dfci.harvard.edu/tgi	To identify and classify transcribed sequences in eukaryotic species through EST and gene sequence data
Swiss-2DPAGE	https://www.expasy.org/ch2d	Data on protein identified through SDS-PAGE and PAGE for human, mouse, *A. thaliana, D. discoideum, S. cerevisiae, E. coli, and S. aureus*
wwPDB (Worldwide Protein Data Bank)	http://www.wwpdb.org	Archive of Protein data for macromolecular structural data that is available free

While seeking sequence homology between an unknown and known sequence, the task would be relatively straightforward and the main task would be merely to assess the statistical significance of the results (*i.e.* to compare with the null hypothesis that a match occurred by chance). However, when we are trying to compare two sequences, they may differ due to mutations, insertions and deletions. Thus the situation becomes complicated. The situation can be best explained through an example showing nucleotide sequences from two DNA fragments compared for sequence homology as follows:

```
TAGCCGTA-CTAT
||   |||| ||||
TA--CGTATCTAT
```

Vertical lines (l) indicate matching, whereas the blanks (-) indicate gaps or mutations. In the absence of gaps, it is possible to compute the Hamming distance between two sequences. Occurrence of gaps creates two problems; the number of possible alignments becomes very large, and it becomes difficult to place the gaps in the nucleotide sequence.

If gaps are not allowed, it is possible to assess and scores the sum of all possible pairs of aligned substrings within the two sequences to be matched. However, if gaps are allowed, there will be $\binom{2n}{n}$ possible alignments of two sequences each of length n[10]. Even with a moderate value of n, there could be several problems while enumerating. These problems can be solved by using dynamic programming algorithms, and by devising a score system with which gaps and substitutions can be assigned numerical values. Thus the essence of sequence alignment is to assign a score for each possible alignment. The score should be given in a scale representing the best and the worst matching. While aligning multiple sequences, degree of kinship can be assigned based on the score in the following manner:

Total score = Score for aligned pairs + Score for gaps

The score will indicate the relative likelihood that pair of sequences is related. It will also indicate the operations (mutations and introductions of gaps) that may be required to edit one sequence onto the other.

The entries in a scoring matrix are numbers related to the probability of a residue occurring in an alignment. Basically they are calculated as (the logarithm of) the probability of the meaningful occurrence of a pair of residues divided by the probability of random occurrence. Probabilities of meaningful occurrence are derived from actual alignments, known to be valid. In the case of gaps, the (negative) score might be a single value per gap, or could have two parameters, one for starting the gap, and the other to be multiplied by the gap length.

Dynamic Programming Algorithms

For sequence alignment, two different methods of dynamic programming algorithms (DPA) are used. First method is known as Needleman-Wunsch (global alignment) algorithm (Needleman and Wunsch, 1970), that builds up an alignment starting with alignments of small subsequences, and the second method is known as Smith-Waterman (local alignment) algorithm (Smith and Waterman 1981) which is similar in concept, but it does not systemically move through the sequences from one end to the other, rather compares subsequences anywhere in the molecule.

The other heuristic algorithms used are known as BLAST and FASTA (Lipman and Pearson 1985), which are faster than DPAs. Usually they look for matches of short subsequences, and then seek to extend. In this case also a scoring system has to be adopted to quantify matches.

Some of the assumptions are quite weak in the sequence alignment methods. Therefore, alternative methods are also being used to evaluate the degree of kinship between sequences, which are not based on symbol-by-symbol comparison.

Intrinsic Methods

In intrinsic method, first a concise descriptions of prototype objects are constructed and then genes are identified by searching for matches to such prototypes. For example, one may look for a short sequence (motif) which is known to interact with a particular drug. The motif can also be defined in terms of amino acid sequence which forms a substructure of a protein that can be linked to protein function or structural stability. This in turn can be connected to a group of evolutionarily related gene sequences. This is based on the consideration that genes which share common sequences are more likely to be evolutionarily related. Basically in this method one or more parameters from the sequence are computed and then compared them with the same parameters computed for sequences of known function, or looking for short sequences that are known to have characteristic of certain functions. One major difficulty that has been encountered while implementing this method is that a gene or an intron will typically be too short to allow a parameter to be estimated precisely to allow its identification.

Through intrinsic method it has been tried to parallel the action of the cell (operations like transcription, splicing and translation) to recognize where the gene expression machinery interacts with DNA. Consideration of the most common base at each position in consensus sequences, the sequences which are well conserved over many species, has played a major role in such analysis. The Hamming distance of an unknown sequence is then computed for the unknown sequence. The closer they are, the more likely that the unknown sequence has the same function as that represented by the consensus sequence. Useful signal include start and stop codons. Other signals include, sequences involved in positioning DNA around histones, intron splice sites, sequences corresponding to ribosome binding sites on RNA etc.

Promoters also are potential target for new drug. In this case, the major problem is the large and variable distance between the promoters and the sequence to be transcribed. Usually relatively well conserved sequences like TATA or CCAAT are used.

Annotation of Sequence Data

With the accumulation of huge amount of sequence data, it became progressively difficult to use them for analysis and interpretation. To develop gene maps out of the sequence data, it is important to identify the structural and regulatory sequences. This was achieved through a process called annotation, which involves large number of software tools.

The first step of annotation is to compare a newly sequenced genomic DNA to the already available sequence in various databases. One of the popular software is called BLAST (Basic Local Alignment Search Tool), provided by NCBI (National Centre for Biotechnology Information) for searching through banks of DNA and protein sequence data. Through BLAST it is possible to compare and identify a sequence of segment of genomic DNA to sequences available in major databases.

Three BLAST programs that are commonly used are BLASTN, BLASTP and BLASTX. BLASTN compares DNA sequences with all the DNA sequences in the non-redundant database (NR). BLASTP compares protein sequences with all the protein sequences in NR, and BLASTX translates nucleotide sequences in all six reading frames and compares the products with NR protein database. Several online tutorials are available which includes, "BLAST Quick star", "Basic Web BLAST" and "Youtube video". Other search engines include; FESTA33, TC-BLAST, MEROPS BLAST, SEARCHGTr, PipeAlign, MPsrch, GOAnno, and COMPASS.

BLAST is considered to be a comprehensive program through which it is possible to align a nucleotide or protein sequence, with that of the available database on nucleotide or protein. The alignment sought is called a "query" and the segment of database of nucleotide or protein sequences is called "subject" sequences. Initially a protein sequence was used as a "query" to scan sequence database of protein through BLAST. Soon thereafter a version operating on nucleotide "query" sequences to nucleotide database was developed. Meanwhile, an intermediate layer was developed through which it was possible to translate the nucleotide sequences into their corresponding protein sequences, and making cross-comparisons of both nucleotide and protein sequences possible. Specialized versions of BLAST allow fast search operations. Both the standalone and web versions of BLAST are available from NCBI (www.ncbl.nlm.nih.gov). Through the web version it is possible to search complete genomes of many model organisms including human.

While searching through BLAST, the alignments found are scored, and assigned a statistical value, called the "expected value". The expected value indicates the number of hits expected to occur by chance while searching a database of a particular size through BLAST. A threshold expected value can be set by the user to discriminate the type of alignments to be recorded. A higher threshold expected value is less stringent in its operation and the BLAST default value of 10 is designated to ensure that no biologically significant ailment is missed. In practice, the "Expect Values" in the range of 0.001 to 0.0000001 are generally used, so that the alignments shown are of high quality.

Sequence homology search tools can be placed into four groups as follows:

1. **Pairwise Searches (e.g., BLAST, Smith-Waterman):** Through these programs it is possible to compare a single sequence (query sequence) against each sequence present in a large database and find significant similarities,
2. **Profile Searches (e.g., HAMMER):** When supplied with several members of a family sequences, these programs can construct a profile of the family, and then a database is searched for sequences that fit the profile,
3. **Automated Searches (e.g., PSI-BLAST):** Through these programs it is first possible to identify close relatives of a single query sequence, and then these close relatives are used to build a profile. They in fact perform through profile search and pairwise tools.

4. **Protein Family Databases (e.g., Pfam, PROSITE, BLOCKS):** Through these programs it is possible to compare both single query sequences and large protein family sequences which may cover the entire database. This approach eliminates the duplications of already well characterized family.

Although each method has its advantages and limitations, the "pairwise searches" has remained the most transparent and fastest of methods. Usually the underlying algorithm used is similar in most of these programs.

Pairwise Searches

In a pairwise search, a comparison is made between a query sequence and a database sequence. A score is generated that indicates the possibility of homology between the two. The comparison is repeated for every sequence present in the database and the high-scoring hits are selected. Various tools available differ in speed and sensitivity. BLAST is most popular and fastest pairwise search tool, whereas the most sensitive algorithm is Smith-Waterman.

The principle on which BLAST program works is based on the regions of homology that are likely to contain indel-resistant segments that are highly conserved. In the alignment analysis, these regions of highly conserved sequence show-up as ungapped blocks. With the advanced versions it is possible to produce gapped alignments. However, gapped alignments are still built from ungapped regions. Therefore for detection of homologous sequence, BLAST is not likely to work, where indels are present randomly throughout any DNA strands. This implies that BLAST will miss the highly divergent sequences. Public interfaces to BLAST can be found on the NCBI and EBI websites:

https://www.ncbi.nlm.nih.gov/BLAST/
http://www2.ebi.ac.uk/
https://www.sdsc.edu/ResTools/biotools/biotools1.html

The BLAST servers mentioned above can be used to search query sequences in the GENBANK, SWISSPROT, TREMBL or non-redundant combination of these databases. Several other local databases e.g. HIV database, and many genome sequence projects, also offer access to their sequences through BLAST interfaces.

Databases such as protein coiled-coils and DNA microsatellites, which are usually low-informative and repetitive, should first be filtered as they produce skewed representation. Normally programs like SEG and DUST are used for the purpose. However these filters are automatically configured into the BLAST server on the NCBI website and filter selectable options are provided in most mirror sites.

Nucleotide BLAST

In nucleotide BLAST, a member of the BLAST suite of programs (e.g. "Blastn") is used to search with a nucleotide "query" against a database of nucleotide "subject" sequences. The "Blastn" is a general purpose search and alignment program for nucleotides and used to align mRNA, rRNA, tRNA, and genomic DNA sequences, which may contain both coding and noncoding regions. An advanced version called MegaBLAST is about 10 times faster than "Blastin", but is designed to align nearly identical sequences. Another program called discontinuous MegaBLAST, which is a refinement of MegaBLAST, and uses a discontinuous template to define an initial "word". In this program, it is essential to match the characters of some positions, e.g. wobble base position of codons. The refined MegaBLAST program overcomes the deficiencies of the original MegaBLAST and allows rapid cross-species mapping involving coding regions.

Protein BLAST

BLASTP, the original member of the BLAST suite of programs is used for searching protein-to-protein sequence homology. During the BLASTP search, construction of the spurious alignments is reduced by filtering the regions with low-complexity of the query sequence. This also speeds-up the search process.

Translated BLAST

In translated BLAST searches, the genetic codes are used to translate either the "query", or the "subject", or both, into protein sequences. Then BLASTP is used for alignment. The translocations are carried out in the three forward and three reverse reading frames, to ensure that all possible translations are covered.

There exist three different versions of translated BLAST search: TBLASTN, BLASTX and TBLASTX. In TBLASTN, a query protein sequence is translated to six reading frame nucleotide sequence and then compared to a nucleotide sequence database. In BLASTX, first a query nucleotide sequence is converted to six reading frames. They are then translated to corresponding six protein sequences, and compared to those present in a protein sequence database. In TBLASTX, the nucleotide sequences of both the query and database are converted to six reading frames. Thereafter 36 (6x6) protein BLASTP comparisons are made. Since comparisons are made at the translated protein level sequences, they are more sensitive compared to comparison at the nucleotide level of sequences. Both TBLASTN and BLASTX programs can be used to identify regions involved in coding within a nucleotide sequence, and in detecting frame-shifts in the coding regions. Through TBLASTX program it is possible to compare transcripts to genomic sequences without any information about the protein translation.

Genome BLAST

Application of any of the BLAST search programs to the transcript of a protein sequence derived from its annotation or to the complete genome sequence of an organism is known as genome BLAST. At NCBI genome BLAST services are available for a number of organisms. For such activities, MegaBLAST and BLASTN searches are used along with other programs like BLASTP and BLASTX.

PSI-BLAST

As an improvement to the original BLAST, another program called PSI-BLAST (Position- Specific Iterative -BLAST) was developed which automates the process profile construction and database searching. PSI-BLAST can construct a profile of the close relatives of a family of a probe after conducting a search in a sequence database. Using the profile, it can search the database again and generate more information. It is possible to repeat the process several times. In general PSI-BLAST is easy, fast and sensitive compared to BLAST.

PSI-BLAST is a method for searching alignment of protein sequencing that is built on the alignments generated through BLASTP. The initial steps of a PSI-BLAST search are similar in nature to the steps involved in BLAST search. Thereafter, a position-specific score matrix (PSSM) is generated from the multiple alignments, on the basis of certain pre-set score or a threshold *e*-value. High score is assigned to the highly conserved sequenced whereas low scores to the weakly conserved sequences. The conservation pattern specified by the PSSM is used to further search the database to identify matching sequences. Another round of searching is carried out by using the fresh sequences that are detected through the second round of search. This helps in refinement of the alignment profile. This process is continued till no new sequences are detected above the threshold value earlier defined. Thus PSI-BLAST is capable to detect distant sequence similarities compared to single query detection by BLASTP. Even after considerable change in their amino-acid sequence, protein molecules can still conserve its three-dimensional sequence. PSI-BLAST has the capability to detect such relationships also.

If an error is committed in the profile early by incorporating a chance similarity, by mistake, and the daughter molecules is picker-up by the subsequent iteration of the search algorithm, it may get fixed at the cost of the query. The best way to overcome such mistakes is to check manually that the sequences reported by the program appear relevant when compared to the query, and not to one another. A site on web interface to PSI-BLAST is available at NCBL at: https://www.ncbi.nlm.nih.gov/cgi-bin/BLAST/nph-psi

Wise2 Programs

The DNA and protein sequences can be compared one to one directly through Wise2 program. For example, comparison between a stretch of genomic DNA (which is untranslated and unspliced) and an amino acid sequence can be made directly through GenWise program. In spite of having intron-exon structure in the genomic DNA, homologies between the two can be obtained. Usually a 'HalfWise' program is included in the Wise2 package which acts as a filter and speed-up the prediction process. A form-based interface to some of the programs is available at: http://www.sanger.ac.uk/Software/Wise2/

Bayesian Alignment Algorithms

The Bayesian Alignment Algorithms search for distant homologies between a pair of sequences by averaging all possible evolutionary relationships. The Bayes aligner is particularly suitable for the sequences which might contain conserved ungapped blocks, e.g. transmembrane protein. It is available at the following site: http://www.wadsworth.org/resnres/bioinfo/software.html

Profile Searches

Profile searches are considered to be more sensitive than simple pairwise searches. They mostly make use of position-specific substitution matrices, but sometimes use position-specific gap penalties. They are much slower than Smith-Waterman, but the speed can be enhanced by the use of accelerated hardware.

The Hidden Markov Models (HMMs) are the most favored profiling tools, and is considered to be the best. The most popular multiple alignment package available is CLUSTAL.

Stochastic Context-free Grammars (SCFGs) are used for modeling similarities between base pairs that are specific for RNA structure. Since they have increased modeling power, they need more computational resources and skill. SCFGs are available at the following site: http://www.genetics.wustl.edu/eddy/software/

Automated Search

For identification of motifs that is present in an unaligned sequence sets, a program called MEMA can be used. It is particularly useful for recognizing motifs which are usually short, e.g. nucleotide binding sites. It starts with a Markov model embedded on a single subsequence of the training set, and then iteratively refines this model till the overall score becomes stable. Repeated motifs can be easily and quickly located through this algorithm. MEME is available in the following site: http://meme.sdsc.edu/meme/website/

Another popular tool used for discovery of short ungapped motifs is called Gibbs sampling. This tool randomly walks through the multiple sequence alignments, and the high-scoring alignments are visited proportionately more often than low-scoring ones, over a long period of time. It is otherwise runs very slowly.

Annotation and Hallmark Characteristics of a Gene

It is important to note that the above approach works only if similar gene sequences are already available in the database. To overcome this limitation, several software on bioinformatics was developed to search for several hallmarks of a typical functional gene (Figure 1). The hallmarks include; regulatory sequence found upstream of gene such as promoters, enhancers and silencers, downstream elements such as termination sequences, triplet nucleotides that are part of the coding region of the gene; 5' and 3' splice sites used to distinguish introns, the noncoding regions of the gene, from exons, the coding regions of a gene; and polyadenylation sites. Since the prokaryotes usually do not have introns, they are comparatively easier for annotation.

For example, if we look into a portion of any human genome sequence it is not possible to ascertain whether this sequence contains any gene(s). But through closer observation and analysis it is possible to find clues to the presence of a protein coding gene sequence. For instance, TATA box, GC box, and CAAT box sequences are often present in the promoter regions of eukaryotic genes. Splice site sequence between exons and introns contain a predictable sequence, most introns begin with GT and end with AG. Similarly, a polyadenylation sequence signals the addition of a ploy (A) tail to the 3' end of an mRNA transcript (Figure 1).

Figure 1. A typical eukaryotic gene consist of coding segments (exons) and noncoding segments (introns). For annotating a genome sequence to determine whether it contains a gene, it is necessary to distinguish between introns and exons, gene regulatory sequences, such as promoters and enhancers, untranslated regions (UTRs), and gene termination sequences.

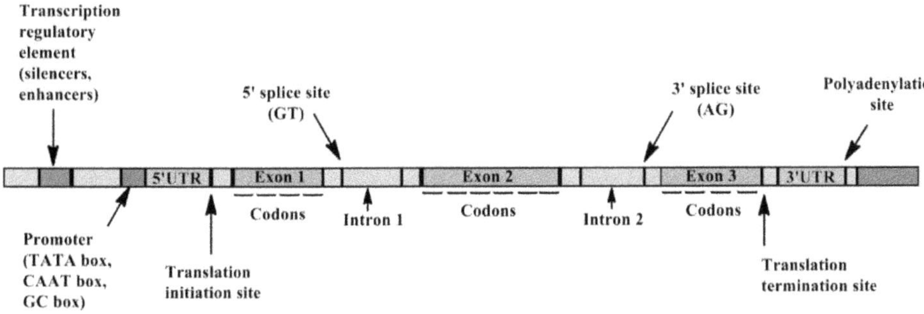

The open reading frames (ORFs) present in the protein coding genes can also be used to identify DNA sequence which represents genes. Sequences of triplet nucleotides present within an ORF, are transcribed to mRNA and after processing translated into the amino acid sequence of a protein. ORFs typically begin with an initiation sequence, usually ATG, which transcribes into AUG start codon in mRNA, and end with a termination sequence, TAA, TAG or TGA, which transcribes to the stop codons of UAA, UAG or UGA in mRNA. It is often difficult to identify the first nucleotide of a triplet code in the DNA sequence data. Typically the sequence adjacent to a promoter is examined for an initiation (start) triplet code. In the absence of an identifiable promoter sequence, ORFs can be used to identify a gene. Software programs can analyze the nucleotide sequence (three at a time) present in ORF. Identification of an ATG sequence followed by a termination sequence at some distance is considered to be a good indication of presence of the coding sequence of a gene. However, utilization of ORFs is difficult in the case of eukaryotes (including human) than in prokaryotes, as eukaryotic genes are not organized as continuous ORFs, rather the gene sequence consist of ORFs (exons) interspersed with introns. Moreover eukaryotic genes are widely spaced and thus the chances of encountering false ORFs in the regions between gene clusters are quite high.

Identification of hallmarks in the DNA sequence through highly efficient and reliable software may reveal presence of say, a promoter sequence, an initiation codon, and exons. Such software can also be used to predict the possible polypeptide sequences encoded by a gene.

Another method of analysis of genes is to look for codon bias. An amino acid can be encoded by more than one codon. For example, alanine can be encoded by GCG, GCC, GCT, and GCA. If the codons were used randomly, each would be used about 25 percent of the time. But in the human genome, GCC is used in 41 percent of the time, whereas GCG only 11 percent of the time. This is called codon bias. Thus codon bias should be present in the exons and not in introns. Such features can easily be identified by the software designed for ORF analysis.

FUNCTIONAL GENOMICS

Functional genomics is the study of gene function, based on the production of mRNAs and proteins thereof. It also deals with regulation of gene expression. Many genes which are sequenced during recent time, have earlier been identified and function assigned through classical methods of mutation and linkage mapping. One approach is to assign function to gene is to look for similar sequence in the database whose function is known, either in the same or different species. Inferring gene function from similarity searches is based on a relatively simple idea. If a genome sequence shows statistically significant similarity to the sequence of a gene whose function is known, then it is likely that the genome sequence encodes a protein with a similar or related function.

A typical example of homology search in the plant *Arabidopsis thaliana* is shown in Table.2.

Table 2. Assignment of Arabidopsis thaliana genes to functional categories based on homology studies. (Genome size: 125 Mb, ORFs: 25,000)

Functional categories	Percentage of genes assigned
Metabolism	11.1
Transcription	9.7
Cell growth, cell division, DNA synthesis	6.6
Cell rescue, defence, cell death, aging	5.8
Cellular communication, signal transduction	5.5
Protein destination	5.3
Intracellular transport	3.9
Energy	2.9
Cellular biogenesis	3.0
Transport facilitation	2.8
Protein synthesis	2.5
Ionic homeostasis	0.6
Unclassified	40.3

Similarity search can be used to find out homologous genes, related evolutionarily. In human genome, many ORFs were identified as protein coding genes based on their alignment with related sequences of genes of known functions in other species. For example, the leptin gene of human (*LEP*) was identified by sequence homology analysis with the mouse leptin gene (*ob/Lep*). The leptin gene is also called obesity (*ob*) in mice. The two genes are over 85 percent identical in sequence.

Homologous genes in the same species are called paralog. For example, α- and β- globin subunits in human are paralogs resulting from duplication of the gene. Homologous genes in different species are called orthologs, which thought to have descended from a gene of common ancestor. Mouse and human α–globin genes have evolved from a common ancestor and thus orthologs (Tufarelli et al. 2004).

Human Genome Project

The largest genomics project completed so far is the Human Genome Project (HGP). Huge amount of information have been generated through this project, and these are still being analyzed and interpreted. Among other things it has been observed that humans and all other species share a common set of genes essential for various cellular functions, and reproduction. This implies that all living organisms evolved from a common ancestor.

One of the important aspects about the human genome revealed through HGP is that less than 2 percent of the genome codes for proteins and that there are only around 30,000 protein coding genes. It was originally estimated to be 50,000 genes present in human to produce about 100,000 proteins. At least half of these genes show sequence similarities to genes present in many other organisms, and majority of the genes are similar in sequence to the genes of chimpanzees, a closely related species. The exact number of genes present in human is yet to be confirmed, as it is not clear whether or not many of these presumed genes produce functional proteins.

It is interesting to note that the number of genes is much less than the number of proteins produced. This can be explained on the basis of alternative splicing of the genes. Alternative splicing pattern of the genes can generate multiple mRNA, which in turn can produce multiple proteins, from a single gene, through different combinations of intron-exon splicing arrangements. It has been estimated that over 50 percent of human genes undergo alternative splicing. HGP also discovered that regardless of racial and ethnic origin, all human genomes are 99.9 percent identical. Most genetic differences between humans result from single nucleotide polymorphisms (SNPs) and copy numbers variations (CNVs). One of contribution of HGP is the development of extensive maps for genes believed to be involved in human disease conditions.

After completion of HGP, a group of research teams from around the world started Encyclopedia of DNA Elements (ENCODE) Project. The goal of this project is to identify and analyze functional elements (transcriptional start site, promoters, and enhancers) that regulate expression of human genes.

Comparison Between Prokaryotic and Eukaryotic Genomes

It has now been possible to compare the organization of prokaryotic and eukaryotic genomes as sequence data from several hundreds of them are available in the database. These include several model organisms of viruses, bacteria, fungi, nematodes, plants and animals.

Features of Prokaryotic Genomes

In the past, bacterial genomes have been thought to be relatively small (less than 5Mb) and contained within a single circular DNA molecule. However, the data generated from sequencing of bacterial genomes challenges this view point (Table 3). Although bacterial genomes are relatively small in size, their sizes vary to a great extent. Number of genes also varies from 500 to more than 5000. In addition, organization of bacterial genome may also vary from single circular molecules to two circular and/or linear molecules. Further, some plasmids found to contain essential genes in certain bacteria (*Borrelia burgdorfera*).

In prokaryotes, the density of the protein coding genes is very high, averaging about one gene per kilo-base of DNA. For example, *E. coli* has genome size of 4.6 Mb which contains 4289 protein coding genes, *Bacillus subtilis* has 4.21 Mb genome and 4779 genes. Thus very small portion of the bacterial genome contains non-coding sequence and often represent regulatory functions.

Table 3. Genome size and number of genes in some bacterial species

Organisms	Genome size (Mb)	Number of genes
Aquifex aeolicus	1.55	1749
Bacillus subtilis	4.21	4779
Escherichia coli	4.64	4289
Haemophilus influenza	1.83	1738
Mycoplasma genitalium	0.58	483
Mycoplasma pneumoniae	0.82	680
Rickettsia prowazekii	1.11	834

Presence of operons has been accepted as a generalization phenomenon in bacterial chromosomes, in which protein products of multiple genes are part of a common biochemical pathway. In *Aquifex aeolicus*, one transcription unit contains six genes involved in several different cellular processes with no apparent common relationship: one for protein synthesis, one for lipid synthesis, one for nucleic acid synthesis, one for protein for cell motility, and two for DNA recombination. This finding has raised interesting question whether the operons encode products control a single metabolic pathway in bacterial cells.

Eukaryotic Genome Organization

Eukaryotic genomes have features which are not found in prokaryotes. First, the gene density in the eukaryotes has a wide range. For example, yeast has about 1 gene per 2 kb of DNA, *Arabidopsis* has 1 gene per 5 kb, *Drosophila* has about 1 gene per 13 kb, and human has about 1 gene per 22 kb (chromosome 22) to 155 kb (chromosome 13). Gene density varies widely from chromosome to chromosome in human. Second, there is wide variation in the number of introns present in the genomes. For example, there exist only 239 introns in the entire yeast genome, whereas in human more than 100 introns may exist within one gene. Thirdly, the average size of the introns is correlated with the size of the genome. Larger the genome, larger is the average size of the introns and vice versa. Lastly, repetitive DNA constitutes a major portion of the genome size. For example, half the human genome is composed of repetitive DNA, and in maize more than two thirds of the genome is repetitive DNA.

More than 100 different angiosperm genome projects are going on around the world. The common denominator among all of these projects is the assembly of genetic maps and the placement of a common set of plant genes on them. Other objectives being comparative genome analysis and QTL mapping.

Commercially important plants that have been sequenced so far include rice, maize, wheat, barley, soybean, sugarcane, cotton, potato, banana, citrus, and sorghum. They contain much larger genome than *Arabidopsis*, but some have about same number of genes. But unlike *Arabidopsis*, genes in these large genome plants are clustered in stretches of DNA separated by long stretches of intergenic spacer DNA. Collectively these gene clusters occupy only about 12 to 24 percent of the genome. In maize, the intergenic DNA is composed of mainly of transposons.

Although the absolute range of genome size is comparable between plants and animals, more plant species have large and complex genome size and structure than do animal species. Information regarding sequencing of all the plants analyzed so far is beyond the scope of this chapter. In 2018, Chen and coworkers reviewed the genome sequencing data of angiosperms (https://doi.

org/10.3389/fpls.2018.00418). Information on some of the important plants sequenced so far is described in the following section.

The *Chlamydomonas* Genome

Chlamydomonas reinhardtii is a unicellular green alga, separated from the higher plants over a billion years ago. It has been used as a model system for studying various aspects plant system including development of eukaryotic flagella (cilia) and its structure and function. It is important to note that *Chlamydomonas* inherited flagella (cilia) from its ancestor, which is common for higher plants, which has lost from its system.

Marchent et al. (2007) reported sequencing of the nuclear genome of *Chlamydomonas reinhardtii* which is of about 120 Mb in size. Analysis of the sequence data could lead to identification of genes which code for proteins that are associated with biogenesis and function of chloroplast found in eukaryotic flagella. Analysis of the *Chlamydomonoas* genome has increased our understanding about the ancestral eukaryotic cells, and could provide information about new genes associated structure and function of flagella.

The Yeast Genome

The first eukaryotic genome sequenced was of yeast (*Saccharomyces cerevisiae*). Yeast genome has 16 chromosomes (12.1 Mb of DNA), and sequencing was done chromosome by chromosome. It has been estimated that yeast has about 5700 genes, out of which about 4000 genes have been characterized to have specific functions.

The *Arabidopsis thaliana* Genome

Arabidopsis thaliana has been recognized as a model plant for study as it has several important features. It is a small flowering plant belonging to the Brassica family, has a short life cycle of about 6 weeks, produce seeds profusely, has large number of mutant lines, has a relatively small genome of 5 chromosomes (140 Mb of DNA), and has extensive genetic and physical maps. It has been estimated that *Arabidopsis* has about 25,498 genes with a gene density of 1 per 5 kb, lacks the repeated sequences, and encodes proteins from 11,000 families. It is expected that systematic studies of this plant will illuminate numerous features of plant biology having significance to agriculture, energy, environment, and human health. Since the plant kingdom evolved independently from animal kingdom, sequence data on *Arabidopsis* can provide vital information about the evolutionary pathway of plants. Although *Arabidopsis* contains genes encoding dozens of protein families unique to plants, about half of its genes are identical or closely related to genes found in bacteria and human.

During evolution of *Arabidopsis* the genome has duplicated once, followed by gene loss and extensive local gene duplications. Many of its 25,000 genes are duplicated and contain about 15,000 different genes. For example, chromosome 2 contain 239 gene duplications (intra-chromosomal gene duplication), several genes on chromosome 4 are also present on chromosome 5 (inter-chromosomal gene duplication).

The Rice Genome

The sequence of japonica rice genome (*Oryza sativa* L. sp. *japonica*) was reported by Goff et al (2002). The genome size of japonica is about 420 Mb, and estimated to have 32,000 to 50,000 protein coating genes. About 98 percent of the proteins found in rice are homologues to maize, wheat, and barley proteins. Although gene homology between rice and *Arabidopsis* is limited, there exists extensive homology with other cereal genomes.

The sequence of *indica* rice genome (*Oryza sativa* L. sp. *indica*) was also reported by Yu et al. (2002). The genome size of *indica* rice was found to be 466 Mb, with an estimated 46,022 to 55,615 protein coating genes. About 42.2 percent of the genome comprises of repeats of 20-nucleotide. The transposons are mostly located in the intergenic regions of the genome. Only 49.4 percent predicted rice genes had a homology in *Arabidopsis thaliana.*

Further analysis of rice genome revealed that it contains about 41,000 genes (389 Mb DNA in 12 chromosomes). About 15 percent of all rice genes are found in duplicated segments of the genome. While about 90 percent of the genes in *Arabidopsis* are found in rice, only 70 percent of genes in rice are found in *Arabidopsis*, indicating that rice contains some genes which are unique to it (may be for other cereals also). Thus analysis of these unique sets of genes in cereals may be critical in improving this crop, say for higher yield.

In 2012, the revised reference genome assembly for *Oryza sativa* Nipponbare (O_s-Nippinbare-Reference-IRGSP-1.0) was released (Kawahara et al. 2012). In early 2000s, to generate 'New rice for Africa' (NERICA), introgressions were carried out by crossing *O. sativa* and *O. glaberrima* cultivars, followed by recurrent backcrossing with *O. sativa*. Wang et al. (2014) sequenced the genome of TOG5681 and CG14, parts of two NERICA generations.

Genome sequences of different *Oryza* cultivars are being added regularly to the rice knowledge base, including several strains of *O. sativa* and *O. glaberrima*. The deeply sequenced O. sativa genomes include IR64, IR8, Swarna, Shu-hui498, DJ123, and N22 (Schatz et al. 2014, Rathinasabapathi 2015, Stein et al. 2018).

Several wild rice species have also been sequenced. These include, *O. nivara, O. rufipogon, O. barthii, O. longistaminata, O. brachyantha, O. punctate* (Chen et al. 2013, Zhang et al, 2014) from Africa, *O. glumaepatula*, from South America, *O. meridionalis* from Australia, *O. granulate* from China (Zhang et al. 202014, Stein et al. 2018, Wu et al. 2018). In addition, two novel perennial wild rice species from tropical Australia were also sequenced (one similar to *O. rufipogon*, and the other similar to *O. meridionalis*) (Brozynska et al. 2017).

The Wheat Genome

Bread wheat (*Triticum aestivum*) is the most widely cultivated and consumed cereal crop in the world. This is hexaploid wheat, having the combination of three genomes A, B, and D. The A genome was contributed by the diploid wild einkorn wheat *Triticum urartu*, the B genome by *Aegilops speltoides* and the D genome by *Aegilopes tauschii*. Due to the complex ploidy nature of the bread wheat genome, it makes genetic and functional analysis extremely difficult.

Ling et al. (2013) reported the draft sequence data of *T. urartu* genome. They estimated the genome size to be 4.94 Gb. About 66.88 percent of the genome was found to be repetitive, including long terminal repeats retrotransposons (49.07%), DNA transposon (9.77%) and unclassified elements (8.04%).

They predicted 34,879 protein-coding genes, with the average gene size of 3,207 bp, and a mean of 4.7 exons per gene. In comparison with the 28,000 genes estimated for the A-genome of hexaploid wheat, *T.urartu* contained 6,800 more genes. Extensive loss of genes in the hexaploid wheat A-genome compared to its diploid progenitor, may have contributed to this difference.

Brenchley et al. (2012) produced a five-fold coverage genome sequence of Chinese Spring wheat, and based on assembles of 5.42 Gb, predicted 94,000 to 96,000 genes. They also assigned two-thirds of these genes to A, B and D subgenomes. International Wheat Genome Sequence Consortium (IWGSC)

(2014) published a chromosome-based draft sequence of Chinese Spring. Compared to the whole genome sequencing strategy, they differentiated the highly conserved gene copies in each chromosome. Clark et al. (2017) published an improved genome sequence of Chinese Spring. Zimin et al.(2017) reported a more complete sequence of wheat genome. By far, this is the most complete wheat genome sequence published. Recently, IWGSC completed a high quality sequence of Chinese Spring which is available in public domain (http://www.wheatgenome.org/News/Latest-news/RefSeq-v1.0-URGI). Avni et al. (2017) reported sequencing of wild emmer, the tetraploid ancestor of common wheat. Decoding of the genome of the emmer will help in understanding the evolution of common wheat.

The Maize Genome

Schnable et al. (2009) reported draft nucleotide sequence of the 2.3 Gb genome of maize. They predicted 32,000 protein-coding genes, and estimated that hundreds of transposable element families are dispersed randomly throughout the genome, which makes-up about 85 percent of the genome. Transposable elements can affect the expression of numerous percent of the genome. Transposable elements can affect the expression of numerous gene and the size and positions of centromeres in the chromosome. They also reported that due to insertions and/or deletions, and uneven gene losses between duplicated regions has converted an ancient allotetraploid to present day maize, a genetically diploid species.

Jiao et al. (2017) reported the assembly and annotation of a reference genome of maize, using single-molecule real-time sequencing and high-resolution optical mapping. Relative to the previous reference genome, they reported a 52-fold increase in contig length and notable improvements in the assembly of intergenic spaces and centromeres. Further, by characterization of the repetitive portion of the genome, they reported more than 130,000 intact transposable elements.

The Barley Genome

Cultivated barley (*Hordium vulgare* L.) derived from its wild progenitor *Hordium vulgare* sp. *spontaneum*, is one of the earliest domesticated crop species. It has a genome size of 5.1 Gb. In 2012, the International Barley Genome Sequencing Consortium has reported nucleotide sequence of barley.

It has been estimated that barley genome contain 26,159 protein-coding genes, and show homology from other plant genomes. Presence of abundant alternative splicing, premature termination codons and novel transcriptionally active regions, suggest that post-transcriptional processing forms an important regulatory mechanism in barley.

The Banana Genome

Domestication of banana (*Musa* species and subspecies) started about 7,000 years ago in Southeast Asia. The present day banana has evolved through hybridization between diverse species and subspecies, and selection of diploids and triploids (seedless). Thereafter, parthenocarpic hybrids were widely dispersed by vegetative propagation. Cultivation of banana mainly involves *Musa acuminate* (A-genome) and *M. balbisiana* (B-genome).

D'Hont et al. (2012) reported draft sequence of *Musa acuminate* (subspecies *malaccensis),* a double-haploid genotype, that contributed one of the three *acuminate* genomes of Cavendish. The genome size of this species is 523 Mb distributed in 11 chromosomes (2n=22). They identified 36,542 protein-coding genes. Almost half of the *Musa* genome sequence is expected to contain transposable elements. Like most of the plants, distal parts of chromosomes have most of the functional genes. However, unlike other plant species, in *Musa*, transposable elements are typically concentrated around centromeres. Remarkably, typical short tandem centromeric repeats are not found in *Musa,* although one long interspersed element (named *Nanica*) is found in the centromeric region of all chromosomes.

The Soybean Genome

Soybean (*Glycine max*) is one of the most commercially important crop plant due to its seed protein and oil content. Schmutz et al (2010) reported sequencing of soybean genome, having a genome size of about 1,115 Mb (2n=40), with an estimated protein-coding genes of 46,430. The estimated number of protein-coding genes is about 70 percent more than *Arabidopsis*, but similar to the poplar genome. About 78 percent of the predicted protein-coding genes occur in chromosome ends. Nearly 75 percent of the genes are present in multiple copies, and it was estimated that genome duplication occurred about 13 million years ago. Genome duplication events were followed by gene diversification, deletion and chromosomal rearrangements. It has also been estimated that there exist 4,991 single nucleotide polymorphisms (SNPs) and 874 simple sequence repeats (SSRs) in the soybean genome.

The *Brassica oleracea* Genome

Liu et al. (2014) presented a draft genome assembly of *B. oleracea* var. *capitata* line 02–12 by interleaving Illumina, Roche 454 and Sanger sequence data. This assembly represents 85% of the estimated 630 Mb genome, and includes >98% of the gene space. The assembly was anchored to a new genetic map to produce nine pseudo-chromosomes that account for 72% of the assembly, and validated by comparison with a *B. oleracea* physical map, a high-density *B. napus* genetic map and complete BAC sequences and . For comparative analyses, identical genome annotation pipelines were used for annotation of protein-coding genes and transposable elements (TEs) for *B. oleracea* and *B. rapa*.

A total of 45,758 protein-coding genes were predicted, with a mean transcript length of 1,761 bp, a mean coding length of 1,037 bp, and a mean of 4.55 exons per gene, similar to *A. thaliana* and *B. rapa*. Publicly available ESTs, together with RNA sequencing (RNA-seq) data generated, support 94% of predicted gene models, and 91.6% of predicted genes have a match in at least one public protein database. Of the 45,758 predicted genes, 13,032 produce alternative splicing (AS) variants with intron retention and exon skipping. Genome annotation also predicted 3,756 non-coding RNAs (miRNA, tRNA, rRNA and snRNA).

The *Brassica napus* Genome

Brassica napus ($2n=4x=38$, AACC) is an important allopolyploid crop derived from interspecific crosses between *Brassica rapa* ($2n=2x=20$, AA) and *Brassica oleracea* ($2n=2x=18$, CC). However, no truly wild *B. napus* populations are known; its origin and improvement processes remain unclear.

Bancroft et al. (2011) sequenced leaf transcriptomes across a mapping population of the polyploid crop oilseed rape (*Brassica napus*) and representative ancestors of the parents of the population. Analysis of sequence variation and transcript abundance enabled them to construct twin single nucleotide polymorphism linkage maps of *B. napus*, comprising 23,037 markers. They used these to align the *B. napus* genome with that of a related species, *Arabidopsis thaliana*, and to genome sequence assemblies of its progenitor species, *Brassica rapa* and *Brassica oleracea*. They also developed methods to detect genome rearrangements and track inheritance of genomic segments, including the outcome of an interspecific cross.

Lu et al. (2019) re-sequenced 588 diverse *B. napus* accessions from 21 countries and obtained 4.03 Tb of clean data. After filtering, we aligned reads to the *B. napus* reference genome. The mapping rate varied from 79.84 to 99.45%, and the effective mapped read depth averaged ~5× and ranged from 3.37× to 7.71×. They generated 5,294,158 single-nucleotide polymorphisms (SNPs; denoted as Bna) and 1,307,151 indels. Validation of 103 randomly selected SNPs in 20 accessions by Sanger sequencing indicated that most of the identified SNPs (95.1%) were authentic. The reliability of SNPs was further confirmed in that most SNPs (93.5 to 96.4%) were repeated in biological replicates of 20 accessions.

Based on their study, Lu and coworkers proposed that the "A" subgenome may evolve from the ancestor of European turnip and the "C" subgenome may evolve from the common ancestor of kohlrabi, cauliflower, broccoli, and Chinese kale. Additionally, winter oilseed may be the original form of *B. napus*. Subgenome-specific selection of defense-response genes has contributed to environmental adaptation after formation of the species, whereas asymmetrical subgenomic selection has led to ecotype change. By integrating genome-wide association studies, selection signals, and transcriptome analyses, they identified genes associated with improved stress tolerance, oil content, seed quality, and ecotype improvement.

The Chickpea Genome

Chickpea (*Cicer arietinum*) is the second most widely cultivated legume after soybean. Varshney et al. (2013) reported draft genome sequencing of chickpea, a *kabuli* chickpea variety. The genome size is about 738 Mb (2n = 16), with an estimated 28,269 protein-coding genes. They further sequenced and analyzed 90 cultivated and wild genotypes of chickpea, and could identify the genes for disease resistance and agronomic traits. The traits that distinguish the two main cultivated varieties- *desi* and *kabuli*, could also be identified.

The Wild Chickpea Genome

Gupta et al. (2017) assembled short-read sequences into 416 Mb draft genome of Cicer reticulatum, a wild progenitor of chickpea (Cicer arietinum) and anchored 78% (327 Mb) of this assembly to eight linkage groups. Genome annotation predicted 25,680 protein-coding genes covering more than 90% of predicted gene space. The genome assembly shared a substantial synteny and conservation of gene orders with the genome of the model legume *Medicago truncatula*. Resistance gene homologs of wild and domesticated chickpeas showed high sequence homology and conserved synteny. Comparison of gene sequences and nucleotide diversity using 66 wild and domesticated chickpea accessions suggested that the *desi* type chickpea was genetically closer to the wild species than the *kabuli* type. Comparative analyses predicted gene flow between the wild and the cultivated species during domestication. Molecular diversity and population genetic structure determination using 15,096 genome-wide single nucleotide polymorphisms revealed an admixed domestication pattern among cultivated (*desi* and *kabuli*) and wild chickpea accessions belonging to three population groups reflecting significant influence of parentage or geographical origin for their cultivar-specific population classification.

The Potato Genome

Potato (*Solanum tuberosum* L.) is the most important non-grain food crop. Potato cultivars are mostly autotetraploid ($2n = 2x = 48$), highly heterogenous, and suffer acute inbreeding depression. Thus genome sequencing is a challenge for the scientific community that will ultimately facilitate advances in breeding.

In 2011, the Potato Genome Sequencing Consortium reported genome sequence of potato. To overcome the issue of heterozygosity, a unique homozygous line of potato called double monoploid, derived through tissue culture techniques was used. The genome size of potato is about 844 Mb, with an estimated 39,031 protein-coding genes. Presence of at least two genome duplication events indicates that potato has a palaeoploidy origin. Presence of addition, deletion and deleterious mutations in the sequence data of heterozygous diploids indicated them as the likely cause for inbreeding depression. Expansion of gene family and incorporation of new genes to carry out new functions contributed towards the evolution of potato tuber development.

The Cotton Genome

Cotton (*Gossypium* sp.) is one of the most economically important fibre crops. The *Gossypium* genus contains five tetraploids (AD_1 to AD_5) and over 45 diploid species. They probably have a common ancestor. The genome of diploid cottons (2n = 2x = 26) varies from about 880 Mb (D-genome) to 2,500 Mb (K-genome). The chromosome number of diploid cotton species are identical (n = 13), and they carry highly similar genotypes. The tetraploid cotton species (2n = 4x = 52, e.g. *G. hirsutum* L. and *G. barbadense* L.) has originated through allopolyploidization of A- and D-genome. Basic knowledge about contributing genomes is essential to understand the nature of cultivate (polyploidy) genome of cotton. Therefore, Wang et al (2012) first prepared the draft sequence of D-genome by utilizing *G. raimondii*, which was brought to near homology by six successive generations of self-fertilization.

The genome was estimated to contain 40,976 protein-coding genes. They identified 2,355 syntenic blocks, and almost half of the genes were found to have been distributed in more than one block. Thus, during its evolution, cultivated cotton genome has undergone substantial chromosome rearrangement. It is important to note that only two plant species (namely, cotton and cacao) having CDN1 gene family for gossypol biosynthesis have so far been sequenced.

The Sorghum Genome

Sorghum (*Sorghum bicolor* L. Moench) is one of important millets. Apart from food, the plant products are used as feed, fibre and fuel. Within the C_4 group of plants, sorghum has a smaller genome size of about 730 Mb, and thus makes it an attractive model for functional genomics of C_4 plants.

Paterson et al. (2009) reported draft sequencing of 730 Mb sorghum genome. Sorghum has gene density and order very similar to those of rice, but genetic recombination is largely confined to about one third of its genome. Larger genome size sorghum (about 75%) compared to rice can be explained on the basis of retrotransposon accumulation in recombinationally recalcitrant heterochromatin. Although gene and repetitive DNA distribution have been preserved since palaeoploypIoidization, one member of the most duplicated genes sets are lost before the divergence of sorghum from rice. About 24 percent of genes are grass-specific and 7 percent are sorghum-specific.

The Pigeonpea Genome

Varshney et al. (2012) published draft pigeonpea (Cajanas cajan) genome. They generated 237.2 Gb of sequence, which along with Sanger-based bacterial artificial chromosome end sequences and a genetic map, they assembled into scaffolds representing 72.7% (605.78 Mb) of the 833.07 Mb pigeonpea genome. Genome analysis predicted 48,680 genes for pigeonpea and also showed the potential role that certain gene families, for example, drought tolerance–related genes, have played throughout the domestication of pigeonpea and the evolution of its ancestors.

The first draft of the pigeonpea (Cajanus cajan (L.) Millsp. cv. Asha) genome with 511 Mbp of assembled sequence information has low genome coverage of about sixty percent. In 2017, Mahato and coworkers presented an improved version of this genome with 648.2 Mbp of assembled sequence of pigeonpea, which has resistance to fusarium wilt and sterility mosaic diseases. With the addition of 137 Mb of assembled sequence information this version has the highest available genome coverage of pigeonpea. They predicted 56,888 protein-coding genes of which 54,286 (96.7%) were functionally annotated. In the improved genome assembly they identified 158,432 SSR loci, designed flanking primers for 85,296 of these and validated them in-silico by e-PCR. The raw data used for the improvement of genome assembly are available in the SRA database of NCBI with accession numbers SRR5922904, SRR5922905, SRR5922906, SRR5922907. The genome sequence update has *been deposited at DDBJ/EMBL/GenBank under the accession AFSP00000000.*

The Sugarcane Genome

Sugarcane (*Saccharum* spp.) is a major crop for the production of sugar and bioenergy. The modern sugarcane cultivar genome is highly polyploidy, aneuploid and heterozygous in nature and thus posse major challenges for producing a reference sequence. Sugarcane breeding is still essentially focused on conventional methods as genetic information in sugarcane lagged behind other major crops. Modern sugarcane cultivars are derived from interspecific hybridization between *S. officinarum* and *S. spontaneum*, both are highly polyploid species. *S. officinarum* (2n=8x=80, x=10) presumed to have derived from *S. robustum* (2n=60, 80 and up to 200), and *S. spontaneum*, a wild species with various cytotypes and aneuploidy forms (2n=5x=40, to 2n=16x=128, x=8). During last 20 years, advancement in molecular genetics has contributed towards understanding of the molecular resources of sugarcane genome. Comparative mapping with other species of Poaceae revealed extensive genome-wide colinearity with sorghum. Thus sorghum has become used as a model for sugarcane to produce a BCA-based monoploid genome sequence.

Sequencing of sugarcane, a highly complex genome poses challenges that have not been encountered in any other sequencing projects. The polyploidy results indicate a total genome size of about 10 Gb for sugarcane cultivars, while the monoploid genome size is about 800-900 Mb, which is very close to sorghum genome (750 Mb). Altogether 25,316 protein-coding gene models are predicted. Out of which 17% was found to have no colinearity with their sorghum orthologs. It has been shown that the two species, *Saccharum officinarum* and *S. spontaneum* differ in their transposable elements and by a few large chromosomal rearrangements. It has also been shown that ploypoidization arose after their divergence in both the species.

The Black Cottonwood Genome

The sequence of black cottonwood (*Populus trichocarpa*) was published in 2006. Sequence analysis has revealed more than 45,000 putative protein-coding genes, and occurrence of a whole-genome duplication event. Nucleotide substitution, tandem gene duplication, and gross chromosomal rearrangement appear to have taken place at a much slower rate in *Populus* than in *Arabidopsis*. *Populus* has more protein–coding genes than *Arabidopsis*, ranging on the average from 1.4 to 1.6 putative *Populus* homologs for each *Arabidopsis* gene. However, the relative frequency of protein domains in the two genomes is similar.

The Grapevine Genome

All grapevine varieties are highly heterozygous. Therefore it was difficult to carry out sequencing from any particular variety. However, the variety PN40024, originally derived from the variety Piont Noir is very close to homozygosity (about 93%), which was achieved by successive selfing. By utilizing this variety, the French-Italian Public Consortium for Grapevine Genome Characterization Project has published the sequence of grapevine (*Vitis vinifera*) in 2007. This was the first sequencing report for a fruit crops (cultivated for both for both fruit and beverages). The 487 Mb sequence include about 30,434 protein-coding genes. This value is considerably lower than the protein coding genes (45,555) in poplars having similar genome size (485 Mb). The genome contains 149,351 exons and 118,917 introns.

It was also revealed that 41.4 percent of grapevine genome is composed of repetitive/ transposable elements, and the distribution of TE along the chromosome quiet uneven. The analysis reveals the contribution of three ancestral genomes to the grapevine haploid genome.

The Date Palm Genome

Date palm (*Phoenix dactylifera*) is a dioecious (separate male and female trees) woody plant. In 2011, sequence of a female of the variety ‘Khalas’ was published. The 380 Mb sequence include about 25,000 protein-coding genes. Sequencing of eight other cultivars, which included both females and males, led to identification of about 3.5 million polymorphic sites, having about 10,000 gene copy number variations. A smaller subset of these polymorphisms can distinguish multiple varieties. A region of the genome has been identified to be linked to gender, which follow a XY system of gender inheritance.

The Moso Bamboo Genome

The *moso* bamboo (*Phyllostachys heterocycla*) is the first non-timber forestry species to be sequenced in 2013. Initially 2.05 Gb of DNA covering 95 percent of the genome was sequenced, which included 31,987 protein-coding genes. Sequence homology analysis revealed that duplication of bamboo genome took place about 7-12 million years ago. Whole genome duplication event generated more duplicate genes involved in bamboo shoot development. Analysis of the RNA sequencing from the bamboo flowering tissue indicate that there could be a positive correlation between bamboo flowering genes and drought condition. This finding should help to understand gregarious flowering of bamboos during draught conditions.

The Sweet Orange Genome

Xu et al. (2013) published a comprehensive analysis of the draft genome of sweet orange (*Citrus sinensis*). The assembled sequence covers 87.3% of the estimated orange genome, which is relatively compact, as 20% is composed of repetitive elements. They predicted 29,445 protein-coding genes, half of which are in the heterozygous state. With additional sequencing of two more citrus species and comparative analyses of seven citrus genomes, they presented evidence to suggest that sweet orange originated from a backcross hybrid between pummelo and mandarin. Focused analysis on genes involved in vitamin C metabolism showed that GalUR, encoding the rate-limiting enzyme of the galacturonate pathway, is significantly upregulated in orange fruit, and the recent expansion of this gene family may provide a genomic basis.

MINIMUM NUMBER OF GENES REQUIRED TO SUPPORT LIFE

After deciphering the genomes of various organisms, a fundamental question arises as to what is the minimum number of genes required to support life. To seek answer for this we shall have to look into the smallest genome studied so far. The bacteria *Mycoplasma genitalium* and *M. pneumonia*, two closely related human parasitic pathogens have been found to be the best candidates to look for the answer. These organisms are among the simplest self-replicating prokaryotes and have the genome size of 580 kb (*M. genitalium*) and 816 kb (*M. pneumonia*).

The *M. genitalium* genome has 483 protein coding genes, whereas *M. pneumonia* has 677 protein coding genes, of which 483 genes are same as that of *M. genitalium*. From the sequence database it is now possible to find out whether similar genes are also present in other organisms and whether they can be identified as essential. Two different approaches namely, comparative and experimental were used to answer this question.

In the comparative method, the sequence of *M. genitalium* (having 483 protein coding genes) was matched with the sequence of *Haemophilus influenza* (having 1783 protein coding genes) it was observed that 240 genes are orthologous for these species. Further it was observed that 16 other genes differ in their sequences, but have identical functions. These represent essential functions performed by nonorthologous genes. Thus it was estimated that 256 genes may represent the minimum genes required for life.

In the second approach, transposons were used to selectively induce mutation in the 483 genes of *M. genitalium*, with the hypothesis that mutation in the essential gene will produce lethal phenotype, whereas mutation in the non-essential gene will not affect viability. Through this experiment it was observed that out of 483 genes, about 265 to 300 genes are essential. The figures on number of essential genes required for life, obtained through two different approaches match closely. Several other experiments conducted using same or other organisms have found the minimum number of genes required for life in similar range.

COMPARATIVE GENOMICS

Through comparative genomics it is possible to study genetic similarities and differences between different organisms and thereby seek information about various aspects of biology including gene and genome evolution. Such information can be used to understand the variations in the phenotypes, life process, and evolution of different organisms. Thus comparative genomics has several goals: 1) to compare and identify the basic process of evolution between related genomes, 2) to relate information derived from model species to other species, and 3) to understand and decipher knowledge on location and expression of genes across species. Table 4 shows a comparative description of chromosome number, ploidy levels, genome size, gene number and gene density of selected prokaryotes and eukaryotes.

Through comparative genomics it has been estimated that the number of genes humans share with other species is very high, ranging from about 30 percent with yeast to about 80 percent with mice and about 98 percent with chimpanzees (Table 4). Interestingly human genome contains about 100 genes which are also found in bacteria. Comparative genomics also revealed that many mutated genes involved in inducing human disease (e.g. colon, prostrate, and pancreatic cancers; cryptic fibrosis; cardiovascular disease) are also present in *Drosophila.* A comparative assessment of size and predicted number of genes for each chromosome of human and rice is presented in Table 5. The table also shows the comparative assessment of the mitochondrial genomes. It is interesting to note that individual chromosomes of rice having much smaller size carries much higher number of genes in each of its chromosomes compared to human. Even the mitochondrial genome of rice carries more number of genes compared to human, although the mitochondrial genome size is much smaller. Obviously the gene density in rice chromosome is much higher than human.

Comparative genomics has also been used to identify members of the multigene families, a group of genes that decent from a single ancestral gene through duplication. These genes are similar but not identical in their DNA sequence, and their products have similar functions. Multigene families are

Table 4 Chromosome number, ploidy level, genome size, gene number and gene density of selected prokaryotes and eukaryotes

Organism	Chromosome number (n) and ploidy	Genome size (Mb)	Number of protein-coding genes (approx.)	Average gene density (kb per gene)	Genes similar with humans (%, approx.)
ARCHAEA					
Archaeoglobus fulgidis	1, haploid	2.17	2437	-	-
Methanococcus jannaschii	1, haploid	1.66	1783	-	-
Methanosarcina acetivoran	1, haploid	5.75	4662	1.23	-
Thermoplasma acidophilum	1, haploid	1.56	1509	1.03	-
EUBACTERIA					
Agrobacterium tumefaciens	1, haploid	5.7	5482	1.04	-
Aquifex aeolicus	1, haploid	1.55	1749	-	-
Bacillus subtilis	1, haploid	4.21	4779	-	-
Bradyrhizobium japonicum	1, haploid	9.1	8322	1.10	-
Escherichia coli	1, haploid	4.64	4289	1.03	-
Haemophilus influenza	1, haploid	1.83	1738	1.00	-
Mycoplasma genitalium	1, haploid	0.58	483	1.11	-
Mycoplasma pneumonia	1, haploid	0.82	680	-	-
Rickettsia prowazekii	1, haploid	1.11	834	-	-
EUKARYOTES					
Fungi					
Neurospora crassa (Bread mold)	7, haploid	43	10,000	3.80	-
Saccharomyces cerevisiae (Yeast)	16, diploid	12	5,700	2.00	30
Algae					
Chlamydomonas reinhardii (unicellular green algae)	17, haploid	120	16,400	-	-
Protozoa					
Tetrahymena thermophile		220	20,000	11.00	-
Plants					
Arabidopsis thaliana (Thale cress)	5, diploid	125	25,498	4.90	-
Arabidopsis lyrata (Rock cress)	5, diploid	230		-	-
Cicer arietinum (Chickpea)	8, diploid	738	28,296	-	-
Glycine max (Soybean)	20, diploid	1100	46,430	-	-
Gossypium raimondii (Diploid Cotton)	13, diploid	880	40,976	-	-
Hordium vulgare (Barley)	7, diploid	5000	26,159	-	-
Musa acuminate (Banana)	11, double haploid	523	36,542	-	50
Oryza sativa sp. *japonica* (Rice)	12, diploid	399	41,000	9.60	-
Oryza sativa sp. *indica* (Rice)	12, diploid	466		-	-
Phoenix dactylifera (Date palm)	18, diploid	658	28,890	-	-
Phyllostachys heterocycla (Moso bamboo)	24, diploid	2050	31,987	-	-
Populous trichocarpa (Black cottonwood, Poplar)	19, diploid	485	45,000	-	-
Solanum lycopersicum (Tomato)	12, diploid	950		-	-
Solanum tuberosum (Potato)	24, dihaploid	844	39,031	-	-
Sorghum bicolor (Sorghum)	5, diploid	770		-	-
Triticum aestivum (Wheat)	21, hexaploid	17000		-	-
Triticum urartu (diploid wheat)	7, diploid	4920	34,897	-	-

continued on following page

Table 4. Continued

Organism	Chromosome number (n) and ploidy	Genome size (Mb)	Number of protein-coding genes (approx.)	Average gene density (kb per gene)	Genes similar with humans (%, approx.)
Vitis vinifera (Grape)	29, diploid	505		-	-
Zea mays (Maize)	10, diploid	2500	32,000	-	-
ANIMALS					
Invertebrates					
Caenorhabditis elegans (Nematode)	6,diploid	97	19,000	5.00	40
Drosophila melanogaster (Fruit fly)	4, diploid	180	13,700	9.00	50
Strongylocentrotus purpuratus (Sea urchin)	21, diploid	814	23,300	-	60
Vertebrates					
Canis familiaris (Dog)	39, diploid	2500	18,400	-	75
Danio rerio (Zebrafish)	25, diploid	1412	26,000	-	70
Gallus gallus (Chicken)	39, diploid	1050	23,000	-	60
Homo sapiens (Human)	23, diploid	2900	30,000	116.00	-
Macaca mulatta (Rhesus monkey)	22, diploid	2870	20,000	-	93
Mus musculus (Mouse)	20, diploid	2500	30,000	90.00	80
Pan troglodytes (Chimpanzee)	24, diploid	3323	31,000	-	98
Rattus norvegicus (Rat)	21, diploid	2750	22,000	91.00	80

present in many genomes, and analysis of these genes provides insight into eukaryotic genome evolution and function. Examples of multigene families include globin, histone, tubulin, actin, immunoglobulin etc.

Comparative genomics can also help in crop improvement programs by transferring information about genes from model species to the species of interest, by identifying genes responsible for expression of traits of interest, and by assessing allelic diversity within a species and thereby help to combine the best alleles to produce superior varieties.

Comparison of in-depth sequence data within and beyond grass family has become possible due to the availability of rice genome DNA sequence data. Comparative sequence analysis between rice and wheat, as reported in 2003 by Sorrels, could interpret the results at much finer details compared to earlier assessments. The increased resolution could reveal several discontinuities in gene order between rice and wheat genome. Such study supports the view that grass genomes are evolving rapidly and that the relationship between structural and functional components of the genes are complex. The sequence-based maps of the model crops shall facilitate their use for locating genes of interest in conserved sequence of other species.

Synteny, or preservation of the gene order on a chromosome, can be used to assess the evolution history of plants and animals. It also provides

Table 5. Size and predicted number of genes for each chromosome of human (Homo sapiens) and rice (Oryza sativa sp. japonica)

Chromosome number	Size (Mb)		Predicted number of genes	
	Human	Rice	Human	Rice
Ch- 1	249	45	2616	4467
Ch- 2	243	36	1733	3011
Ch- 3	198	37	1388	3197
Ch- 4	191	36	1076	2679
Ch- 5	180	30	1185	2426
Ch- 6	170	32	1328	2342
Ch- 7	159	30	1250	2507
Ch- 8	146	29	920	2286
Ch- 9	141	24	1101	1618
Ch-10	135	24	1001	1724
Ch-11	135	31	1618	1834
Ch-12	133	28	1338	1870
Ch-13	115	-	466	-
Ch-14	107	-	890	-
Ch-15	102	-	916	-
Ch-16	90	-	1075	-
Ch-17	81	-	1462	-
Ch-18	78	-	401	-
Ch-19	59	-	1598	-
Ch-20	63	-	733	-
Ch-21	48	-	325	-
Ch-22	51	-	601	-
X	155	-	1129	-
Y	59	-	140	-
Chloroplast	-	0.135	-	159
Mitochondria	1.66	0.491	37	96
Total	3089.66	382.626	26327	30216

information about the functional relationships between genes. Chromosomal rearrangements take place during evolution, and, therefore, the degree of synteny can provide information about shared ancestry. In organisms with known shared ancestry, prediction of the presence of genes in different species can be made through synteny. This implies that closely related species should carry similar genes in their genome. For example, high degree of synteny between rice and other cereals implies that genes present in rice may well be present in other cereals.

Goff et al. (2002) reported that 98 percent of known proteins from maize, wheat and barley are homologous in rice genome. Mere existence of the similar genes in the same order on a chromosome in rice and wheat does not necessarily mean that they have same gene sequences or produce identical proteins. However, several conserved sequences of genes can tolerate high degree of diversity, while others show virtually identical sequence across a wide evolutionary range. Homology is usually expressed in terms of "percent identity" or 'percent similarity'. If we consider 'percent identity' between rice and wheat, it will be evident that they share about 80 percent nucleotide homology between them. With maize also similar results was obtained.

APPLICATIONS OF THE PLANT SEQUENCED GENOMES

Availability of sequencing data from model plant such as *Arabidopsis thaliana* and several other crop plants such as rice, wheat, maize, potato, barley, banana, sugarcane, cotton etc. has opened new avenue for their utilization in crop improvement programs. Some of the possible applications of sequence data obtained from plant genome are described in the following section.

Translational Biology: from Models to Crop

Sequencing of *Arabidopsis* genome revolutionized understanding of plant biology by unraveling basic mechanisms in plant development, tolerance to abiotic and biotic stresses and adaptation. Since many of these basic pathways are common to all plants, *Arabidopsis* genes can be used either directly in heterologous systems or as candidate genes for identifying orthologs in crops. The most successful translations of a gene from *A. thaliana* to improve a trait in crop were achieved with genes involved in abiotic stress tolerance, and primarily with transcription factors, because of their central role in controlling cellular processes. For example, CRT binding factor (CBFs) from *Arabidopsis* were expressed in tomato, rapeseed, strawberry, rice and wheat, providing evidence of improved freezing, salt and draught tolerance.

Because of the complexity of disease resistance mechanisms in plants, such translational biology is difficult for adoption in disease resistance.

The two resistance mechanisms, pathogen-associated molecular pattern-triggered immunity (PTI) and effector-triggered immunity (ETI) are usually superimposed on each other in plants. ETI corresponds to classic race-specific disease resistance that has evolved rapidly and more specifically in each plant species, thereby making it a difficult subject for direct transfer from model to crops. On the other hand, the PTI translation affords more opportunities, as was demonstrated by the successful engineering of broad-spectrum disease resistance in tomato and tobacco by the expression of an *Arabidopsis* elongation factor Tu receptor (EFR).

In several other cases, the candidate genes underpinning basic traits have been identified in crops by screening with sequences of functional or structural orthologs for closely related or model species. For example, drought tolerance in maize was obtained after transforming a maize ortholog (*ZmNF-YB2*) of the *Arabidopsis* transcription factor *AtNF-YB1*, which was identified through a screening for drought tolerance in *Arabidopsis*. But such technique has limitations as one-way translational biology cannot explain all the differences in species. Therefore, such techniques can be applied most effectively if the knowledge available about the biological processes from the model plants can be matched with the crop genome sequence.

Comparison of one crop genome sequences with other crop genomes, or model species, may be useful in identifying species-species differences that can underlie essential traits, particularly so for the conserved sequences. For example, with the help of flowering-time genes in rice it was possible to determine that rice and *Arabidopsis* share common regulatory pathways but functional analysis demonstrated differential regulation that results from reverse function of key central regulator. These studies have refined our understanding about the triggering the events in short-day and long-day plants.

Application of knowledge gained from model species to crop plants has various other limitations, which include role of genotype x environment interaction in crop plants, selection and adoption of species, nature of polyploidy in crop plants etc.

Application of Sequences From Crops to Crop Improvement

Rice being the first crop genome sequenced has provided excellent opportunity for crop improvements. Rice researchers were rapidly able to integrate and apply genome sequence information to understand rice genome structure and evolution. They also could discover and mine genes, having agricultural importance. Through genome-wide survey it was possible to discover members of large gene families such as, transcription factors, peptide transporters, kinases, nucleotide binding leucine-rich repeats (NB-LRRs), microRNAs and germins. Subsequently, their cellular functions and their roles in plant growth and development were also elucidated. The molecular changes that took place during domestication of crops (*e.g.* seed-shattering) were also identified.

One of the most important outcomes of rice genome sequence is the identification of high-throughput molecular markers to assist genetic analysis, gene discovery and breeding program. Rice is now rich in tools for mapping and breeding, including high density simple sequence repeats (SSRs) (about 15 SSR Mb^{-1}), comprehensive single nucleotide polymorphisms (SNPs), insertion-deletion polymorphisms (IDPs) and custom design (candidate gene) markers. Genome sequencing has also made it much easier to identify and clone quantitative trait loci (QTLs), having agronomic importance.

Carillo et al. (2009) compared sequences of alleles of an oxalate oxidase gene family that contributes to a disease-resistance QTL and discovered an indel in the promoter of one gene family member from the QTL donor that was lacking in varieties. Association and functional analysis of the oxalate oxidase gene demonstrated that presence of the indel was associated with QTL effectiveness and that the polymorphism created by the indel is a useful marker for the QTL. In the same year, Fukuoka and coworkers used sequenced-based markers from the blast disease resistance gene *pi21* region to identify recombinants between *pi21* and another gene located 37 kb apart that confers poor eating quality. With this information they were able to select for rice varieties that combined durable blast resistance and good eating quality. This happens to be the first successful report on selection of resistance with quality parameters in rice.

Recently genome sequencing of several crop species have been released, which include sorghum, maize, barley, wheat and grapevine. Sorghum genome sequence has now been coupled with dense molecular marker maps that were previously impossible to link because of the lack of common markers across populations. This enabled, for the first time, the integration of major effect genes into a single map, thereby providing a foundation for breeders to link these easily recognizable landmarks to QTL studies for enhancing breeding program.

Gore et al. (2009) used two datasets comprising 3.3 million SNPs of maize to produce a first haplotype map (HaploMap) and to analyze the distribution of recombination and diversity along maize chromosomes. This Maize "HaploMap" and comparative genome hybridization (CGH) experiments enabled the identification of >100 low-density regions that are possibly associated with the domestication and geographical differentiation of maize. In 2009, Springer and coworkers reported extensive structural variations, including hundreds of CNVs (copy number variations) and thousands of PAVs (presence-absent variants), among maize lines. Many of the PAVs contain intact, expressed, single-copy genes that are present in one haplotype but absent from another, including hundreds of expressed genes potentially involved in heterosis. Thus maize genome sequence has the potential to unveil the molecular mechanism of heterosis by identifying the underlying genes and/or small RNAs.

METAGENOMICS

Sequencing the genomes from entire community of microbes in environmental samples of soil, water, and air through whole genome shotgun approaches is called metagenomics or environmental genomics. One of the major objectives of metagenomics is to know about the millions of species of bacteria, which are yet to be characterized. Metagenomics is providing new information about genetic diversity in microbes. It also has potential to identify genes with novel functions, having potential applications in medicine and environmental pollution control.

It is well established that only about 10 percent of the bacteria that exist on earth are actually culturable in known media. Therefore, the rest 90 percent has remained uncharacterized. In metagenomics DNA from different microbes were sequenced directly from the environmental samples without culturing them. For example, water samples from different layers of water column were passed through high density filters of various sizes to capture the bacteria. DNA was then isolated directly from these microbes and sequenced through shotgun method and genome assembly. In one such experiments conducted in Sargasso Sea off Bermuda yielded 1.2 million novel DNA sequences from 1800 microbial species, including 148 previously unknown bacterial species. Subsequent experiments have generated billions of sequences of uncharacterized microbes. If we compare the sequences in the publicly available databases for predicted protein sequences, eukaryotic sequences comprise the majority (63 percent) of predicted proteins (Table

6). However, comparison for novel predicted proteins has indicated that the bacterial sequences (90.8 percent) dominate the databases (Table 6).

Table 6. Frequency of predicted proteins and novel predicted proteins identified in different kingdoms using publicly available databases of sequence genomes

Kingdom	Predicted proteins (%)	Novel predicted proteins (%)
Eukaryotes	63	2.8
Bacteria	28	90.8
Viruses	7	3.7
Archaea	2	2.7

TRANSCRIPTOMICS

Although sequencing of any genome is considered to be an achievement, the finding is not complete without understanding how genes are expressed. Study on expression of genes is called transcriptome analysis or transcriptomics.

Although all cells in an organism possess the same genome, all the genes are not expressed in all the cells or tissues. Certain genes will be highly expressed in some tissue, while other will expressed at low levels, and some not expressed at all. Transcriptome analysis provide information on normal pattern of gene expression in different cells and tissues during development, how gene expression controls the physiology of differentiated cells, and how altered gene expression lead to disease development.

Identification and quantification of all the mRNA produced in a cell is known as transcriptomics. This can be done by using microarrays or 'gene chips'. The principle of this technique is to coat a flat surface with spots of DNA complimentary to the expressed mRNA. Hybridization between the DNA and RNA base pairs shall lead to capture of the mRNA and their quantification. The basic information about mRNA can be derived from Northern blot (see Chapter 12). The detailed protocol for microarray assay is described in Chapter 12. Some features of microarray are presented in Table 7.

Analysis of Transcriptomcs

The analysis explores the data on the basis of correlations and similarities. First the groups of genes that have correlated expression profiles are found

Table 7. Some features of microarrays

Application	Sample	Captors
Genomics	DNA	ESTs
Transcriptomics	mRNA	Cdna
Proteomics	Proteins	Antibodies
Metabolomics	Various	Various

out, from which it can be inferred that they participate in the same biological process. In the second approach, the tissues are grouped according to their gene expression profiles, on the assumption that the tissues with the same or similar expression profile belong to the same clinical state.

If a set of experiments comprising samples prepared from cells grown under m different conditions is carried out, then the set of normalized intensities (i.e transcript abundances) for each experiment defines a point in m-dimensional expression space, whose coordinates give the normalized expressions. Distance (D) between the points can be calculated by the Euclidean distance metric as:

$$d = \left[\sum_{i=1}^{m} \left(a_i - b_i \right)^2 \right]^{1/2}$$

For two samples a and b subjected to m different conditions. Clustering algorithms can then be used to group transcripts. The closest pair of transcripts forms the first cluster, the transcript with the closest mean distance to the first cluster forms the second cluster, and so on. This is the un-weighted pair-group method average. Other methods include single linkage clustering, in which the distance between two clusters is calculated as the minimum distance between any members of the two clusters.

For large and complex data sets, fuzzy clustering algorithms have been found to be better and more successful. Fuzzy schemes allow points to belong to more than one cluster. The degree of membership is defined by:

$$U_{r,s} = \frac{1}{\sum_{j=1}^{m} \left\{ \frac{d\left(x_r, \theta_s\right)}{d\left(x_r, \theta_j\right)} \right\}^{\frac{1}{(q-1)}}} \text{ where, } r = 1, \ldots\ldots, N; s = 1, \ldots\ldots, m$$

For N points and m clusters (*m* is given at the start of the algorithm), where $d(x_i, \theta_j)$ is the distance between the point x_i and the cluster represented by θ_j, and $q > 1$ is the fuzzifying parameter. The cost function:

$$\sum_{i=1}^{N}\sum_{j=1}^{m} u_{r,s}^{j} d\left(x_i, \theta_j\right)$$

is minimized, subject to the condition that the $u_{i,j}$ sum to unity and, clustering converges to cluster centers corresponding to local minima or saddle points of the cost function. The procedure is typically repeated for increasing numbers of clusters until some criterion for clustering quality, i.e the partition coefficient:

$$\left(\frac{1}{N}\right)\sum_{i=1}^{N}\sum_{j=1}^{m} u_{i,j}^{2}$$

becomes stable. Closer the partition coefficient to unity, it becomes harder for clustering.

The dimensionality of expression space can also be reduced by principle component analysis (PCA), in which the original dataset is projected onto a small number of orthogonal axes. The original axes are rotated until there is maximum variation of the points along one direction. This becomes the first principal component. The second is the axis along which there is maximal residual variation, and so on.

Microarrays depend on exposure duration, temperature gradients, flow conditions etc. Thus variations in any of these conditions may lead to variations in the results. Microarrays also depend on pre- and post-processing of mRNA, matching of mRNA fragment size distribution with the probe, and quantitative interpretation of data. Thus it is important to take precautions on the above aspects to obtain reproducible results.

Another technique called serial analysis of gene expression (SAGE) is also used in transcriptomics. In this technique, a short but unique sequence tag is generated from the mRNA of each gene using the PCR and joined together. The joint molecule is then sequenced. The representation of each tag in the sequence will be proportional to the degree of gene expression.

DNA MICROARRAY

Although several techniques can be used for transcriptome analysis, DNA microarray (also known as gene chips) analysis is widely used as it enables to analyze all the expressed genes in a sample simultaneously. Gene chips consist of a glass microscopic slide onto which single-stranded DNA molecules are attached, or 'spotted', using a computer controlled high speed robotic arm called an arrayer. Arrayers are fitted with a number of tiny pins, and each pin is immersed in a small amount of solution containing millions of copies of a different single-stranded DNA molecules. The single-stranded sequences can be complimentary DNA (cDNA) or expressed sequenced tags (ESTs). The array fixes the DNA onto the slide at specific locations (spots of points) that are recorded by a computer. A single microarray can have more than 20,000 different spots of DNA, each containing a unique sequence for a specific gene. These way entire genomes can be translated into microarrays. Thus microarrays can be used to compare patterns of gene expression induced due to different conditions in a tissue, gene expression patterns in normal and diseased tissues and also to identify pathogen(s).

For transcriptome analysis through microarrays, first mRNA should be extracted from cells or tissues, and then mRNA is used to synthesize cDNA through reverse transcriptase. The nucleotides of the cDNA are tagged with fluorescently labeled nucleotides. Usually cDNA prepared from one tissue is labeled with one color dye, say red, and from another tissue labeled with a different color, say green. Labeled cDNAs are then denatured and incubated overnight with the microarray. cDNAs will hybridize to the complimentary DNA sequences present at specific spots in the microarray. The microarray is then washed, and scanned by a laser that causes the hybridized cDNA to fluoresce. The pattern of fluorescent spots indicates which genes are expressed, and the intensity of inflorescence indicates the relative level of expression.

Microarrays can yield variable results from one experiment to another identical experiment. Some of these differences are due to real differences in gene expression, but such differences can also be due to variability in chip preparation, cDNA synthesis, probe hybridization, or washing conditions. Use of commercially available microarrays can reduce such variability. Cluster algorithm programs can be used to retrieve spot intensity data from different locations on a microarray and to group gene expression data from one or multiple microarrays into cluster images incorporating results from many experiments. It can also be used to group genes according to whether they show increased or decreased expression under the specific experimental conditions.

PROTEOMICS

Proteomics involves identification, determination of amount, locations and interactions of all the proteins expressed in a cell. Although every cell of an organism contains an equivalent set of genes, all cells do not express the same genes and proteins. Eukaryotic gene is a mosaic of introns (Ys) and exons (Xs), and can form different mRNAs after processing. For example, let us consider a gene having the exons (X) and introns (Y) as: $X_1\ Y_1\ X_2\ Y_2\ X_3\ Y_3\ X_4\ Y_4\ X_5\ Y_5\ X_6\ Y_7$, which can form different mRNA having the composition of $X_1\ X_2\ X_3\ X_4\ X_5\ X_6$, $X_1\ X_2\ X_4\ X_6$, $X_1\ X_3\ X_5\ X_6$, $X_1\ X_4\ X_6$ etc. The ensemble of these transcripts is called the transcriptome, and its study is called transcriptomics. As apparent from the above example, the transcriptome is considerably large in number than the gene or genome.

Although the number of different proteins far exceeds the number of genes present in an organism, the actual number of proteins present in a cell at any given stage may be much smaller, as only a small fraction of genes are likely to be expressed within a cell of a particular tissue. Although proteome is a term that represents the complete set of proteins encoded by a genome, but most often used to mean the entire complements of proteins in a cell. Thus cell type in an organism has a markedly different proteome. Proteome is usually highly dynamic as it is likely to depend on the environment within a cell type (Komatsu et al. 2013).

Proteomic Analysis

Form the comparative data available on transcriptome and proteome, it is interesting to note that the amount of the mRNA and the corresponding protein produced are very different. Thus transcriptome has lost some of its importance, as it is an intermediate step and does not contribute to the phenotype directly. Further transcriptome contains no information about the numerous post-transcriptional modifications of proteins.

For their identification and quantification, proteins must first be isolated and separated. The detailed procedure for isolation, separation, and identification are presented in Chapter 12.

Protein Expression Pattern

Both transcriptome and proteome usually generate huge amount of data on the expressed object (mRNA and protein). Therefore, it is required to reduce the quantum of these data for carrying out meaningful analysis. One approach is to group proteins into blocks whose expression tends to vary in the same way i.e. increase, decrease or remain same. The second approach is to search for global parameters characterizing the proteome. It has been observed that the distribution of abundance of protein follows the same canonical law as the frequency of words in literary texts. The canonical law has two parameters, the informational temperature, and the effective redundancy. The informational temperature is low for limited expression of the potential gene repertoire and high for extensive expression, whereas effective redundancy is high when many alternative pathways are active and low when alternative pathways are limited.

Regulatory Networks

In prokaryotes and some eukaryotes, genes are organized in operons, where a promoter controls the expression of several genes, positioned successively downstream from the promoter. In most eukaryotes, a similar, but less clearly delineated arrangement also exists; the same transcription factor may control the expression of several genes, but they may be quite distant from each other along the DNA in the same chromosome or in different chromosomes. Genes observed to be close to each other in expression space are likely to be controlled by the same activator. Each gene can have its own promoter sequence, and coexpression can be achieved by binding of the transcription factor to a multiplicity of sites. Thus, a potentially interconnected network can be established from the information on which gene code for which protein, and which in turn control the expression of other genes (Eldakak et al. 2013).

Let us consider a network model to explain the process of the analysis. Let gene A activates the expression of gene B, B activates C, and C inhibits A. Thus, A, B and C form a network, which can be represented through a Boolean weight matrix as:

	A	B	C
A	0	1	-1
B	1	0	0
C	0	1	0

if one reads from top to bottom it gives the cybernetic formalization; and if one read horizontally it gives the Boolean rules: A=B Not C, B=A, C=B can be transformed to produce a stochastic matrix and the evolution of transcription given by a Markov chain. Alternatively the system can be modeled as a neutral net in which the evolution of the expression level a_i of the *i* th protein in time *t* is:

$$T\frac{da_i}{dt} = F_i\left(\sum_j \omega_{ij} a_j - x_i\right) - a_i$$

where, ω is an element of the weight matrix, *F* a nonlinear transfer function, x is an external input and the negative term at the extreme right represents degradation. The Boolean network can be extended to hundreds of genes.

Interactions

Proteins are known to be highly interactive molecule, within the cell, between them. If the proteins are considered as the nodes of a graph, a pair of proteins will be joined by a vortex if the proteins associate with each other. This is in contrast to metabolic network, in which two metabolites are joined if there is a chemical reaction leading to formation of a third metabolite. On this basis, a set of interactions in which a protein could participate, would be characterized by such a graph, or an equivalent list of all the proteins in a cell, each associated with a sub-list of the proteins with which they interact.

It is important to note that this graph shall be very large and complex. Even if we confine to N expressed proteins in a cell, there are ~ N^2 potential binary interactions, and vastly higher order ones. Even if a small fraction of these interactions actually occur, it can safely be around 10^7 interactions, assuming that about 10^4 expressed proteins in a eukaryotic cell. In a prokaryotic cell, the situation should be better as around 1000 proteins are expressed. The other challenge being that many of these proteins are expressed in extremely low concentration. At present the interactome has mostly been assembles on the basis of dichotomous inquiry, i.e. whether the protein interact or not. With the advancement of technology this approach is obvious to change, and it will become important to assign gradations of affinity to the interactions. The influence of other factors like compartmentalization of cytoplasm, resulting in restriction of protein interaction, and interaction of proteins at the internal surfaces of cells (lipid membranes) should also be considered. Several *in vitro* analysis have been developed to understand protein-protein interactions. But all of them have several limitations.

PROTEIN FAMILY DATABASES

Several protein databases are available, but most of these databases are inaccurate or incomplete. Therefore, most of them contain cross-reference to one another, and provide links to databases on protein structure and literature. Brief descriptions of the most widely used databases are given in the following section. Some other important databases on proteomics are presented in Table 1.

PROSITE contains annotated information on protein families and domains, and also provide links to scientific literature and experts. Domains are constructed manually and quantitatively less flexible compared to HMM statistical methods. This can be found at the following site: http://www.expasy.ch/prosite

Pfam contains information on protein domains which are non-overlapping. Entries of each domain contains information about a seed alignment from which an HMM profile has been created. The database was built with the help of HMM-profile and is linked inextricably to the HMMER. In fact there exist two databases, the curated database (Pfam-A), and putative families database (Pfam-B), generated automatically. Usually more families are upgraded from Pfam-B to Pfam-A, with the release of each families. Pfam is available at the following sites: http://www.cgr.ki.se/Pfam/, https://www.sanger.ac.uk/Software/Pfam/, http://www.pfam.wustl.edu/

PRINTS contains information on protein fingerprints which are well annotated. It facilitates using of a single motif by multiple fingerprints. The curators of PRINTS suggest that it can be used for recognizing larger structural patterns. PRINTS is available at the following site: http://www.biochem.ucl.ac.uk/bsm/dbbrowser/PRINTS/PRINTS.html

ProDom contains information on protein sequences, generated automatically from a cluster of the protein sequence database using PSI-BLAST algorithm. Although it is slightly less tidy, its coverage is usually more complete. ProDom is available at the following site: http://www.toulose.inra.fr/prodom/doc/prodom.html

InterPro has been constructed as a composite database by combining the four databases mentioned above namely, Pfam, PROSITE, PRINTS, and ProDom. Query for the sequences can be made simultaneously with all the four databases, and the results are also displayed simultaneously. InterPro is available at the following site: http://www.ebi.ac.uk/interpro/

Blocks is a database of ungapped multiple alignments. Since gaps are excluded in this database, it will not include some motifs which are present in other databases, and it will tend to represent highly conserved sequences. The Block database is available at the following site: http://blocks.fhcrc.org/

SMART database contains HMM profiles of protein families, but much lesser number compared to other databases. However, for each domain it provides a significantly enriched structural, functional and phyletic annotation, and thereby compensates its deficiency. SMART is available in the following site: http://smart.embl-heidelberg.de/

PROTEIN MICROARRAYS

Protein microarrays are designed by following similar basic concept as in DNA microarrays (gene chips). These are often constructed with antibodies that specifically recognize and bind to different proteins. Protein microarrays allow assessment of expression levels for thousands of genes across various treatment conditions and time. However, it is difficult to place thousands of protein capture agents on the array, as capture does not depend on simple hybridization, but on certain arrangement of amino acids in the three-dimensional space. The proteins may also lose their original structure due to immobilization on the chip surface. Alternatively, it may be possible to use nucleic acid immobilization by using aptamers, oligonucleotides binding specifically to proteins, for protein capture. This approach may be useful to determine the expression levels of transcription factors through microarrays (Eldakak et al. 2013).

Non-specific adsorption of proteins is a major problem with protein microarrays. Pretreatment with "blocking protein" like seralbumin, can eliminate the non-specific adsorption sites, although they may interfere with specific binding sites as well.

Statistical analysis of protein microarray data focuses on ascertaining similarities of gene expression profile through clustering or calculating the differential expression between treated and control samples.

METABOLOMICS AND METABONOMICS

Metabolomics is defined as the measurement of the amounts (concentrations) and locations of all the metabolites in a cell. Metabolites are the products formed, due to transformation in the process of metabolism, mostly the substrate and products of enzymes. Metabolomics is essentially an extension of proteomics, which basically examine correlations between expression data and metabolite data (Adamski 2012).

On the other hand, metabonomics is defined as the quantitative measurement of the multiparametric metabolic responses of living systems to pathophysiological stimuli, genetic modification, and environmental stress. In the case of many non-genetic diseases, and adverse effects due to exposure to toxic substances continuously, metabolites are the most revealing markers for their identification. Metabolites are also used to study the effect of drugs.

In many cases, concentrations of a fairly small number of metabolites have been shown to be well correlated with a pathological state of the organism. Thus, metabolomics is being integrated with genomics and proteomics in order to create new systems biology. For example, a toxin ingested by an organism may induce expression of a gene, which in turn produces a protein (enzyme), which may get involved in a metabolic pathway involved in production of other proteins and so on (Wen et al. 2014).

Data Collection and Analysis

Metabolomics has to deal with a diverse set of metabolites, and therefore, require techniques which can separate and identify them. Usually different chromatographic techniques are used to separate them, and mass spectrometry to identify them. Alternately, high resolution nuclear magnetic resonance spectroscopy can be used directly to analyze bio-fluids, tissues and organs.

For development of metabolic microarrays, usually large numbers of small molecules are synthesized, using combinatorial or other chemistry for generating high diversity. The array is then exposed to the target, whose components of interest are usually labeled. Binding of the macromolecule with the metabolite can easily be identified and further analyzed. Several other advanced methods like Probes Encapsulated by Biologically Localized Embedding (PEBBLES), High Resolution Scanning Secondary Ion Mass Spectrometry (nanoSIMS) are also used to determine spatial variations in selected metabolites.

While analyzing the metabolomics the first task is to correlate the presence of metabolites with gene expression, which means correlating two datasets, containing vast information. Usually two categories of algorithms called supervised and unsupervised, are used for this task.

In unsupervised methods, it was determined whether there exists any intrinsic clustering within the dataset. Principle Component Analysis (PCA) is a widely used unsupervised technique. In this method the original dataset is projected onto a space of lower dimensions. For example, let us consider a set of metabolomics data which contains one hundred metabolites, is a point in a space of one hundred dimensions. The original axes are rotated to find a new axis along which there is the highest variation in the data. This axis becomes the first principle component. The second one is orthogonal to the first, and has the highest residual variation, the third axis is again orthogonal, and has the next highest residual variation, and so on. The first two to three axes are usually sufficient to account for most of the variations in the original data. The principle components remain uncorrelated (zero covariance), as they are orthogonal.

In supervised methods, the classes are non-overlapping and sequences of parameter values characterizing the object of the class are previously known. In the first stage, decision functions are elaborated enabling new objects from a dataset to be recognized, and during the second stage, those objects are recognized. In these methods neural networks are commonly used.

Interpretation of the Results

After analysis of the data, it is important to interpret the irregularities or patterns. The chemical relationship between regulatory effector molecules, and their immediate effect like feedback inhibition of enzyme activity or repression of enzyme biosynthesis, is described as simple regulation. On the other hand, intracellular effector molecules that accumulate whenever the cell is exposed to a particular environment are called complex regulation. They are effective within the metabolic process controlled by them. For example, hormones are synthesized and secreted by specific tissue, circulated through the blood, execute specific metabolic pathways, and finally degraded into basic ingredients.

Metabolic Control Analysis (MCA)

Metabolic control analysis is a used to determine the relationship between properties of the network of a biochemical reaction (as a whole), and that of its basic component. Since it is useful for both theoretical and experimental aspects of any cellular activity, it helps to understand the mechanism of control and regulation of gene expression. Essentially, it is a sensitivity analysis of a dynamical system, and due to the stoichiometric structure of reaction networks it generates a character of its own.

Metabolic control analysis is the application of systems theory to metabolism. If we consider say,

$$X = \{ x_1, x_2, \ldots\ldots\ldots, x_m \}$$

where, x_i is the concentration of the *i*th metabolite in the cell. These concentrations vary in both time and space.

Let $v = \{ v_1, v_2, \ldots\ldots, v_r \}$

Where v_j is the rate of the *j*th process. To a first approximation, each process corresponds to an enzyme, then

$$dx / dt = Nv$$

where the N specifies how each process depends on the metabolites. Metabolic control theory (MCT) is concerned with solutions to the above equation and their properties. The dynamerical system is generally too complicated to difficult to obtain a solution, and numerical solutions are of little use without detailed information about enzyme rate coefficients. Therefore, currently metabolic centers are discussed on qualitative features only.

Network

It is possible to construct a network of the metabolism, in which the nodes are the enzymes and the edges connecting them are the substrates and products of the enzymes. It is then possible to investigate further by following two different independent ways. In the first approach a series of consecutive enzyme-catalyzed reactions producing specific products are defined. The intermediate products in the biochemical pathway are defined as substances with a sole reaction producing them and a sole reaction consuming them. The biggest challenge of this approach is to identify different products in the biochemical pathway and correlate to the living system. The second approach is to focus on the distribution of the density of connections between the nodes. The number of nodes of degree k appears to follow a power law distribution, i.e the probability that a node has k edges ~ k^{-r}.

One of the major challenges in metabolomics is to establish the relationship between the physical structure (the nodes and their connecting edges) and the state structure. Integration of the metabolic network into expression network is another challenge.

APPLICATIONS OF GENOMICS AND PROTEOMICS

Medical Science

Investigating the physiological disorder in the organism is one of the primary concerns of medicine. Bioinformatics allows us to establish correlations between those disorders and variations in the genome and proteome of a patient.

Variations in the nucleotide sequence in the gene(s) between normal and diseased individuals have been established to be the cause of many diseases. With the completion of human genome sequencing, it has become much easier to check any variation in the DNA sequence of the diseased individuals. Millions of single nucleotide polymorphisms (SNPs) have been documented, and most of the genetic variability across human populations can be accounted for by SNPs, and most of the SNP variation can be grouped into a small number of haplotypes. This database is extremely useful for elucidating the genetic basis of disease, or susceptibility to disease. This implies that one can look for preventive treatments for the known genetic diseases.

Genes responsible for a particular phenotype can be identified through ESTs obtained from the cDNA library of the patient's tissue. These ESTs can then be used to find out any variation in the nucleotide sequences by comparing with the ESTs of normal individuals. This can be done by using BLAST. It can then be verified whether a variant gene is known to cause a different (diseased) phenotype with the help of Mendelian Inheritance in Man (OMIM) database. Most of the ESTs are developed and annotated from known genes. But there could be ESTs developed from new (mutated) genes, which are yet to be annotated and put on the Genes_seq map. For identification of new genes Model map are very useful. Any new SNPs found in ESTs, can be deposited in SNP database (dbSNP). It is possible to determine whether the new SNPs are associated with the disease phenotype by comparing the DNA sequence of the patient with the normal individual.

Microarrays are extensively applied to screen the mutations. In effect each bead corresponds to one spot on a microarray. The beads are individually tagged with the help of fluorophores. Several hundred different types of beads can be mixed and discriminated in a single experiment. However, this technique has certain limitations, as it is not possible to discriminate between completely match and slightly mismatch sequences.

It is possible that a disease may develop due to combined effect of more than one gene. The problem of correlation then becomes a combinational aspect and much more difficult to solve. Thus the role of genomics in medicine will expand once gene syntax is understood.

Another approach called functional cloning, in which the gene is cloned into cells and examined for pattern of expression of certain proteins linked to the disease symptoms, in cell culture system.

DNA sequencing has also contributed towards various applications in forensic medicine. Repeated sequences like variable number of tandem repeats (VNTRs) and short tandem repeats (STRs) have potential to be used in forensic medicine, as they appear to be uniquely different for each individual. However, degradation of the DNA samples before collection may limit the use of VNTRs. Amplification by PCR can extend the use of degraded DNA samples for STR analysis.

Gene chips can also be used to identify unambiguous foreign DNA in a patient due to an invading microorganism. Such analysis shall make the conventional techniques of culture and identification of disease causing microbes redundant.

Drug Discovery and Testing

Bioinformatics has opened new avenues for drug discovery. Unlike traditional methods which mainly sought to bind the drugs to the enzymes, thereby blocking their activity, bioinformatics driven drug discovery focuses on control points. Once a gene or set of genes are found to be associated with a disease, they can be cloned into the cells and their protein products can be investigated as possible targets for the drugs (functional cloning).

Based on the tissue and environmental factors, proteins can have different structural forms. However from the point of view of the drug discovery it is important to identify the discrete domains within the subunits of proteins, which can be used as targets for the drugs. Structural genomics can be used to predict the three dimensional structure of a protein suspected to be at a control point from the corresponding gene sequence. These structures can be effectively used to study the binding sites of new drugs and its consequence.

Proteins often bind with other proteins to form multi-protein structures and thereby control functions such as the response to hormones, allergens, growth signals etc. During disease development all or some of these responses may go wrong. Therefore, knowledge of the network of interactions is needed to understand which proteins are the best targets for the drug.

Advances in genomics and proteomics have triggered development of 'targeted' therapeutics. The recent healthcare approach is directed towards development of personalized therapy, with the help of individual's genetic makeup.

Metabolic Pathway Elucidation

Mathematical modeling techniques have enabled researchers to identify interactions between components. Data from feedback inhibition and competitive binding studies have been analyzed using Gauss and Stoke's theorems. Evolution optimization of metabolic pathway has been studied using game theory. One of the examples where the accuracy of the mathematical models has been experimentally verified is the insulin signaling pathway. It is possible to find out the presence or absence of an interaction in a pathway by formulating equations for all the components and checking whether the theoretical values obtained hold true for experimental findings. In the case of any discrepancies, it is possible to create additional interactions and formulate equations until the values obtained theoretically matches with the experimental value.

Disease Pathway Identification

Information on metabolic pathway enables identification of any defect in the regulatory proteins causing disorders. It is possible to obtain experimental data in the case of anomalous interactions it is possible to obtain experimental data on concentration of metabolites from the affected individuals through protein detection techniques. By extending the experiments, the regulatory characteristics of the pathway and different interactions can be studied.

Target Protein Identification

The structure of the target protein can be accurately determined by advanced techniques like laser desorption ionization. Identification of regulatory proteins causing metabolic disorder can be effectively utilized for drug discovery.

Bioactive Molecule Identification

Usually the bioactive molecules are low molecular weight compounds that may cause changes in the biochemical pathway to overcome metabolic disorder. The usual mode of action of such molecules is to bind to the target proteins, and/or to the receptor and transcription factors that are involved in the expression of the target proteins. With the help of high content screening (HCS) and High content analysis (HCA), it is now possible to observe system level changes of potential molecule at an early stage of drug development. This has resulted in increased in the number of failures in the clinical trials, as normally exposure of new molecule into the system is done at the last stage of the clinical trials. Further, safety of the molecule can be assessed at every stage of the drug discovery.

Translational Genomics

Ability to utilize functional test systems (cell lines) that mimic the features involved in actual genetic investigation is called translational genomics. With the availability of most precise and flexible genome-editing solutions, it is now possible to insert, delete or substitute a known DNA sequence at any target locus in any cell line. However, for such manipulations it is essential to have accurate laboratory-based models that are capable of recapitulating various genetic events known for disease initiation and its progression in human. There exist ample opportunity for utilizing such models for identification and validation of targets, screening for repositioning of drugs and development of personalized medicines.

Through translational genomics it is possible to: i) identify persons who are susceptible to particular disease conditions, ii) find out the response of patients against a specific therapy, and iii) determine the response of any newly developed to patients on the basis of their genetic background. Application of translational genomics should eliminate treatments which are not necessary, and thereby minimize the effect of any adverse treatment. Ultimately this will lead to improved treatment, which will benefit both the drug developers and patients.

Systems Biology

The term systems biology has been coined to describe the frontier of cross-disciplinary research in biology. It basically encompasses the discipline that incorporates and analyzes data from genomics, transcriptomics, proteomics, metabolomics and other areas of biology and engineering.

Proteins usually function in complex interconnected networks under regulation and control of other proteins and metabolites. Therefore, to understand the complex cellular processes such as signal transduction pathways, metabolic pathways, and regulation of cell division, DNA replication, and gene expression a more comprehensive and integrated approach is required. As genomics and proteomics have advanced, the discipline of systems biology has emerged as a more holistic approach to analyze various cellular interactions among the molecular components of an organism. Systems biology tries to develop models, based on the information derived from genomics, proteomics and metabolomics that can be used to better understand the biological functions.

Biological systems are composed of many diverse components which are functionally different from one another. They usually interact selectively to produce coherent behaviors. These components may be individual molecules, network of interacting complexes, or sets of physical factors that are involved in the development of an organism. For example, the mitotic cell cycle of an organism depends on production of say 20 compounds, whose interactions can be approximately described with the help of 10 differential equations and about 30 kinetic parameters. The dynamic behavior of this network of interaction is possible to understand only with the help of computer simulation and dymural systems theory.

By studying relationships between all components in an organism, biologists are trying to build a "system" through which it is possible to understand how organisms function. It combines the traditional knowledge about gene and protein structure and function, with the advanced knowledge on genomics and proteomics, while developing a "system". Much of this data is retrieved from databases such as PubMed, GenBank etc. While constructing the models various interactions like protein-nucleic acid interactions, protein-protein interactions, protein-metabolite interactions are considered. The interacting components of a cell have been termed as interactome. With the development of the biochemical systems biology networks, databases on properties and processes in those biochemical networks have been developed. These include: basic metabolic pathways, human biological pathways, protein-protein interactions, and interactions between small molecules and proteins. The most important bioinformatics tools that are required should be applicable for analysis of network structures and for simulation of experimental data. One such tool includes yEd graph editor networks, and tools for visualization of data generated through biochemical networks include Cytoscape, and Pathway Tools Omics Viewer.

Although several types of models are used by system biologists, the most common one is called network map, which shows the interacting proteins, genes and other molecules involved in the "system" through a sketch. Network maps are static diagrams and do not show when and where these interactions occur. But such diagrams provide important information for development of computer simulated models to understand how signaling events occur.

Areas that are of interest and fruitful for systems biology includes: network on application of biochemical pathways in microbes, network on protein-protein interactions in microbes, network on gene regulation, and network on metabolic pathways. Several factors have contributed towards utilizing microorganisms as convenient models for systems studies. These include: unicellular in nature, a convenient boundary created by the cell membrane that delineates the "system" for genome-wide studies, generation of immense information on biochemical and metabolic pathways, development of sophisticated molecular biological techniques, possibility of culturing them in inexpensive media, possibilities to conducting ample controlled experiments, many of them are pathogens to humans, plants and domestic animals and thus have medical and environmental importance, and many of them have industrial applications.

In the past, three types of biochemical network have mostly been studied, namely network on gene regulation, metabolic network and network on protein interactions. Some of the examples of such networks include: mitotic cell cycle modeling in yeast, and modeling of specific metabolic and signal transduction pathways in microbes. Among all the biological pathways, metabolic networks have so far been the best understood networks. So far, different components and topologies of metabolic networks of only two model microbes namely, *Saccharomyces cerevisiae* and *E. coli* are available. In the case of *E. coli*, involvement of different enzymes, coenzymes, and substrates, in most of the metabolic reaction are known. However, how these three components interact in different situations, and their control mechanisms are yet to be understood. While the topologies of metabolic pathways are well understood, it is not clear how these interactions control metabolism.

In general systems biology approaches involve mathematical and computational modeling, but it will be required to overcome many challenges before such tools can be utilized in real life situations. Data repositories and their standards, simulation tools, identification of different components of the reaction and accuracy in data analysis shall contribute towards success of modeling. To overcome these problems and create a platform foe exchange of such information the Systems Biology Markup Language (SBML) was developed. The SBML project basically develops computer-readable format for the representative of biological processes. It provides a format for application of different software tools for development and exchange of biological models with high fidelity. The graphical notation to be used for different biological processes has been developed as Systems Biology Graphical Notation (SBGN). SBGN consists of three complimentary languages to describe biological processes and to establish relationships between different biological components.

Several software tools are available to assist in the analysis of biochemical networks, which are mostly freely available. One of this is called Systems biology Workbench (SBW). Through this it is possible communicate between different components for systems biology, exchange models between users via SBML. Another useful tool is CellDesigner, a Java based program which can be used for constructing and editing of biochemical networks. Through advanced versions of CellDesigner it is possible to import models via SBML and to display of biochemical networks developed SBGN language. Through CellDesigner, models can be simulated either with in-built simulator or through external simulators.

Biological systems are composed many biochemical networks which in turn are composed of various independent components. These networks are integrated into a system to make whole organism. Thus, it is possible to conceptualize many biochemical processes as complex dynamic network at molecular level. By altering the level of one component of a metabolic pathway it is possible to observe the corresponding change in the metabolic chain. Thus, it is important to know the different communication forms operating within a metabolic pathway to understand the overall behavior of the system.

Systems biology models basically tries to develop biochemical networks between genome, transcriptome, proteome, and metabolome, highlighting their complexities and mutual dependence. In future, for accurate and comprehensive measurements of the experimental results, it will be essential to develop advance and sophisticated models. Deeper understanding of diverse biochemical processes has already become possible through systems biology approaches. These include understanding of individual metabolic pathways, to signaling networks, to genome-scale metabolic networks. It is believed that systems biology shall be able to unveil many new facts of living systems in due course of time.

CONCLUSION

The DNA found within each cell contains the genetic blueprint for the entire organism. Each gene contains the information necessary to instruct the cellular machinery how to make mRNA, and in turn the protein encoded by the order of bases constituting the gene. Each one of these proteins is responsible for carrying out one or more specified molecular functions within the cell. Differing patterns of gene expression (i.e., different mRNA and protein levels) in different tissues explain differences in both cellular function and appearance.

Genomics deals with the discovery and noting of all the sequences in the entire genome of a particular organism. Once this is done, the genomic sequence is used to study the function of the numerous genes (functional genomics), to compare the genes in one organism with those of another (comparative genomics), or to generate the 3-D structure of one or more proteins from each protein family, thus offering clues to their function (structural genomics).

Large-scale DNA sequencing has stimulated the development of proteomics by providing a sequence infrastructure for protein analysis. Rapid and automated protein identification can be achieved by searching protein and nucleotide sequence databases directly.

The original central dogma of molecular biology posited that each gene encodes the information for the synthesis of a single protein. In recent years, however, it has become apparent that the primary RNA transcript of eukaryotic genes can be processed in more than one way—one gene can produce more than one protein. In one dramatic case, it is speculated that a single gene in *Drosophila* has the potential to encode more than 30,000 closely related but distinct proteins. Furthermore, once a protein is produced through a process called translation, it can be further modified by the covalent attachment of substances such as sugars, fats, phosphate groups, and other so-called post-translational modifications that affect the function that the protein performs for the cell. Together, these various possibilities constitute the proteome—the entire set of proteins made by a cell or, in the case of multicellular organisms all the cells of the body.

An application of proteomics is known as protein "expression profiling" where proteins are identified at a certain time in an organism as a result of the expression to a stimulus. Proteomics can also be used to develop a protein-network map where interaction among proteins can be determined for a particular living system. Proteomics can also be applied to map protein modification to determine the difference between a wild type and a genetically modified organism. It is also used to study protein-protein interactions involved in plant defense reactions.

Metabolomics can be used to determine differences between the levels of thousands of molecules between a healthy and diseased plant. The technology can also be used to determine the nutritional difference between traditional and genetically modified crops, and in identifying plant defense metabolites.

In crop agriculture, the main purpose of the application of genomics is to gain a better understanding of the whole genome of plants. Agronomically important genes may be identified and targeted to produce more nutritious and safe food while at the same time preserving the environment.

During last decade, advances in genomics and proteomics in combination with technical advances in molecular biology, image analysis, liquid-handling robotics, miniaturization, and computing platforms has transformed the way in which biologists approach the study of cells and even entire organisms.

REFERENCES

Adamski, J. (2012). Genome-wide association studies with metabolomics. *Genome Medicine*, *4*(4), 34–38. doi:10.1186/gm333 PMID:22546499

Avni, R., Nave, M., Barad, O., Baruch, K., Twardziok, S. O., Gundlach, I., ... Distelf.eld, A. (2017). Wild em.mer genome architecture and diversity elucidate wheat evolution and domestication. *Science*, *357*(6346), 93–97. doi:10.1126cience.aan0032 PMID:28684525

Bancroft, I., Morgan, C., Fraser, F., Higgins, J., Wells, R., Clissold, L., ... Trick, M. (2011). Dissect.ing the genome of the polyploid crop oilseed rape by transcriptome sequencing N*ature. Biotechnology (Faisalabad)*, *29*, 762–766. PMID:21804563

Barth, D., Khan, M. S., & Davis, E. (Eds.). (2015). *PlantOmics: the omics of plant science*. Springer.

Brenchley, R., Spannagl, M., Pfeifer, M., Barker, G. L., D'Amore, R., Allen, A. M., ... Hall, N. (2012). Analysi.s of bread wheat genome using whole-genome shotgun sequencing. *Nature*, *491*(7426), 705–710. doi:10.1038/nature11650 PMID:23192148

Brozynska, M., Copetti, D., Furtado, A., Wing, R. A., Crayn, D., Fox, G., Ishikawa, R., & Henry, R. J. (2017). Sequenc.ing of Australian wild rice genomes reveals ancestral relationships with domesticated rice. *Plant Biotechnology Journal*, *15*(6), 765–774. doi:10.1111/pbi.12674 PMID:27889940

Carrillo, M. G. C., Goodwin, P. H., Leach, J. F., Leung, H., & Cruz, C. M. V. (2009). Phylogenomic relationships of rice oxalate oxidases to the cupin superfamily and their association with disease resistance QTL. *Rice (New York, N.Y.)*, *2*(1), 67–79. doi:10.100712284-009-9024-0

Cen, J., Huang, Q., Gao, D., & Wang, J., Lang. Y., Liu, T., … Chen, M. (2013). Whole genome sequencing of *Oryza brachyantha* reveals mechanisms underlying *Oryza* genome evolution. *Nature Communications*, 4.

D'Hont, A., Denoeud, F., Aury, J., Baurens, F. C., Correel, F., Garsmeur, O., ... Wincker, P. (2012). The ban.ana (*Musa acuminata*) genome and the evolution of monocotyledonous plants. *Nature*, *488*(7410), 213–217. doi:10.1038/nature11241 PMID:22801500

Eldakak, M., Milad, S. I. M., Nawar, A. I., & Rohila, J. S. (2013). Proteomics: A biotechnology tool for crop improvement. *Frontiers in Plant Science*, *4*, 35–42. doi:10.3389/fpls.2013.00035 PMID:23450788

Goff, S. A., Ricke, D., Lan, T. H., Presting, G., Wang, R., Molly Dunn, M., ... Briggs, S. (2002). A draft. sequence of the rice genome (*Oryza sativa* L ssp jap.onic.a). *Science*, *296*(5565), 92–100. doi:10.1126cience.1068275 PMID:11935018

Gore, M. A., Chia, J.-M., Elshire, R. J., Sun, Q., Ersoz, E. S., Hurwitz, B. L., Peiffer, J. A., McMullen, M. D., Grills, G. S., Ross-Ibarra, J., Ware, D. H., & Buckler, E. S. (2009). A first.-generation haplotype map of maize. *Science*, *326*(5956), 1115–1117. doi:10.1126cience.1177837 PMID:19965431

Gupta, S., Nawaz, K., Parween, S., Roy, R., Shahu, K., & Pole, A.K., … Chottopadhyay. (2017). Draft genome sequence of *Cicer reticulatum* L., the wild progenitor of chickpea provides a resource for agronomic trait improvement. *DNA Research*, *24*, 1–10. PMID:27567261

International Wheat Genome Sequencing Consortium (IWGSC). (2014). A chromosome-based draft sequence of the hexaploid bread wheat (*Triticum aestivum*) genome. *Science*, *345*, 1126–1131. PMID:25035500

Jiao, Y., Peluso, P., Shi, J., Liang, T., Stitzer, M. C., Wang, B., ... Ware, D. (2017). Improved maize reference genome with single-molecule technologies. *Nature*, *546*(7659), 524–527. doi:10.1038/nature22971 PMID:28605751

Kawahara, Y., de la Bastide, M., Hamilton, J. P., Kanamori, H., McCombie, W. R., Ouyang, S., ... Matsumo.to, T. (2013). Improve.ment of the *Oryza sativa* Nipponbare reference genome using next generation sequence and optical map data R*ice, N. Y (Dayton, Ohio)*, *6*, 4–6. PMID:24280374

Komatsu, S., Mock, H. P., Yang, P., & Svensson, B. (2013). Application of proteomics for improving crop protection/artificial regulation. *Frontiers in Plant Science*, *4*, 522–532. doi:10.3389/fpls.2013.00522 PMID:24391656

Kusha, B. (1998). Beer, bethesda, and biology: How "genomics" came into being. *Journal of the National Cancer Institute*, *90*, 91. PMID:9450566

Ling, H., Zhao, S., Liu, D., Wang, J., Sun, H., Zhang, C., Fan, H., Li, D., Dong, L., Tao, Y., Gao, C., Wu, H., Li, Y., Cui, Y., Guo, X., Zheng, S., Wang, B., Yu, K., Liang, Q., ... Wang, J. (2013). Draft g.enome of the wheat A-genome progenitor *Triticum urartu*. *Nature*, *496*(7443), 87–90. doi:10.1038/nature11997 PMID:23535596

Lipman, D. J., & Pearson, W. R. (1985). Rapid and sensitive protein similarity searches. *Science*, *227*(4693), 1435–1441. doi:10.1126cience.2983426 PMID:2983426

Liu, S., Liu, Y., Yang, X., Tong, C., Edwards, D., Parkin, I. A. P., ... Paterso.n, A. (2014). The B*ra.ssica oleracea* genome reveals the asymmetrical evolution of polyploid genomes N*ature. Communication*, *5*, 3930–3932. PMID:24852848

Lu, K., Wei, L., Li, X., Wang, Y., Wu, J., Liu, M., Zhang, C., Chen, Z., Xiao, Z., Jian, H., Cheng, F., Zhang, K., Du, H., Cheng, X., Qu, C., Qian, W., Liu, L., Wang, R., Zou, Q., ... Li, J. (2019). Whole-genome resequencing reveals *Brassica napus* origin and genetic loci involved in its improvement. *Nature Communications*, *10*(1), 1154–1157. doi:10.103841467-019-09134-9 PMID:30858362

Merchant, S. S., Prochnik, S. E., Vallon, O., Harris, E. H., Karpowicz, S. J., Witman, G. B., ... Grossman, A. R. (2007). The *Chlamydomonas* genome reveals the evolution of key animal and plant functions. *Science*, *318*(5848), 245–250. doi:10.1126cience.1143609 PMID:17932292

Needleman, S. B., & Wunsch, C. D. (1970). A general method applicable to the search for similarities in the amino acid sequence of two proteins. *Journal of Molecular Biology*, *48*(3), 443–453. doi:10.1016/0022-2836(70)90057-4 PMID:5420325

Paterson, A., Bowers, J., Bruggmann, R., Dubchak, J., Grimwood, J., Gundlach, H., Haberer, G., Hellsten, U., Mitros, T., Poliakov, A., Schmutz, J., Spannagl, M., Tang, H., Wang, X., Wicker, T., Bharti, A. K., Chapman, J., Feltus, F. A., Gowik, U., ... Rokhsar, D. S. (2009). The S*or.ghum bicolor* genome and the diversification of grasses. *Nature*, *457*(7229), 551–556. doi:10.1038/nature07723 PMID:19189423

Schatz, M. C., Maron, L. G., Stein, J. C., Hernandez, W. A., Gurtowski, J., Biggers, E., ... McCombi.e, W. R. (2014). Whole g.enome de novo assemblies of three divergent strains of rice, *Oryza sativa,* document novel gene space of *aus* and *indica Genome. Biology (Basel)*, *15*, 506–510. PMID:25468217

Schmutz, J., Cannon, S., Schlueter, J., Ma, J., Mitros, T., Nelson, W., ... Jackson, S. A. (2010). G.enome. sequence of the palaeopolyploid soybean. *Nature*, *463*(7278), 178–183. doi:10.1038/nature08670 PMID:20075913

Schnable, P. S., Ware, D. D., Fulton, R., Stein, J., Wei, F. F., Pasternak, S. S., ... Wilson, R. K. (2009). The B73. Maize Genome. *Complexity, Diversity, and Dynamics Science.*, *326*, 1112–1115. PMID:19965430

Smith, T. F., & Waterman, M. S. (1981). Identification of common molecular subsequences. *Journal of Molecular Biology*, *147*(1), 195–197. doi:10.1016/0022-2836(81)90087-5 PMID:7265238

Stein, J. C., Yu, Y., Copetti, D., Zwickl, D. J., Zhang, L., Zhang, C., ... Wing, R. A. (2018). gGenomes of 13 domesticated and wild rice relatives highlight genetic conservation, turnover and innovation across the genus *Oryza. Nature Genetics, 50*(2), 285–296. doi:10.103841588-018-0040-0 PMID:29358651

Tufarelli, C., Hardison, R., Miller, W., Hughes, J., Clark, K., Ventress, N., ... Higgs, D. R. (2004). Compara.tive Analysis of the alpha-like globin clusters in mouse, rat, and human chromosomes indicates a mechanism underlying breaks in conserved synteny cristina G*enome. Research, 14*, 623–630. PMID:15060003

Varshney, R., Chen, W., Li, Y., Bharti, A. K., Saxena, R. K., Schlueter, J. A., ... Jackson, S. A. (2012). Draft g.enome sequence of pigeonpea (*Cajanus cajan*), an orphan legume crop of resource-poor farmers N*ature. Biotechnology (Faisalabad), 30*, 83–89.

Varshney, R., Song, C., Saxena, R., Azam, Y., Wu, S., Sharpe, A. G., ... Cook, D. R. (2013). Draft g.enome sequence of chickpea (*Cicer arietinum*) provides a resource for trait improvement N*ature. Biotechnology (Faisalabad), 31*, 240–246. PMID:23354103

Wang, K., Wang, Z., Li, F., Ye, W., Wang, J., Song, G., ... Yu, S. (2012). The dra.ft genome of a diploid cotton *Gossypium raimondii Nature. Genetics, 44*, 1098–1103. PMID:22922876

Wang, M., Yu, Y., Haberer, G., Marri, P. R., Fan, C., Goicoechea, J. L., ... Wing, R. A. (2014). The gen.ome sequence of African rice (*Oryza glaberrima*) and evidence for independent domestication. *Nature Genetics, 46*(9), 982–988. doi:10.1038/ng.3044 PMID:25064006

Wen, W., Li, D., Li, X., Gao, Y., Li, W., Li, H., Liu, J., Liu, H., Chen, W., Luo, J., & Yan, J. (2014). Metabolome-based genome-wide association study of maize kernel leads to novel biochemical insights. *Nature Communications, 5*(1), 3438–3445. doi:10.1038/ncomms4438 PMID:24633423

Wu, Z., Fang, D., Yang, R., Gao, F., An, X., Zhao, X., ... Luo, Q. (2018). De novo genome assembly of *Oryza granulate* reveals rapid genome expansion and adaptive evolution. *Communications Biology, 1*(1), 84–92. doi:10.103842003-018-0089-4 PMID:30271965

Xu, Q., Chen, L., Ruan, X., Chen, D., Zhu, A., Chen, C., ... Ruan, Y. (2013). The dra.ft genome of sweet orange (*Citrus sinensis*). *Nature Genetics, 45*(1), 59–66. doi:10.1038/ng.2472 PMID:23179022

Yu, J., Hu, S., Wang, J., Wong, G., Li, S. Z., ... Yang, H. (2002). A draft. sequence of the rice genome (*Oryza sativa* L ssp ind.*ica). Science, 296*(5565), 79–92. doi:10.1126cience.1068037 PMID:11935017

Zhang, Q. J., Zhu, T., Xia, E. H., Shi, C., Liu, Y. L., Zhang, Y., ... Gao, L. Z. (2014). Rapid d.iversification of five *Oryza* AA genome associated with rice adaptation. *Proceedings of the National Academy of Sciences of the United States of America, 111*(46), 4954–4962. doi:10.1073/pnas.1418307111

Zimin, A. V., Puiu, D., Hall, R., Kingan, S., & Salzberg, S. L. (2017). The first near complete assembly of the hexaploid bread wheat genome, *Triticum aestivum. bioRxiv, 6*(11). Advance online publication. doi:10.1093/gigascience/gix097 PMID:29069494

ADDITIONAL READING

Adam, J. (2008). Transcriptome: Connecting the genome to gene function. *Nature Education, 1*, 1–4.

Alvarez, S., Marsh, E. L., Schroeder, S. G., & Schachtman, D. P. (2008). Metabolomic and proteomic changes in the xylem sap of maize under drought. *Plant, Cell & Environment, 31*(3), 325–340. doi:10.1111/j.1365-3040.2007.01770.x PMID:18088330

Ansong, C., Purvine, S. O., Adkins, J. N., Lipton, M. S., & Smith, R. D. (2008). Proteogenomics: Needs and roles to be filled by proteomics in genome annotation. *Functional Genomics Proteomics, 7*(1), 50–62. doi:10.1093/bfgp/eln010 PMID:18334489

Arif, I. A., Bafee, S. O., Alfarhan, A. H., Ahamed, A., Thomas, J., & Bakir, M. A. (2014). Nucleotide based validation of the endangered plant *Diospyros mespiliformis* (Ebenaceae) by evaluating short sequence region of plastid *rbc*L gene. *Plant Omics Journal, 7*, 102–107.

Baginsky, S., Henning, L., Zimmermann, P., & Gruissem, W. (2010). Gene expression analysis, proteomics, and network discovery. *Plant Physiology, 152*(2), 402–410. doi:10.1104/pp.109.150433 PMID:20018595

Baxevanis, A., & Ouellette, F. (2001). *Bioinformatics: a practical guide to the analysis of genes and proteins.* John Willey & Sons. Inc. doi:10.1002/0471223921

Bernardo, R., & Charcosset, A. (2006). Usefulness of gene information in marker-assisted recurrent selection: A simulation appraisal. *Crop Science*, *46*(2), 614–621. doi:10.2135/cropsci2005.05-0088

Bohme, K., Calo-Mata, P., Barros-Velazquez, J., & Ortea, I. (2019). Recent applications of omics-based technologies to main topics in food authentication. *Trends in Analytical Chemistry*, *110*, 221–232. doi:10.1016/j.trac.2018.11.005

Bohra, A. (2013). Emerging paradigms in genomics-based crop improvement. *TheScientificWorldJournal*, *17*, 584567–584577. PMID:24348171

Bohra, A., Sahrawat, K. L., Kumar, S., Joshi, R., Parihar, A. K., Singh, U., Singh, D., & Singh, N. (2015). Genetics and genomics-based interventions for nutritional enhancement of gain legume crops: Status and outlook. *Journal of Applied Genetics*, *56*(2), 151–161. doi:10.100713353-014-0268-z PMID:25592547

Chen, F., Dong, W., Zhang, J., Guo, X., Chen, J., Wang, Z., ... Zhang, L. (2018). The sequenced angiosperm genomes and genome databases. *Plant Science*, *9*, 418–432. doi:10.3389/fpls.2018.00418 PMID:29706973

Church, G. M. (2006). Genomes for all. *Scientific American*, *294*(1), 47–54. doi:10.1038cientificamerican0106-46 PMID:16468433

Davey, P. A., Pernice, M., Sablok, G., Larkum, A., Lee, H. T., Golicz, A., Edwards, D., Dolferus, R., & Ralph, P. (2016). The emergence of molecular profiling and omics techniques in seagrass biology; furthering our understanding of seagrasses. *Functional & Integrative Genomics*, *16*(5), 465–480. doi:10.100710142-016-0501-4 PMID:27443314

Fernie, A. R., & Schauer, N. (2008). Metabolomics-assisted breeding: A viable option for crop improvement? *Trends in Genetics*, *25*(1), 39–48. doi:10.1016/j.tig.2008.10.010 PMID:19027981

Ferri, E., Galimberti, A., Casiraghi, M., Airoldi, C., Ciaramelli, C., Palmioli, A., ... Libra, M. (2015). Towards a universal approach based on omics technologies for the quality control of food. *Journal of Biomedicine & Biotechnology*, *7*, 1–14. PMID:26783518

Ford, K. L., Cassin, A., & Bacic, A. (2011). Quantitative proteomic analysis of wheat cultivars with different draught stress tolerance. *Frontiers in Plant Science*, *2*, 44–51. doi:10.3389/fpls.2011.00044 PMID:22639595

Hossain, Z., & Komatsu, S. (2014). Potentiality of soybean proteomics in untyping the mechanism of food and drought stress tolerance. *Proteomes*, *2*(1), 107–127. doi:10.3390/proteomes2010107 PMID:28250373

Ingvarsson, P. K., & Street, N. R. (2011). Association genetics of complex traits in plants. *The New Phytologist*, *189*(4), 909–922. doi:10.1111/j.1469-8137.2010.03593.x PMID:21182529

Jorrin, J. V., Maldonado, A. M., & Castillejo, M. A. (2007). Plant proteome analysis: A 2006 update. *Proteomics*, *7*(16), 2947–2962. doi:10.1002/pmic.200700135 PMID:17654459

Koller, A., Washburn, M. P., Lange, B. M., Andon, N. L., Deciu, C., Haynes, P. A., Hays, L., Schieltz, D., Ulaszek, R., Wei, J., Wolters, D., & Yates, J. R. (2002). Proteome survey of metabolic pathway in rice. *Proceedings of the National Academy of Sciences of the United States of America*, *99*(18), 11969–11974. doi:10.1073/pnas.172183199 PMID:12163647

Kosova, K., Vitamvas, P., & Prasil, I. T. (2014). Proteomics of stress response in wheat and barley: Each for potential protein markers of stress tolerance. *Frontiers in Plant Science*, *5*, 711–718. doi:10.3389/fpls.2014.00711 PMID:25566285

Kumar, P., Gupta, V. K., Misra, A. K., Modi, D. R., & Pandey, B. K. (2009). Potential of molecular markers in plant biotechnology. *Plant Omics Journal*, *2*, 141–162.

Minh-Thu, P. T., Hwang, D. J., Jeon, J. S., Nahm, B. H., & Kim, Y. K. (2013). Transcriptome analysis of leaf and root of rice seedling to acute dehydration. *Rice (New York, N.Y.)*, *6*(1), 38–44. doi:10.1186/1939-8433-6-38 PMID:24341907

Rafalski, J. A. (2010). Association genetics in crop improvement. *Current Opinion in Plant Biology*, *13*(2), 174–180. doi:10.1016/j.pbi.2009.12.004 PMID:20089441

Rahman, M., Shaheen, T., Rahman, M., Iqbal, M. A., & Zafar, Y. (2016). *Bioinformatics: a way to explore "Plant Omics". Bioinformatics – Updated Features and Applications*. Retrieved from., doi:10.5772/64043

Ramsden, J. J. (2009). *Bioinformatics: an introduction*. Springer. doi:10.1007/978-1-84800-257-9

Rios, R. O. (2015). *Plant breeding in omics era*. Springer. doi:10.1007/978-3-319-20532-8

Singh, B., Bohra, A., Mishra, S., Joshi, R., & Pandey, S. (2015). Embracing new-generation 'omics' tools to improve draught tolerance in cereal and food-legume crops. *Biologia Plantarum*, *59*(3), 413–428. doi:10.100710535-015-0515-0

Tyagi, S., Singh, R. U., Kalra, T., & Munjil, K. (2010). Applications of metabolomics- a systematic study of the unique chemical fingerprints: An overview. *International Journal of Pharmaceutical Sciences Review and Research*, *3*, 83–86.

Vassilev, D., Leunissen, J., Atanassov, A., Nenov, A., & Dimov, G. (2005). Application of bioinformatics in plant breeding. *Biotechnology, Biotechnological Equipment*, *19*(sup3), 139–152. doi:10.1080/13102818.2005.10817293

APPENDIX

1. What are the differences between functional genomics and comparative genomics?
2. What features of bacterial genomes are similar to eukaryotic genomes?
3. What are the possible practical applications of large genomic sequence data?
4. What are the basic features of BLAST? Explain with the help of an example how BLAST can be used to generate useful genetic information.
5. What is DNA microarray? Describe the technique of analyzing through DNA microarray.
6. With the help of microarray analysis, how it is possible to demonstrate that although all cells of an organism have the same genome, some genes show cell- and tissue specific expression?
7. Based on the sequence data, how can you predict that the organization of genetic material is more complex in eukaryotes than prokaryotes?
8. It has been discovered that in humans of all races and nationalities approximately 99.9 percent of the sequences are the same. But different individuals can still be identified by DNA fingerprinting techniques. Explain how human genomes from different individuals can be distinguished?
9. For identification of genes and regulatory sequences annotation of the genome is required. What are the characteristics of a genome that are hallmark for identifying genes in an unknown sequence in eukaryotes and prokaryotes?
10. What is metagenomics? What genetic information can be obtained through metagenomics?
11. Describe the different BLAST search machines available? What are the other important search machines used in bioinformatics?
12. Proteomics analysis indicates that human cells are capable of synthesizing more than 300,000 different proteins. Whereas annotation of human genome sequence revealed that human genome has about 30,000 protein-coding genes. Explain why there exist such discrepancies and how it can be resolved?
13. How the minimum number of genes required for life has been determined? Is there any possibility of reducing in this minimum number of genes? Explain.

14. How it is possible to determine whether a genomic sequence of DNA contains a protein-coding gene?
15. How with the help of proteomics it is possible to identify differences between the number of protein-coding genes predicted for a genome and the number of proteins expressed by a genome?
16. Explain how sequence databases are contributing towards drug designing?
17. What are isogenic cell lines of human? How isogenic cell lines are being used for personalized drug development and delivery?
18. What is the difference between metabolomics and metablonomics? How metabolomics and metabonomics are being increasingly used to understand biochemical pathways in living systems?
19. Describe how completion of human genome project (HGP) has revolutionized the science of bioinformatics.
20. What do you understand by the term "systems biology"? Describe the applications of systems biology.

Chapter 12
Molecular Biology Techniques

ABSTRACT

The development of vast array of laboratory methods and their applications provided great leaps in the ability of the researchers to discover new features and functions of macro-molecules. Most of them represent procedures for measuring or visualizing ever-smaller quantities or tinier features of molecules, or part of molecules. Especially when applied in combination, these methods have led to enormous advances in understanding the structural features of proteins and nucleic acids. New techniques have been regularly introduced and the sensitivity of older techniques greatly improved upon. The originators of several of those breakthrough methods were awarded Nobel Prizes. Basic principles of some of most important techniques invented and applied in molecular biology research are described in this chapter.

INTRODUCTION

Recent advances in vast array of laboratory methods and their applications provided great leaps in research scientist's ability to discover new features and functions of macro-molecules. Most of them represent procedures for measuring or visualizing ever-smaller quantities or tinier features of molecules, or part of molecules. When applied in combinations, the application of these methods has led to enormous advances in understanding the structural features of proteins and nucleic acids. Newer techniques have been regularly introduced and the sensitivity of older techniques greatly improved upon.

DOI: 10.4018/978-1-7998-4312-2.ch012

FLUORESCENCE IN SITU HYBRIDIZATION (FISH)

Fluorescence *in situ* hybridization (FISH) is a technique used to analyze chromosomes at the gene or DNA level. Both metaphase (dividing) and interphase (no-dividing) cells are used to identify structural abnormalities in the chromosomes through FISH. FISH is usually applied to cytogenetic preparations on microscopic slides, but it can be used on slides of formalin-fixed tissue, blood or bone marrow smears, and directly fixed cells or other nuclear isolates. The basic principle on which FISH works is that a dsDNA molecule can be denatured to form ssDNA, and a complimentary strand (other than the original complimentary strand) can bind to the ssDNA thus generated. Thus, a specific DNA sequence in the metaphase or interphase chromosomes can be identified by using a DNA probe having complimentary sequence. However both the chromosomal DNA sequence and the probe should be in single-stranded conformation, which can be achieved by heating them in a solution containing formamide. The advantages of FISH over ISH (*in situ* hybridization) are faster detection, higher resolution, sensitivity and speed.

The target DNA sequence and the probe get hybridized to form dsDNA under ideal conditions. An excess of the repetitive sequence DNA is added to the hybridization mixture to avoid non-specific binding. Depending on the nature of the probe and target DNA, hybridization process should be completed in about 2-18 hr at 37^0 C. Thereafter the excess non-bound probe is removed by washing the slides with formamide-saline citrate solutions. The probe should be previously labeled directly with a fluorescent tag so as to locate them on the target DNA. The probes can also be labeled indirectly first by joining with a haptan molecules (biotin or digoxigenin) and then binding the haptan with a fluorescent tag. The target DNA is counterstained with another fluorochrome of a complimentary color.

With the help of a fluorescent microscopy and specific filters it is possible to observe the flourochrome labeled probe bound to the target DNA. With the help of special filters it is possible to observe several flourochromes simultaneously. To increase the sensitivity of detection of the probes, digital cameras capable of detecting low light intensity along with computer imaging software are used. The FISH preparations tend to fade over time (photobleaching) and therefore should be stored in the dark. However it is possible to improve the longevity and documentation of the FISH preparations by application of anti-fad solution like phenylenediamine.

When total genomic DNA (consisting of the entire nuclear DNA of a species) is used as a probe in hybridization experiments to chromosomal DNA *in situ*, the technique is called *GISH* (genomic *in situ* hybridization). GISH permits characterization of the genome and chromosomes of hybrid plants, allopolyploid species, recombinant breeding lines, and phylogenetic relationship. Multicolor FISH (mFISH) using genomic DNA probes are a promising approach for simultaneously discriminating each genome in natural and artificial amphidiploid.

The normal FISH approaches are based on air-drying procedure of chromosome preparation, which leads to well spread metaphase chromosome preparations. But when this procedure is adopted to study spherical interphase nuclei, the nuclei becomes flattened, and thus may lead to questionable results. To overcome this problem, a procedure called suspension FISH (sFISH) was developed, which allows 3D analysis. In this technique suspended cells are placed in polished concave slides as the final step of the procedure, just before evaluation.

Three different types of probes used in FISH studies are as follows:

1. Probe that bind to specific chromosome structures typically recognize repetitive DNA sequence, such as within centromeres (α satellite DNA) or telomeric sequences. Within the repeated sequence, the nucleotide probes patterns are unique for specific chromosomes. Specific centromere probes are now available that can identify most of the individual chromosome homologues. Similarly, there are probes that specifically recognize the telomeres (chromosome ends) of the long and short arms of many of the homologues, and more are rapidly being developed. Most alpha satellite centromeric probes give a large, bright signal and are useful for both chromosome identification in metaphase preparations and chromosome enumeration in interphase nuclei. Chromosome-specific telomere probes can be used for the above, for detection of cryptic translocations, and to define interstitial and terminal deletions.
2. Unique sequence probes hybridize to single copy DNA sequences in a specific chromosomal region or gene. In clinical cytogenetics these probes are usually referred to as cosmids, named for their cloning vector. These are the probes used to identify the chromosomal critical region or gene associated with microdeletion syndromes. On metaphase chromosomes, they hybridize to each chromatid, usually giving two small, discrete signals per chromosome.

3. Whole chromosome paints are cocktails of unique sequence probes that recognize the unique sequences spanning the length of a particular chromosome. At metaphase, both chromosome homologues are 'painted' or fluoresce brightly. Among other applications, paint probes are used to define the chromosomal origin of derivative segments on translocation chromosomes or supernumerary markers.

FISH for RNA

For detection and quantification of long RNA molecules including mRNAs, single molecular RNA FISH or Stellaris FISH method is used. The target molecule is probed by multiple short singly labeled oligonucleotides. It is possible to detect and localize RNA molecules containing at least 48 oligos (fluorescent labeled) to a single molecule of mRNA, as they provide sufficient fluorescence. The unbound probes remain scattered and therefore cannot produce sufficient fluorescence to produce background effect. The method has potential applications in gene expression, diagnosis of cancer, neuroscience, and companion diagnostics.

Automation of FISH

With the help of microfluidic chips, the interphase FISH procedure has been automated and used as a diagnostic tool for detection of chromosomal abnormalities on cell by cell basis. Automation of FISH technique has not only reduced the time required (only few minutes compared to several hours to days required in conventional FISH) for setting-up of the complex procedure but also become cost effective. Since microchannels permit sophisticated level of fluid control (up to picolitres), these devices can reduce analysis time, lower reagent consumption, minimize human intervention, and provide reproducible results with accuracy. Compared to conventional FISH methods, microfluidic chip method could be 10-20 times cost effective and about 10-fold higher throughput, which enables simultaneous assessment of several chromosomal abnormalities or patients.

Metaphase FISH had continued to be difficult to integrate with the microfluidic chips, owing to the complex sample preparation protocol. However, with the development of novel lab on chip device, it has been possible to integrate the entire sample preparation protocol for metaphase FISH called FISHprep.

Quantitative measurement of intensity of fluorescence can be made by combining FISH with PANs and computer software, which is known as Q-FISH. Usually this method is used to quantify fluorescence intensity in the analysis of telomeric regions. The Flow-FISH technology uses flow cytometry to perform FISH automatically through measurement of per-cell fluorescence.

Fluorescent *In Situ* RNA Sequencing (FISSEQ)

Knowledge about the location of functional gene product, called mRNA, within the living tissue often helps to understand how cells and tissue grow and develop. But it is difficult to analyze several mRNAs simultaneously, as the tissue containing cells have to be crushed as pulp, thereby making it impossible to place any mRNAs it its original site of operation. To overcome this problem a new technique called fluorescent in situ RNA sequencing (FISSEQ) was developed by Lee (2015). Through this technique it is possible to locate the exact position of thousands of mRNAs within intact cells. It is also possible to simultaneously determine the sequence of the bases and understand their functions.

The method combines the RNA FISH and RNA sequencing. The sequencing of RNA is done through multiple single-base-extensions. Addition of bases in serial sets (e.g. C A T G C A T G …) is interrupted by scanning (for acquisition of data) and then the slides are treated chemically to remove signal, before extending further.

The expressed mRNA transcripts should be localized *in situ* within the cell for identification through single-nucleotide resolution technique. Stable cross-linkage is established with cDNA amplicons through FISSQE for sequencing.

At any given time, about half of the human genes, of the nearly 30,000 genes present, are expressed by its cells, to carryout various functions. Cells can also adjust gene expression process to produce anywhere from few to several thousand copies. Simultaneous pinpointing of the cellular locations of all those mRNAs is extremely difficult. Using 30-base reads from 8742 genes *in situ*, it has been possible to carry out RNA expression and localization in human primary fibroblast with a simulated wound-healing assay.

FISSQE can be carried out on whole mount embryos and tissue sections and problems of reduced optical resolution and noisy signal associated with single-molecule detection technique can be considerably reduced. The technique has the potential to use for large scale detection of genetic elements simultaneously. This include transcription of genes, DNA barcodes, and also to investigate cellular phenotype, gene regulation *in situ*.

The method has the potential to early detection of cancer by identifying molecular changes that are linked to the development and proliferation of cancer cells. It can also be used to detect cancer mutations and how they respond to new therapies. It also has the potential to use in developmental biology to reveal the changes during embryonic development and to map the size of neurons present in the human brain.

FLOW CYTOMETRY

Flow cytometry (FC) is a powerful technique for analyzing multiple parameters of individual cells within heterogeneous populations. FC is used for a range of applications such as; cell counting, immunophenotyping, ploidy analysis and GFP expression analysis. FC performs this by analyzing thousands of cells per second, as they move in a liquid stream through laser beam and capturing the light that emerges from each cell as it passes through. These characteristics are determined with the help of an optical-electronic coupling system which can records the incident laser light or fluorescence scattered by the cells or particles. The data gathered can be analyzed statistically by flow cytometry software to find out cellular characteristics such as: relative size, internal complexity, relative fluorescence intensity, phenotype and health. The basic characteristics components of FC are as follows (Figure 1).

1. The fluidics system – presents the sample into the point of interrogation and removes the waste,
2. The lasers – the source of the light for scattering and fluorescence,
3. The optics – gathers and detects the light,
4. The detectors – receives the light, and
5. The electronic and peripheral computer system – converts analogue signals to digital signal and analyze the data.

In the flow cytometer, particles or cells are carried to the laser intercept, one at a time, in a liquid stream. Suspended cells or particles having diameter ranging from 0.2–150 μm can be analyzed through this system. Therefore, cells or particles of bigger aggregates should be broken down to smaller size before analysis. The particles are located in the portion of the fluid stream called sample core. Scattered laser light beams are produced by the particles when passed through the laser intercept. The particles having fluorescence property shall fluoresce. With the help of appropriately placed lenses the scattered and fluorescent lights can be collected. The scattered light and fluorescent beams are guided to the appropriate detectors with the help of a combination of beam splitters and filters. Electronic signals are produced by the detectors which are proportional to the strength of the optical signals produced. For each event or particle the list mode data are recorded. Based on the strength of scattering light and fluorescent properties the characteristics or parameters of each event are prepared. The collected data are stored in the computer. Data thus generated can be analyzed to obtain information on subpopulations which may be present within the sample.

Among other applications, flow cytometry is used for DNA analysis. The DNA content in the nucleus of the cells can be measured to know about their ploidy level. There exist a direct relationship between the DNA content and the ploidy level of a cell. Such measurements are useful in tumour analysis, particularly to know the distribution of aberrant cells in a tissue. The somatic cells contain diploid number of chromosomes (2n) whereas the germ cell (gamete) contains haploid number (n). In most of the tumour, the chromosome number in the cells is greater than 2n (hyperdiploid), although sometimes chromosome number may be less (hypodiploid). Such aberrant cells are called aneuploids and shall have either more or less DNA content than the diploid cells. When change in the chromosome number (aneuploidy) is measured as change in the DNA content, it is called DNA aneuploidy. The ratio between the amount of DNA present in a tumour cell and that of a normal diploid cell is called DNA index (DI), which is used to reflect the status of the tumour development.

For measurement of the DNA content, a fluorescent dye is used to stain the chromosomes. In aqueous solution most of these dyes fluoresce weakly, but once they bind to the DNA they fluoresce strongly due to the hydrophobic nature in the surrounding environment. Propidium iodide (PI) is the most commonly used dye, which fluoresce red colour at 488 nm wave length. Before treating with PI the cells should be fixed with the help of a fixative (70 per cent ethanol) to make the plasma membrane permeable for the dye (PI) and also treated with RNase to remove all RNA molecules. Alternatively cells can be enucleated by treating with a detergent.

Ultraviolet or violet laser can be used for the measurement of DNA. High quality DNA histograms can be generated if the cells can be permeabilized. DNA histograms can also be generated without permeabilization by treatment with either DRAQ5 or Hoechst 33342. Knowledge about the correct time

Figure 1. Flow cytometers use the principle of hydrodynamic focusing for presenting cells to a laser (or any other light excitation source). The sample is injected into the center of a sheath flow. The combined flow is reduced in diameter, forcing the cell into the center of the stream. Thus the laser is passed through one cell at a time. Scattered and emitted light signals are converted to electronic pulses that can be processed by the computer.

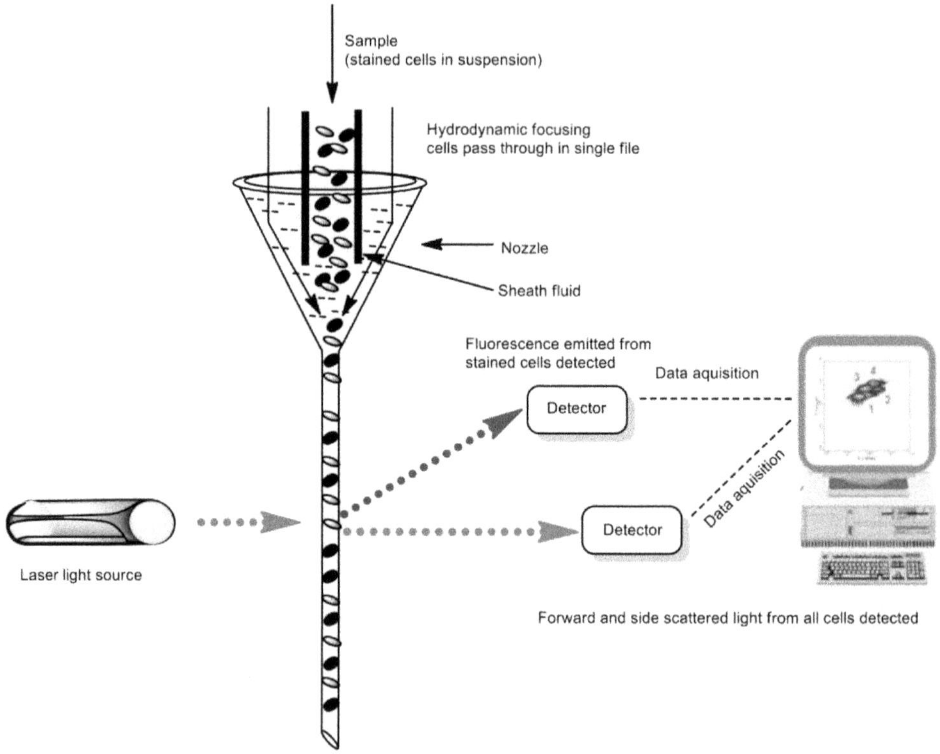

of incubation and concentration of the dye to be used is essential to obtain satisfactory DNA histogram for each cell type.

Nuclei extracted from processed clinical specimens (*i.e.* formalin-fixed, paraffin embedded) can also be studied. In this technique, nuclei are extracted from 50 μm thick sections (paraffin removed) of the histological blocks, by treatment with pepsin. The quality of the DNA histograms can be improved by incubating the specimen at 80°C before treating with pepsin.

ISOLATION OF PROTEINS

While general methods for isolation and purification of proteins are applicable to all organisms, it is invariably necessary to develop unique strategies for isolation of the target protein of interest. Unlike research with DNA, no manuals or standard protocols or "recipes" are available, outlining a stepwise approach applicable to all proteins. Furthermore, there are no organism-specific procedures that can allow one to plan a course of action with a predictable outcome. The design of an appropriate procedure for isolation of a given protein should be tailored in accordance with the objective(s) of the research project, which may require relatively pure product in modest amounts for analytical purposes (e.g. enzyme kinetics) or a highly purified, homogeneous preparation for physicochemical or structural studies. Isolation and purification of a single protein from cells containing a mixture of thousands of unrelated proteins is achievable because of the remarkable variation in the physical and chemical attributes of proteins. Characteristics unique to each protein—amino acid composition, sequence, subunit structure, size, shape, net charge, isoelectric point, solubility, heat-stability, hydrophobicity, ligand/metal binding properties and post-translational modifications—can be exploited in formulation of a strategy for purification. Based on these properties a combination of various methods can be used for separation of cellular proteins.

In general, protein purification entails essentially five types of steps: 1) efficient extraction from biological material, 2) separation from non-protein components (nucleic acids and lipids), 3) precipitation steps, initially to recover the bulk protein from a crude extract, followed by preliminary resolution into manageable fractions, 4) use of ion-exchange chromatography/size fractionation or hydrophobic chromatography columns to further separate the target protein-containing fraction from the bulk protein, and 5) a more refined set of steps including an "affinity" matrix to enable recovery of the target protein in a highly purified state along with a high yield. A variety of agarose-based matrices with immobilized reactive dyes, covalently bound nucleotides, metals and numerous other ligands are commercially available.

In order to evaluate the progress of purification, a convenient assay procedure based on enzymatic activity or some other easily monitored property specific to the protein are available. A spectrophotometric or colorimetric method for enzymatic activity measurement is most convenient and a progressive increase in specific activity (for enzymes, activity in units /mg protein) is an excellent indicator of the efficacy of the purification step. For proteins lacking a readily measurable biological activity, it may be feasible to use an immunochemical procedure such as western blotting or ELISA (Enzyme-Linked-Immunosorbent Assay), provided suitable antibodies are available. In this case, electrophoretic resolution of the protein population in samples at each stage of purification will be required. Detailed procedures for isolation of proteins from various sources are beyond the scope of this book.

ELECTROPHORESIS OF PROTEINS

Electrophoresis is one of the most popular and efficient method used for studying macromolecules like proteins and nucleic acids. The method basically takes advantage of the different electrical charges being carried by the macromolecules. When exposed to an electric field the positively charged molecules will move towards the negative pole (cathode) and the negatively charged molecules towards the anode (positive pole). Depending upon the characteristics of the macromolecules the method of electrophoresis may vary. Some of the most important electrophoresis techniques used in molecular biology is described.

Protein and nucleic acids are charged molecules. Consequently, when an electric field is applied to such charged molecules dissolve in a liquid, they will move. In simple liquid containing water and salts, all the molecules will move at nearly same speed. Under such conditions, it will not be possible to distinguish molecules from one another. However, instead of a solution, a gel having pores is used, then different molecules will travel at different speeds through the pores. Usually smaller molecules move faster than large ones. Thus the molecules can be separated according to their size. When the molecules of similar size accumulate at a site in the gel, they will form bands at different locations within the gel. Each of these bands contains molecules of a specific size.

Polyacrylamide Gel Electrophoresis (PAGE)

This is one of the earliest techniques used for separation of protein molecules. In this technique, proteins are applied to a porous polyacrylamide gel and are exposed to an electric field. When electrical current is applied, the molecules will separate on the basis of their net charges and size. The polyacrylamide gel will act as the molecular sieve and allow the molecules to move through the pores. Small molecules will move faster than the large molecules.

Polyacrylamide is chemically inert and the gel is formed by polymerization of acrylamide. It is possible to regulate the pore size of the gel by mixing appropriate quantity of methylene bisacrylamide, a crosslinking agent. The higher the concentration of acrylamide, the smaller will be the pore size. The gel is usually casted between two glass plates of 7-20 cm^2 placed with a gap of 0.5-1.0 mm. The protein samples are poured into the wells, at the top of the gel. The wells are created by placing a plastic or Teflon comb in the gel before it sets (Figure 2). A dye called bromophenol blue should be mixed with the protein samples to trace the movement of the molecules during electrophoresis. Because of its small molecule, the bromophenol blue will move quickly through the pores of the gel during electrophoresis. Thus progression of the dye through the gel over time can be trace easily by looking into the movement of the blue color.

The buffer in the two tanks (upper and lower) and in the gel should be same, with a pH of 9.0, and as most proteins have net negative charges, they migrate towards the anode. Usually an electric current of ~300 V, is applied across the gel. When the blue indicator dye reaches the bottom end of the gel, the electric current is stopped. Then the gel is removed from the apparatus, stained and protein bands can be visualized (Figure 2).

SDS-Polyacrylamide Gel Electrophoresis (SDS-PAGE)

A modification of PAGE is SDS-PAGE (sodium dodecyl sulphate- PAGE). In this technique, the protein samples are treated with reducing agents like 2-mercaptoethanol or dithiothreitol to break all disulfide bonds. Thereafter, by treatment with sodium dodecyl sulphate (SDS), a strong anionic detergent, the protein is denatured, by disrupting all the noncovalent interactions. This will make the protein samples negatively charged. Since approximately one SDS molecule binds to every two amino acids residues, it will generate negative charge to the denatured protein, which will be proportionate to its mass. Bromophenol blue dye is then added to the SDS-protein mixture and poured into the wells of the gel as described earlier for PAGE. Since all the proteins now have an identical charge to mass ratio, they will be separated on the basis of their mass. The mass of unknown protein can be known by comparing with the mass of known protein, by running them parallel in the gel.

Figure 2. Important steps of SDS –PAGE technique (For details see text)

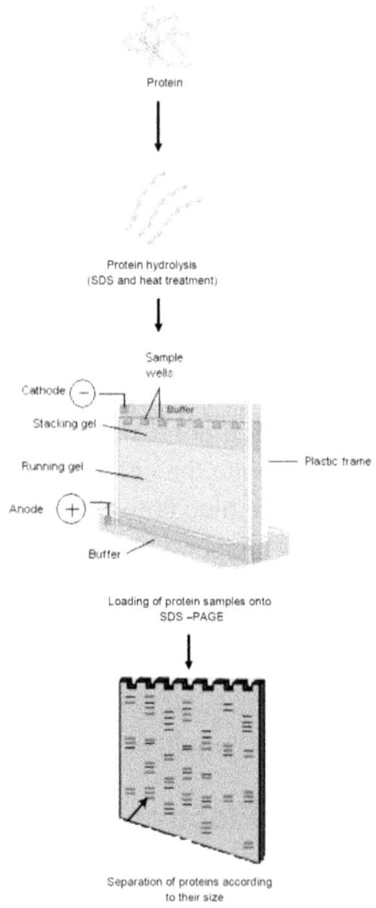

Proteins can be separated on the basis of their ability to move when an electrical current is applied. The rate of the movement of the protein will depend on the length of the polypeptide chains and/or their molecular weight. This property is used in SDS-PAGE technique, by removing the secondary and tertiary structures in proteins through treatment with the detergent SDS. After denaturation, proteins are coated proportional to their molecular weight by SDS. The original negative charge of all polypeptides is also maintained by SDS. Movement of the glycosylated proteins may not take place at the expected rate as their movement depends on the mass of their polypeptide chains, and not on the attached sugars molecules.

Laemmli system is the most widely used SDS-PAGE gel system for separating a wide range of proteins. In this system tris-glycine gels is used as a stacking gel and varying concentration of acrylamide gel to separate the proteins on the basis of their mass weight. Stacking gel helps the proteins to form sharp bands before electrophoretic run. A discontinuous buffer system is used, for example, Tris at pH 8.3 for running gel, Tris at pH 6.8 for stacking gel, and Tris at pH 8.8 for resolving gel.

Since the Laemmli system operates in a highly alkaline environment it may affect proper band development and their resolution. The poor resolution of the bands is due to: 1) shortening of the self-life of polyacrylamide due to hydrolysis, 2) alkylation and deamination of proteins, 3) reoxidation of the reduced cysteine disulphide present in the proteins, and 4) cleavage of the bonds in the protein at the Asp-Pro site.

Isoelectric Focusing (IEF)

Isoelectric focusing (IEF) is technique used for separation of proteins on the basis of their isoelectric point (pI). The pH at which a protein has no net electric charge is called its pI. Therefore in an electric field, a particular protein does not move further from this point. Isoelectric point of a protein can be determined by using specific IEF gels. Minor changes that may occur in the protein due to phosphorylation and glycosylation can also be determined through this technique.

In this technique polyacrylamide gels or IPG strips containing fixed pH gradient is used. The mixed protein samples are applied to this gels/ strips and an electrical field is applied. The mixed protein sample will migrates through the pH gradient. As the protein molecules approach their specific pI, they will be immobilized. The gels are then stained and documented. The immobilized proteins are then separated through 2D gel electrophoresis.

Two-Dimensional (2-D) Gel Electrophoresis

Two-dimensional (2-D) gel electrophoresis is a method for separating mixtures of complex protein obtained from various biological materials (cells, tissues and organs). The advantage of 2D protein electrophoresis is that it combines two methods of separation and allows the resolution of up to several thousand proteins at a time. The first dimension of a 2D gel separates proteins according to their iso-electric point using an immobilized pH gradient (IPG) strip. Proteins migrate in the strip until their charge becomes neutral and then stop their migration. The second dimension gel is typically an SDS gel which separates proteins based on their mass (molecular weight) (Figure 3). After separation, individual proteins resolved into small circular regions or spots. These gels are often used to analyse samples in proteomics research since the individual spot can be excised, digested and used for mass spectrometry.

In this technique samples are placed in gel with a pH gradient and then an electrical current is applied. The proteins will migrate along the pH gradient till they reach their pI point. There are two alternative methods by which the pH gradients can be created, either through carrier ampholites or by immobilized pH gradient gels (IPG).

In 2-D electrophoresis the most critical step is isoelectric focusing (IEF). Usually a high concentrated urea solution, reducing agent and chaotraphs are used to dissolve the proteins to maintain their charges. High data quality can be obtained by maintaining low ionic strength before isoelectric focusing. For each type of samples, it may be necessary to adjust the buffer and the electrical profile, as different types of samples may differ in their ion content.

The separation in the second dimension is done by SDS-PAGE. In a standard apparatus twelve parallel gels can be run simultaneously, which helps is reducing the time for separation.

WESTERN BLOTING

When biological samples are transferred from a gel to a nylon or nutrocellulose membrane and subsequently used for their analysis is referred to as blotting. Although, Towben et al (1979) developed the Western blotting technique for analysis of proteins, the term Western blotting was given by Burnette (1981). The technique is based on the principle that antibody and antigen binds specifically and therefore the target protein can be easily identified from a complex mixture of proteins. Both quantitative and semi-quantitative analysis of proteins can be done through Western blotting.

In Western blotting, first the proteins have to be separated through gel electrophoresis. After separation, the proteins are transferred (blotted) to a nitrocellulose or polyvinylidene difluoride (PVDF) membrane. Transfer of the protein molecules can be done by several methods, such as, capillary transfer, diffusion transfer, vacuum blotting, heat-accelerated convectional transfer, and electro-elution. However, the most commonly used transfer method is electro-elution, because of its speed and transfer efficiency. The electrophoretic mobility property of proteins is used in this method. In this procedure the polyacrylamide gel containing the protein is placed in direct

Figure 3. Two dimensional gel electrophoresis of proteins. A protein solution (e.g. crude extract of cells) is first subjected to electrophoresis in a tube with a semisolid gel inside it. The gel contains a pH gradient. This first step separates proteins according to their net charges (isoelectric focusing). Next, that gel is set on top of a rectangular slab containing an acrylamide gel that includes the anionic detergent sodium dodecyl sulphate (SDS). SDS molecules interact with amino acids (one SDS molecule/amino acid) to give all proteins a similar charge-to-mass ratio. In this second dimension, once an electric field is applied proteins are separated during electrophoresis solely according to their molecular size. Using this method, several hundred of the proteins in a crude extract of cells can be identified as individual spots on the rectangular second dimension gel.

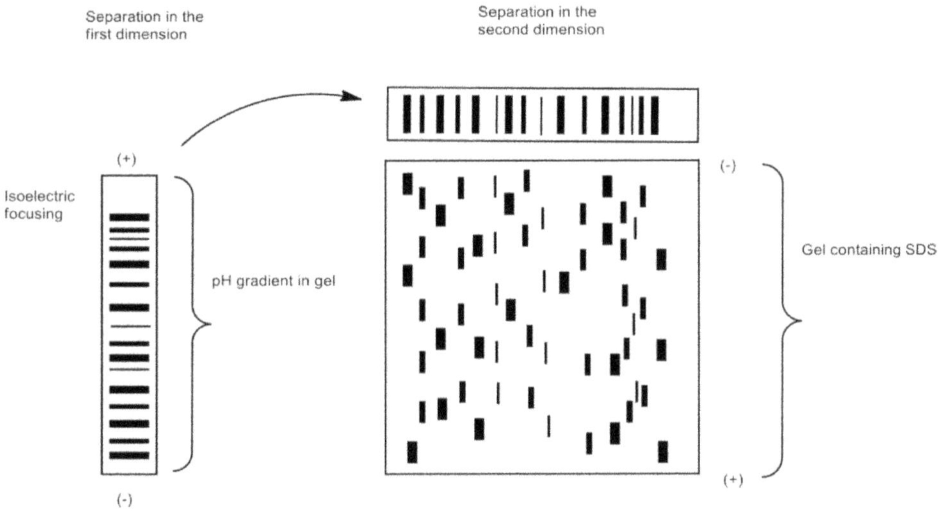

contact with a nitrocellulose or PVDF membrane, and two electrodes are placed connecting the gel-membrane sandwich and the conducting solution. The proteins from the gel will move to the membrane after application of an electrical current.

The efficiency of transfer may vary in different proteins according to their ability to migrate and binding propensity to the membrane. The factors which determine the efficiency of transfer include: composition of the gel, contact of the gel with the membrane, position of the electrodes, transfer time, size and composition of protein, field strength and characteristics of the buffer. Optimal efficiency can be obtained by using low ionic strength buffers and low electric current. Usually, several dyes (e.g. Ponceau S or Amido black 10B) are used to confirm transfer of protein to the membrane. The dye used should be easily removable, so that it cannot interfere in further analysis such as antibody binding.

Usually high affinity exists between the proteins and the membrane used in Western blotting. Therefore the portion of the membrane not involved in blotting should be protected from binding to nonspecific antibodies. For this purpose a variety of blocking buffers are used. These include highly purified protein, milk protein and normal serum. However, no single blocking agent is recommended for every occasion as each antibody-antigen binding has unique characteristics. To reduce background effect, it is essential to follow several washing steps for the Western blots.

Identification of protein is done with the help of a primary antibody that can recognize a specific protein or its epitope. The choice of a primary antibody will depend on the characteristics of the antigen to be detected and availability of antibodies for that antigen. Primary or secondary antibodies can be conjugated by many different tags. Radioisotopes were used extensively. But they are expensive, have a short self-life and require careful handling and disposal. Alternatively, enzymes or fluorophores can be used. Two most extensively used enzymes for detection and labeling of proteins are Alkaline phosphatase (AP) and Horseradish peroxidase (HRP). The flurophore-conjugated antibodies require special equipment to detect and document the fluorescent signals. Recent advances in digital imaging and availability of new flurophores (e.g. near-infrared, infrared and quantum dots) has increased its sensitivity and wide adoptability in Western blotting and other immunoassay. Steps involved in Western blotting are presented in Figure 4.

ENZYME LINKED IMMUNOSORBENT ASSAY (ELISA)

Enzyme linked immunosorbent assay (ELISA) technique is used to determine both quantitative and qualitative aspects of a particular protein is present in a sample. This method has two major variations: to determine how much antibody is present in a sample, or how much protein is bound by the antibody. The distinction is whether quantify an antibody or some other protein. In the ELISA test there exist two components, an enzyme and an antibody or antigen. ELISA is mainly applicable to detect substances having antigenic properties, such as proteins. The substances such as hormones, bacterial antigens and antibodies can be detected by ELISA.

For example, following procedure has to be followed to determine the amount of a particular antibody present in a blood sample. ELISAs are usually

Figure 4. Steps involved in Western blotting technique. (a) Proteins are separated by gel electrophoresis (through SDS-PAGE), (b) Separated proteins are transferred to nitrocellulose membrane, (c) The membrane containing the proteins is incubated with a generic protein (such as milk proteins) so that remaining sticky places on the nitrocellulose membrane are blocked. An antibody attached with an enzyme (such as alkaline phosphatase or horseradish peroxidase) is then added to the reaction mixture, which binds to its specific protein, (d) A colorless substrate is then added which is converted to a colored product by the attached enzyme and thereby the location of the antibody can determined.

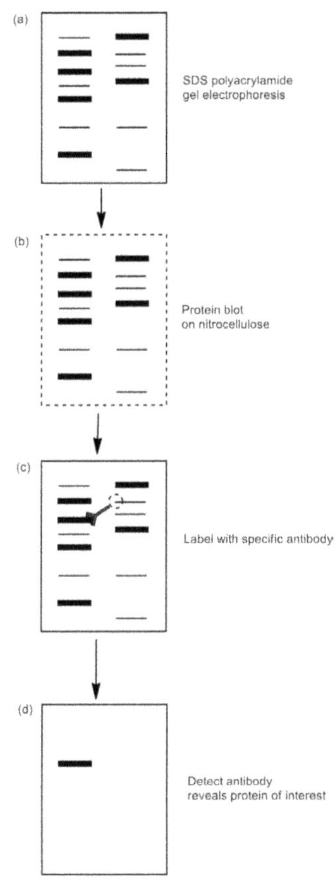

performed in 96-well plates which permit high throughput results. A protein to which the antibody (to be measured) can bind is used to coat the bottom of each well (Figure 5). The clotted whole blood is centrifuged to obtain the serum with antibodies (primary antibodies). In one of the wells, the serum is

Figure 5. Basic protocol for traditional double antibody sandwich ELISA

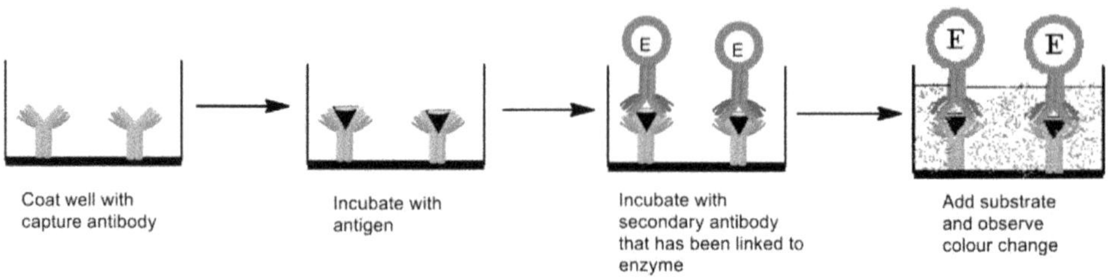

Figure 6. Detection of antigen-antibody binding (ELISA and IFA tests). First, the antigen is immobilized on a 96-well plate and the test serum is added. It is then incubated in the presence of another antibody labeled with an enzyme, and washed to remove the unbound antibodies. When a colorless substrate is added, it gets converted colored product by the attached enzyme and the location of the bound secondary antibodies can be detected.

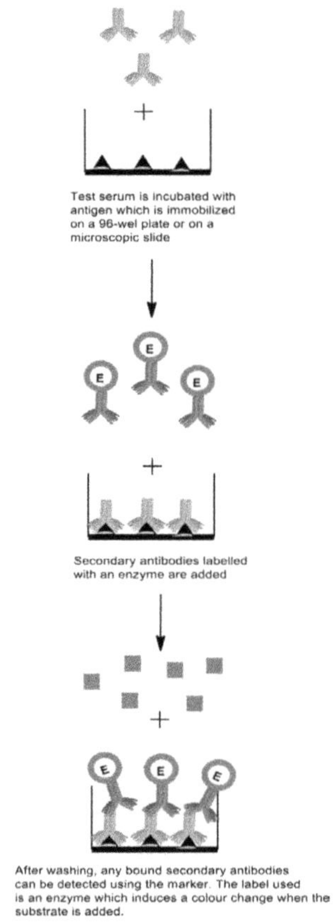

incubated. Different serum is put into each well for incubation. The 96-well plate shall contain a positive control serum and a negative control serum.

ELISA exploits the specific interactions of antibodies with antigens. The basic procedure for ELISA involves coating the wells of a microtitre plate with antibody specific for the target antigen. After washing the wells to remove any unbound antibody, non-specific binding sites on the well are blocked with a protein solution, usually a solution of bovine serum albumin. After washing the wells again the test sample is added. If the target antigen is present in the test sample it will bind to the antibody that has coated the wells (Figure 6). The wells are again washed to remove unbound material and another antibody specific for the target antigen is added. The antibody is labeled with an enzyme and is chosen so that it binds to a different site on the antigen than does the 'coating antibody'. In practice this labeled antibody is usually a polyclonal antibody and so it will bind to multiple sites on the

Figure 7. Different types of ELISA (For details see text)

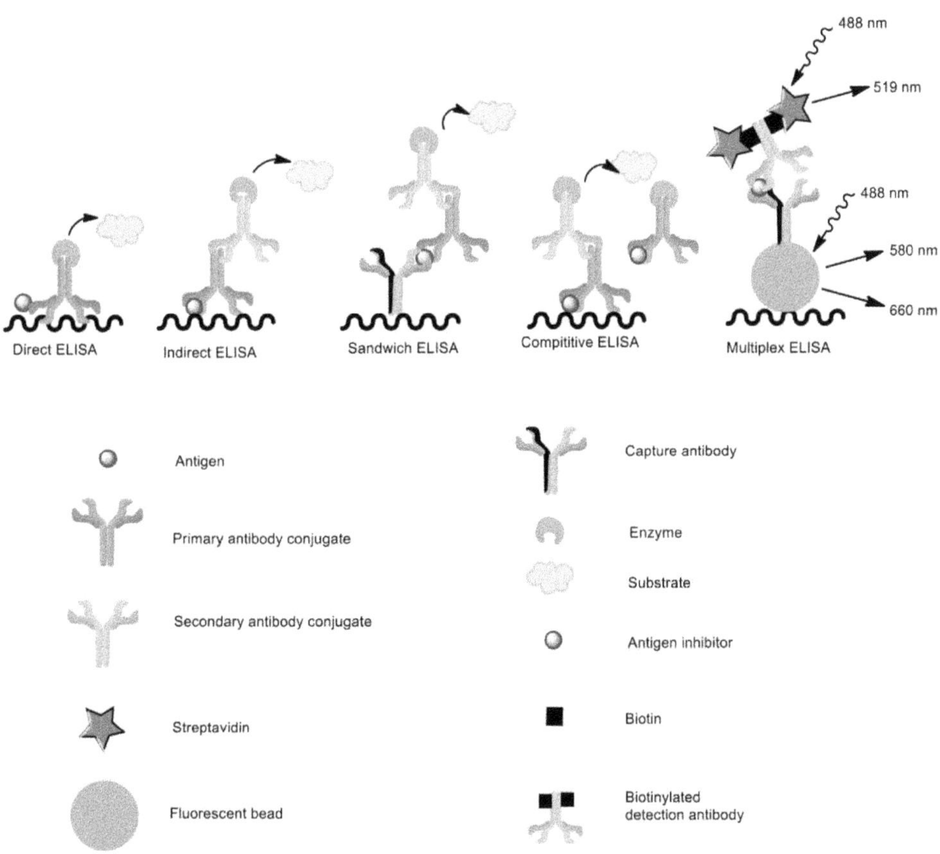

antigen allowing a greater signal to be obtained. After washing again to remove unbound labeled antibody, the enzyme substrate is added. The substrate is chosen so that the enzyme will convert it to a soluble product that can be quantified spectrophotometrically. By comparing the absorbance values of test wells with those of wells containing known quantities of target antigen, the amount of antigen in the test sample can be determined. This form of ELISA is called 'sandwich ELISA'. Variations of this basic procedure include: direct ELISA where the target antigen is immobilized on to the plate, indirect ELISA which uses a labeled secondary antibody, competitive ELISA where the concentration of antigen is calculated based on competition with a known amount of labeled antigen, and multiplex ELISA which allows the simultaneous detection of many different antigens in the same sample (Figure 7).

Direct ELISA

In direct ELISA method labeling of the antibody is done directly. The wells are coated with the target antigen and the antibodies (previously labeled) are added for binding. The bound antibodies are measured through colorimetric, chemilluminescent, or fluorescent end-point. The direct ELISA is relatively quick as the secondary antibody step is omitted. Further, possibility of cross-connection with the secondary antibody is eliminated. However, since every antibody to be tested should be labeled, it can be expensive and time-consuming. Moreover, certain antibodies may not be suitable for direct labeling. This method also lacks the possibility of having additional signal amplification that can be achieved through use of a secondary antibody.

Indirect ELISA

Indirect ELISA is a two-step process in which a second antibody (previously labeled) is used for detection. After coating the wells with the antigen, they are incubated in presence of a primary antibody. Thereafter a labeled secondary antibody that recognizes the primary antibody is added and further incubated. It is important to note that the antibody enzyme conjugate should be of high specific activity. This can be achieved by purification of antibody through affinity chromatography and by preserving antibody specificity.

Sandwich ELISA

Sandwich ELISA is designed to measure the amount of multivalent antigen present between two layers of antibodies. The antigens to be quantitatively measured should be capable of binding to two different antibodies. The two antibodies act as the sandwich. Therefore sandwich assays are restricted to the quantization of multivalent antigens like proteins or polysaccharides. Sandwich ELISAs is useful when the antigen concentration is very low.

In this method, a purified antibody (called "capture" antibody) is placed in the well and incubated in presence of the antigen. The unbound antigens are removed by washing, and a secondary labeled antibody (called "detection" antibody) is added. The second antibody binds to the antigen at a different site forming a "sandwich". The amount of labeled second antibody is measured through colorimetric method. Since the antigen to be measured is not required to be purified the technique is fast and very specific. However, not all antibodies can be used in this technique.

Competitive ELISA

In competitive ELISA a purified primary antibody, without being labeled is used to coat the wells. Then the unlabeled standards and unknowns are added and incubated. Once the reaction reaches equilibrium, conjugates antigen is added. The antigen binds to the primary antibody in those sites which are free (not occupied by unlabeled antigen). Thus, if the site of unlabeled antigens in the sample or standard is more, then the amount of conjugated antigen bound to the antibody will be lower and vice versa. The substrate is then added and the change in color is measured. The main advantage of this technique is that the primary antibodies can be used without being purified.

Multiplex ELISA

In multiplex ELISA, the microtitre plate ELISA is used to detect multiple analytes simultaneously at multiple array addresses within a well. Several types of multiplex ELISA is available. In one type, a sandwich ELISA is constructed, where antigen to be measured is placed between two different antibodies within single well. Multiplex ELISA can be carried out through an array of antibodies, in which primary antibodies are coated on glass plate to arrest corresponding antigens present in biological samples such as tissue extract, cell lysates, or plasma. Depending upon antibody array technologies, any of the detection methods such as direct or indirect, labeling or non-labeling, sandwich or competitive, can be used.

Microtitre plate ELISA uses very small quantities both of reagents and of sample and many of the steps in the assay can be automated through the use of automatic plate washers, reagent dispensers and plate readers. Robotic ELISA is used in most routine diagnostic immunology laboratories for the testing of patient blood for the presence of IgE antibodies to allergens such as pollen, nut etc.

ISOLATION OF NUCLEIC ACID

The procedure for isolation of nucleic acids (DNA and RN) is described in the following section.

Isolation of DNA

Isolation of DNA is needed for a variety of applications in agriculture, horticulture, medical, forensic, microbiology, taxonomy, fishery, animal science etc. The main purpose of DNA isolation include: genetic analysis, introduction of foreign DNA into cells of microbes, animals or plants, diagnostic purposes, and identification of individuals, plant or animal etc.

Based on the requirements the sources for DNA isolation can be very diverse. Isolation of DNA can be done from both living and dead organism. For various experimental purposes DNA is isolated from: plant and animal tissues, whole blood, buccal swabs, hair, bones, sperm, nails, saliva, epithelial cells, urine, bacteria and viruses.

Depending on the age, source, and size of the sample the isolation methods may vary. Although a wide variety of methods are available, some similarities do exist among them. In general, the goal is to isolate DNA from the nucleus without being contaminated by other cellular components. Analysis of DNA may be affected due to presence of proteins, polysaccharides, lipids, and other organic or inorganic compounds in the DNA sample. In experiments requiring high purity of DNA e.g. polymerase chain reaction (PCR), special precautions are required. Presence of impurities can also reduce the life of the DNA molecule.

Size of the sample is considered to be an important factor for isolation of DNA. For small samples such as sperm, or a single hair, the method is different from the method used for large samples such as tissue (few milligrams) or blood (few millilitres). The method of isolation will also vary depending upon whether the sample is fresh or stored. Stored samples are usually derived from frozen tissue or blood, archived tissue samples, exhumed bones or tissues, and fossilized plants, animals and humans.

The first step of isolation of DNA starts with breakdown or lysis of cells or tissue. This helps in the release of DNA from the nucleus by destroying the protein structure. Cells are usually broken down by treatment with a salt solution containing detergents. This helps to denature proteins or protein digesting enzymes and in dissolving membrane structures. For soft tissue this treatment works well. For hard tissues, such as various plant materials, the sample is required to be frozen in liquid nitrogen before pulverizing the tissues to a fine powder. In the case of bones, removal of the ions from the samples is an important step before extraction, as they are highly mineralized. The samples are then homogenized mechanically in lysis buffer.

A number of DNA isolation kits are available commercially. These kits usually contain the common lysis solutions which include: sodium chloride, Tris (trimethamine), EDTA (ethylenediaminetetraacetic acid, which binds metal ions), SDS (sodium dodecyl sulphate, a detergent), Proteinase K (an enzyme), and a buffer (which maintains constant pH).

Various organic solvents are used for purification of DNA samples. A mixture of chloroform, phenol, and isoamyl alcohol is used for separation of protein from DNA. The organic mixture denatures the proteins. Centrifugation of the sample treated with the organic mixture leads to separation of the DNA from proteins. The denatured proteins forms a cloudy interface between the aqueous (water) layer containing DNA at the top and phenol is at the bottom of the tube. Some of the disadvantages of this method are: i) it works efficiently only if the starting material is reasonably large, ii) the organic solvents used are health hazards, and iii) the quality of the DNA may not be suitable for sensitive analytical techniques such as DNA sequencing.

In a modified method high salt (sodium chloride) concentration is used. After denaturation of cellular proteins (through treatment with organic mixture for a few hours or overnight), NaCl is added to the solution. Salts of nucleic acid is formed which can be recovered through centrifugation after treating with ethanol.

In the case of buccal swabs and blood stains, DNA can be released by alkaline denaturation of the samples. Cells obtained from such source are put into small eppendorf tubes and treated with NaOH for denaturation. The pH of the solution is then brought to neutral by adding acidic buffer solution. The procedure is simple and quick but the quality of DNA produced may not be applicable for all analytical techniques.

Denaturation of the dsDNA can be done by heating or boiling the samples. When the sample containing dsDNA is heated to 100° C, the two strands of DNA are separated and released from the sample. Although in some cases DNA thus released can be amplified by PCR, in most of the time they are contaminated by proteins, other organic compounds, or ions.

In a modified method, used commonly in forensic laboratories, Chelex ion exchange resin is used. The resin can bind multivalent metal ions and other organic compounds from DNA, while the DNA remains in the solution. Such resin can be used with wide variety of sample, such as buccal swabs, hair, blood stain, whole blood, seminal stains etc. The resin containing the contaminants can be separated by centrifuging. In another method instead of Chelex, paramagnetic beads are used to bind DNA. In this method, the paramagnetic beads are added to the sample after lysis. The magnetic beads containing the impurities are separated from the sample on a magnetic stand and DNA is eluted at 65°C.

In some other methods various columns are used for DNA purification. These columns are usually packed with various matrices or resins (ion exchange, or silica). These matrices are positively charged and therefore can bind the negatively charged DNA. Unbound materials are removed from the columns by washing with salt solutions. The columns are then treated with neutral pH salt solution (which breaks resin-DNA bonding) to recover DNA. Application of columns has several advantages such as: i) require less time for isolation, ii) yield of recovered DNA is increased, and iii) quality of DNA recovered is improved. Liquid resins are also used in certain cases. In such cases, the DNA is separated from the resin by centrifugation.

The above methods are applicable for simple, single samples. However, sample consists of a mixture of cells has also to be analysed in several situations. For example, analysis of sperm cells and non-sperm epithelial cells (which exist together) may be required. In such situations, isolation of DNA is carried out on the basis of differential properties exhibited by the two cell types. Sperm cells are resistant to Proteinase K digestion. Therefore, the non-sperm cells can first be digested with Proteinase K. The two cell types can be separated by centrifugation, the epithelial cells will be present in the solution, and the sperm cells in the pellet. The separated sperm cells can be lysed with DTT (dithiothreitol) along with Proteinase K. Thereafter, isolation of DNA can be done by adopting any of the methods described above.

Analysis of DNA from plant cells has many applications. Presence of cell wall makes it more difficult to isolate DNA from plants. Plants often have high levels of carbohydrates (sugars and starch) in their tissues and other organic compounds such as polyphenols and pigments, which interfere with the isolation procedure. Freezing the samples in liquid nitrogen and grinding thereafter leads to breaking of the cell wall and release of the nucleus. The polysaccharides are removed by treatment with chloroform-octanol mix, hexadecyltrimethylammonium bromide (CTAB) and high salts, whereas phenolic compounds are removed by treatment with polyvinylpyrrolidone (PVP). For optimization of the quality and yield of the DNA extract, specific method has to be selected based on the objective of the experiment and the available material.

Isolation of RNA

Rapid inactivation of the endogenous RNase released during the disruption of cells is a key factor for isolation of intact RNA from the cells. The chaotropic agent Guanidinium Isothiocyanate (GTC) is used for disruption of cells and for dissolving cellular protein. RNase remains inactivated when dissolved in GTC along with reducing agents, and the disintegrated cell suspension maintains the integrity of RNA molecules. The RNase is then removed by treatment with phenol. The treatment is done at a pH of 4.5. At this pH the DNA partitions into the organic phase, and can be removed along with enzymes and proteins. Precipitation of the RNA is done with ethanol and through centrifugation they can be collected.

AGAROSE GEL ELECTROPHORESIS OF DNA AND RNA

Usually extremely large molecules of DNA and RNA are required to be purified for various molecular biological experiments. Molecular weight of a piece of single stranded DNA or RNA of 1 kilobase shall be of 330,000 dalton, which is larger than vast majority of proteins. An extremely open matrix structure is required for separation of such large biological molecules. Agarose, became the choice for separation of DNA and RNA as it forms gels having sufficient strength even at 0.5% concentration and can be used to separate molecules having over 1000 bp.

Agarose is derived from seaweed and is a natural polysaccharide. Agarose has to be purified from its crude precursor material (agar), as it contains a number of contaminants, which affect quality separation. Mainly contamination with sulphonated polysaccharides creates strong negative charges to the gel matrix. These charges on the matrix induce flow of water through the gel thereby balances the effect of osmotic created due to migration of counter ions. This is called electroendosmosis (EEO). This causes the smearing and broadening of the bands. Sulphonated polysaccharides contamination can also act as effective DNA mimics, and can affect later processing steps by inhibiting enzyme action, such as restriction analysis or ligation. Since agarose has low EEO, the banding resolution is excellent, and can be further processed with different enzymes, although this may not be true always.

With the help of hydrogen bonding the 3-dimentional structure of an agarose gel is maintained. Since covalent bonds are not present in this network, gels can be easily melted by heating. The melted agarose is poured to gel moulds to create new gels after cooling. According to the nature of the nucleic acid (native or denatured) to be analyzed, the buffer can be selected to facilitate analysis of dsDNA, ssDNA and RNA on agarose gels. Usually horizontal apparatus is used to run the agarose gels, where the gel lies beneath a thin layer of buffer (submarine gels). However, a vertical apparatus can also be used to run the agarose gels, particularly in those cases where discontinuous buffer systems or thin gels are required.

Analysis of RNA through electrophoresis is rather tricky as it occurs in a single stranded form, without complementary sequences. However, certain RNA molecules can form a complex resulting into secondary structures, which are usually very stable and difficult to denature. Usually a denaturing agent is included in the gel during agarose electrophoresis of RNA, otherwise RNA molecules forms a compact secondary structures, which disturbs the electrophoretic separation. Application of urea a denaturant can cause disruption of the hydrogen bonds in the agarose gel, whereas alkaline conditions will hydrolyze RNA molecules. Therefore, none of these chemicals can be used. On the other hand most of the denaturants which can be used for RNA analysis are toxic. Therefore, among the various types of denaturants available it is he RNase enzymes are present in the sample or in the processing environment can cause extensive degradation to the RNA molecules thereby affecting their isolation and analysis. Therefore in all effective procedures, contamination of RNase should be avoided. RNases are thermostable small enzymes which are found to be widely distributed in nature. Interestingly they are found on the surface of human skin, where acts as a defense mechanism against retroviruses. Therefore gloves should be worn while carrying out any experiment on RNA, and any surface touched with bare hands should be cleaned before using. Certified RNase free glassware or disposable plastic can also be used. Although RNases can be denatured by boiling, they renature upon cooling. Therefore, they cannot be eliminated by boiling the contaminated solution. Glassware can be decontaminated from RNase by dry heating to 250°C for 4 hours.

Solutions can be decontaminated from RNase by several methods. Treatment with diethylpyrocarbamate (DEPC) is the most commonly used practice. RNase is inactivated by DEPC and can be removed by autoclaving. It is important to note that DEPC is extremely toxic and volatile and therefore must be used only inside a fume hood. DEPC has to be removed as it would otherwise covalently modify the RNA and react with amines. For removal of DEPC the solution has to be heated, which may affect the heat labile RNA. Thus, this is a major constraint of using DEPC. To overcome this problem, RNA containing samples are decontaminated by guanidinium salts treatment. RNase inhibitors such as RNasin or Vanadyl Ribonucleosides along with small volume of Tris buffer can also be used to deactivate RNase.

PURIFICATION OF DNA AND RNA FROM AGAROSE GELS

After separation of the DNA and RNA fragments through electrophoresis it is usually essential to purify them from the agarose gel to carry out subsequent operations. A variety of techniques are available to carry out such operations. In cases where agarose and buffer components will not interfere, low melting agarose gels are used for electrophoretic separation. The gel is stained and the band(s) of interest excised. The excised material is then melted at $60\text{-}65^{0}$ C, which is much below the melting temperature of any DNA longer than 30 bp. The challenge of recovering DNA from a matrix arises from the fact that nucleic acid molecules are caged in a 3-dimentional network of matrix molecules. Removing the DNA from this cage requires forcing the DNA through the matrix pores. The DNA can be released easily by melting the low melt agarose without disturbing the structure of DNA molecules.

DNA purified from low melt agarose is sufficiently clean for many purposes. When further purification is necessary, glass power elution is an effective method. In this technique, the DNA is bound to finely powered glass or microscopic glass beads in a high salt suspension. Agarose and other contaminants do not bind to the glass and can be washed away. The DNA is then eluted in water or a low salt buffer. A selection of glass powder elution kits is commercially available. DNA can also be separated from agarose gel directly through this technique by using sodium iodide (NaI), which disrupts the gel matrix and thereby the DNA gets released.

The alternative to glass elution for purification of DNA from agarose is electroelution. The electroelution process can be easily carried out as the agarose gels are run in horizontal apparatus. However, electroelution is not possible in vertical gels as they are performed in encased glass plates.

In the simplest form of electroelution, the portion of the gel containing specific band is excised and put inside a dialysis membrane bag. Then electrophoresis buffer is put inside the bag and run through an electric field. The DNA molecules shall come out from the gel slice and remain in the buffer, but shall not be come out from the bag because of its large sizes. DNA can be easily recovered from the buffer inside the bag. Alternatively, a 'trench' is made by removing the portion of the gel just ahead of the band of interest, and then electrophoresis is continued till the band is migrated to the trench. Care should be taken to monitor the exact timing so that the band dose not migrates past the trench. To overcome monitoring of the exact timing and a piece of DEAE ion-exchange paper is inserted into the gel just ahead of the band through a slit. If electrophoresis is continued thereafter, the band will run into the paper and bind tightly to the paper. Once the band is bound to the paper it will not move any further, and DNA can be recovered by washing the paper in a high salt buffer. The elution buffer can be removed by precipitating the DNA by ethanol.

SOUTHERN BLOTTING

Edward Southern (1970) developed the Southern blotting technique. The name of the technique was given after the developer. With the help of this technique it is possible to locate a specific DNA fragment within a mixture of complex sample. For example, the technique can be used to locate a particular gene (represented in a DNA fragment) within a genome.

For Southern blotting, the requirement of quantity of DNA depends on the characteristics of the probe, such as its size and specific activity. Short probes are more specific. Under optimal conditions, it is possible to detect as low as 0.1pg of DNA through specific probe. For Southern blotting, genomic

Figure 8. Steps involved in Southern blotting technique. Genomic DNA is digested with a single restriction endonuclease resulting in a complex mixture of DNA fragments of different sizes, that is, molecular weights. Digested DNA is arrayed by size using electrophoresis through a semisolid agarose gel. Because DNA is negatively charged, fragments will migrate toward the anode, but their progress is variably impeded by interactions with the agarose gel. Small fragments interact less and migrate farther; large fragments interact more and migrate less. The arrayed fragments are then transferred to a sheet of nitrocellulose or nylon based filter paper by forcing buffer through the gel as shown. The DNA fragments are carried by capillary action and can be made to bind irreversibly to the filter. Now the DNA fragments, still arrayed by size on the filter, can be probed for specific nucleotide sequence using a radiolabelled nucleic acid probe. The probe will hybridize to complementary sequence in the DNA, and the position of the fragment that contains these sequences can be revealed by exposing the filter to x-ray film.

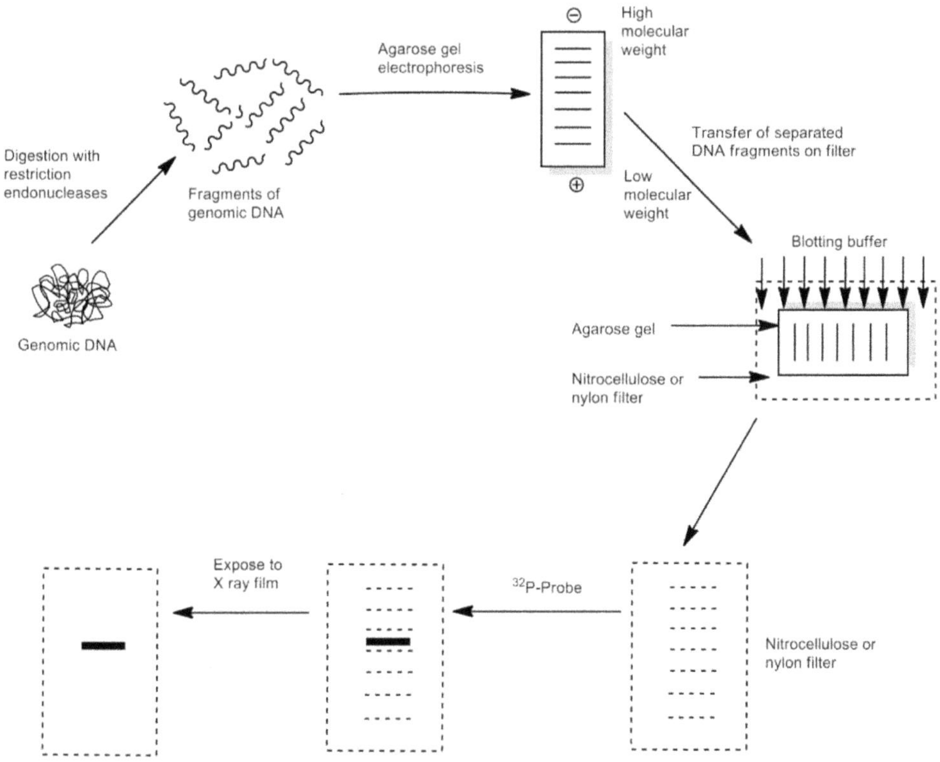

DNA is digested with one or more restriction enzymes, and the resulting fragments are separated by agarose gel electrophoresis, as described earlier. The separated double-stranded DNA fragments through electrophoresis are denatured into single-stranded DNA by soaking the gel in about 0.5M NaOH. It is important to note that only single-stranded DNA can be transferred to a blotting membrane. Fragments greater than 15Kbp are difficult to transfer to the blotting membrane. Therefore, in such situations depurination of the DNA molecule, with HCl (about 0.2M HCl for 15 min.), is required. Depurination takes the purines out thereby cutting the DNA into smaller fragments. It is essential to neutralize the gel after NaOH and/or HCl treatments.

For blotting DNA fragments, a nitrocellulose or nylon membrane can be used. Binding capacity of nitrocellulose membrane is about 100 μg/cm, and that of nylon membrane is about 500 μg/cm. Nylon membranes is also less fragile. Therefore they are preferred over nitrocellulose membrane. Transfer to the membrane takes place through capillary action, which is a time consuming process (usually overnight). With the help of a vacuum blot apparatus, the process can be enhanced and completed in about an hour. After transfer of DNA, the membrane should be treated with UV light. This cross links (by covalent bonds) the DNA to the membrane.

Probing or hybridization is usually done with ^{32}P labeled ATP, biotin/streptavidin or bioluminescent probe. Non-specific sites are blocked through prehybridization. This ensures that single-stranded probe does not bind anywhere else on the membrane. To prehybridize, non-specific ssDNA such as sonicated salmon sperm DNA is commonly used. The procedure is shown in Figure 8.

For labeling the probe with ^{32}P, the dsDNA should first be treated with mild DNase, to induce double-stranded nicks in DNA. Then ^{32}P, dATP, dNTPs and DNA polymerase should be added. The DNA polymerase has 5' to 3' polymerase activity as well as 3' to 5' exonuclease activity. In nick translation, while the DNA polymerase continues to nick down the DNA strand through its polymerase activity, its exonuclease activity fill-in the nick continuously. In the process, ^{32}P gets incorporated and DNA becomes labeled. By raising the temperature to about 90^0 C the DNA should be made single stranded, and then by immediately placing it on ice the two strands are kept separated, without being reannealing to each other. When the DNA is placed on ice, the DNA will pass the reannealing temperature too quickly, and thus they fail to rehybridize into dsDNA. After hybridization the radioactivity can be visualized by autoradiography. Biotin/streptavidin detection is done by colorimetric method, and bioluminescent visualization uses luminescence.

Figure 9. Steps involved in Northern blotting technique. Northern blots allow investigators to determine the molecular weight of an mRNA and to measure relative amounts of the mRNA present in different samples. (a) RNA (either total RNA or just mRNA) is separated by gel electrophoresis, usually an agarose gel. Because there are so many different RNA molecules on the gel, it usually appears as a smear rather than discrete bands. (b) The RNA is transferred to a sheet of special blotting paper called nitrocellulose, though other types of paper, or membranes, can be used. The RNA molecules retain the same pattern of separation they had on the gel. (c) The blot is incubated with a probe which is single-stranded DNA. This probe will form base pairs with its complementary RNA sequence and bind to form a double-stranded RNA-DNA molecule. The probe cannot be seen but it is either radioactive or has an enzyme bound to it (e.g. alkaline phosphatase or horseradish peroxidase). (d) The location of the probe is revealed by incubating it with a colorless substrate that the attached enzyme converts to a coloured product that can be seen or gives off light which will expose X-ray film. If the probe was labelled with radioactivity, it can expose X-ray film directly.

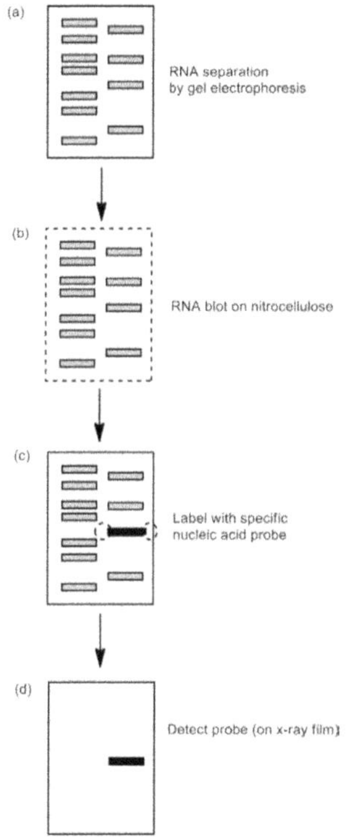

NORTHERN BLOTTING

Alwine et al. (1977) developed the Northern blotting technique to measure the amount and of mRNA present in different samples. It is also used to determine its molecular weight. Before discussing the technique, it is appropriate to consider the question of why one should wish to measure an mRNA. There are two main reasons. The first is to know which tissues express a particular gene, and the second is to determine the factors which regulate the expression of a given gene, *i.e* by nutritional, hormonal, or environmental.

Measurement of mRNA can be done by three techniques. The first is Northern blotting, the second is RNase protection assay and the third is reverse transcriptase polymerase chain reaction. Although the second and the third methods provide considerable increase in sensitivity, Northern blotting is still the most preferred method for mRNA analysis. The Northern blotting technique is outlined in Figure 9.

The underlying principle of Northern blotting is that the RNAs are separated by agarose gel electrophoresis by their size and detected on a membrane using a hybridization probe. The RNA molecules on the gel will appear as a smear rather than discrete bands, because there will be many different RNA molecules on the gel. The separated RNA fragments are then transferred to a nitrocellulose or nylon membrane by following the procedure described in Southern blotting. Then the RNA must be immobilized on the membrane, either by exposure to UV light or by baking. This results in covalent linkage of RNA to the membrane, which prevents the nucleic acid from being washed away during the subsequent processing. The blot is then incubated with a single-stranded DNA probe. This probe will form double-stranded DNA-RNA molecule. The probe is previously labeled with a radioactive isotope or by a fluorescent bound enzyme, such as horseradish peroxidase or alkaline phosphatase. The probe can be identified in specific locations by exposing to X-ray film. In the case of enzyme labeling, the membrane should be incubated with a colorless substrate which can be converted to a colored product by the attached enzyme, and which can be seen or gives off light that can expose X-ray film.

Figure 10. Analysis of DNA by dot blotting technique. The steps involved are: isolation of double-stranded DNA from the given sample, immobilization of single stranded DNA onto membrane (dot blotting, hybridization (binding) with the labeled probe, washing to remove unbound probe and visualization of labeled DNA.

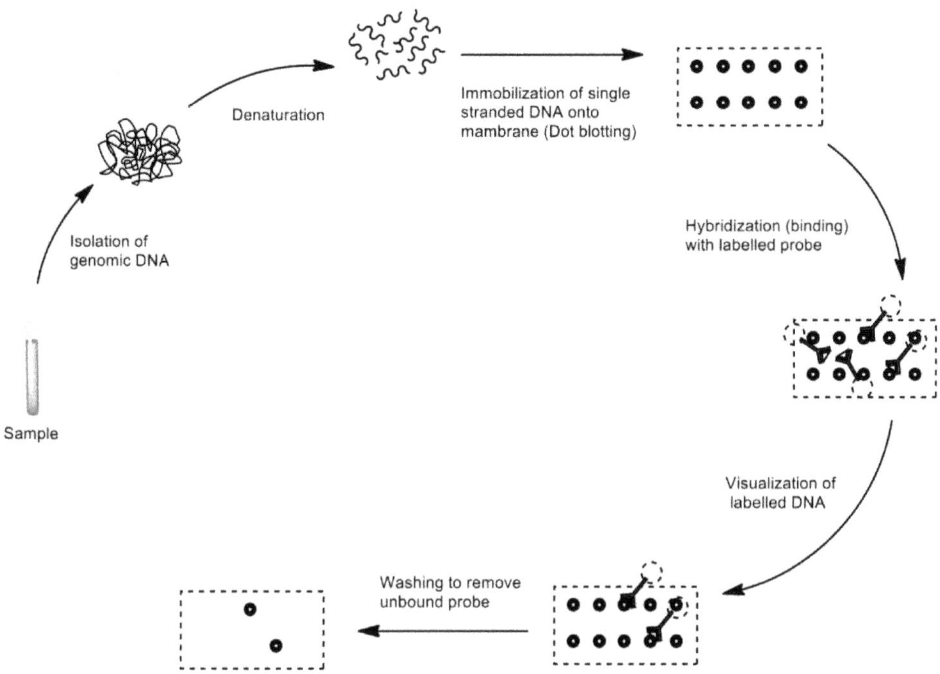

NUCLEIC ACID HYBRIDIZATION PROBE

Nucleic acid hybridization requires that a probe is complimentary to all, or part, of the sequence of the mRNA (in Northern blotting) and DNA (in Southern blotting) of interest. It depends upon the strict base pairing between A-T (A-U) and G-C. In general, the minimum size for a probe to ensure specificity is approximately 25 bases, provided there is a complete match between the probe sequence and target sequence. This can however be modulated by the stringency conditions. With a probe of approximately thirty bases in length, the probability that same sequence occurs in the mammalian genome is of the order of 1 in 1 billion.

Basically there are two main forms of nucleic acid hybridization probe; the complimentary DNA (cDNA) or anti-sense oligonucleotide (about 30-40 base length). The anti-sense oligonucleotide probe can be developed from sequence data. Oligonucleotide probe is comparatively simple to develop, as plasmid isolation is not required for it. For Northern blotting, 'riboprobes' based on RNA can also be used. 'Riboprobes' may increase sensitivity compared to DNA probes, but they are less stable as they are easily broken-down by RNase.

Detection

Detection is achieved either through the use of radioactivity or by non-radioactive strategies. Most laboratories prefer radioactively-labeled probes with ^{32}P. The procedure for using radioactively-labeled probe is well established and it provides high level of sensitivity. However interest in non-radioactive probing is growing, as it is comparatively safe and stable. Use of radioactive labeling has also been discouraged due to short half-life of the isotopes and their disposal problem. At present, the main non-radioactive approach is based on chemiluminescence. Detection may also be based on fluorescence, but this requires dedicated instruments.

In the case of chemiluminescent detection, the breakdown of specific chemiluminescence substrate is catalyzed by alkaline phosphatase or horseradish peroxidase, with the emission of light. The enzymes may be conjugated directly to a probe, or a probe is labeled with a ligand (e.g. digoxigenin, fluorescein, biotin), which is then localized by an antibody (or avidin or streptavidin if biotin is the ligand), to which alkaline phosphatase or horseradish peroxidase is attached.

Hybridization signals (both radioactive and chemiluminescence) are collected by means of X-ray film. Quantification is achieved by densitometry. Alternatively, phosphor storage screens together with a molecular imager can also be used. This offers reduced exposure time, and can quantify over several orders of magnitude.

DOT BLOTTING

All the blotting techniques described above (Southern, Northern and Western blotting) require extensive purification (electrophoresis and blotting of the gel) of nucleic acid, which is time consuming and expensive. Dot blotting technique is a simplified version of all the blotting techniques described so far (Figure 10). Sometime it may be required to simply detect and quantify a given sequence. In such case, the process may be shortened by avoiding the purification steps. The purified sample, either nucleic acid (DNA or RNA) or protein is applied directly to nitrocellulose or nylon membrane with the help of a pipette. A dissolved protein sample is usually pulled through the membrane either by applying vacuum, absorption or intrusion. The nucleic acids or the proteins bind to the membrane and the other sample compounds pass through the membrane. Subsequently detection is done by hybridization with nucleic acid probe (for nucleic acid)) or antibodies (for protein) and detection procedure is similar as described for Southern and Western blotting techniques. For semi-quantitative estimate of a specific protein within a mixture crude protein, both purified protein and specific antibody against it should be available. Since double stranded DNA molecule does not bind efficiently to the membrane, it is essential to denature the DNA by treating with NaOH (0.4M). The intensity of hybridization of nucleic acid sample is quantified by autoradiography. As the sample is applied in a circular form (either as a drop or oblong shape), and visualized as 'dot' or 'slot', the technique is known as dot or slot blotting.

The technique is applied for quick detection of specific sequences of nucleic acid and for determination of the relative amount of any given sequence (RNA or DNA) in a complex sample. It offers no information on the size of the target biomolecules. Furthermore, if two molecules of different sizes are detected, they will still appear as a single dot. Dot blot, therefore, only confirm the presence or absence of a biomolecule(s), which can be detected by nucleic acid probe or by the antibiotics.

PULSE FIELD GEL ELECTROPHORESIS

With the help of homogenous field agarose gel electrophoresis it is possible to separate and analyze DNA fragments of size less than 50-70 Kbp. But resolution of megabase (mb) fragments is required in many genomic DNA analyses. With the help of Pulse Field Gel Electrophoresis (PFGE) it is possible to separate and resolve larger DNA fragments up to ~10Mbp. The principle of PFGE is based on the fact that upon application of an electric field the DNA molecules elongates, and reverts to the un-elongated state after withdrawal of the electric field. The rate of relaxation depends on the size of the DNA molecule. During electrophoresis, by changing the orientation of the electric field, it is possible to make the DNA molecule get relaxed prior to reorientation, which affects the migration rate. For migration of DNA through the gel there must be pores large enough for the molecules to pass through. The requirement of the pore size shall depend on length of the DNA

Figure 11. Pulse field electrophoresis. (a) Voltage clamping by the CHEF mapper system in the mode: (i) Relative electrode potentials when the +60° field angle is activated, (ii) Relative electrode potentials when the -60° field angle is activated, (b) Voltage clamping by the CHEF mapper system in the FIGE mode: (i) relative electrode potentials when the 0° field vector is activated, (ii) Relative electrode potentials when the 180° field vector is activated.

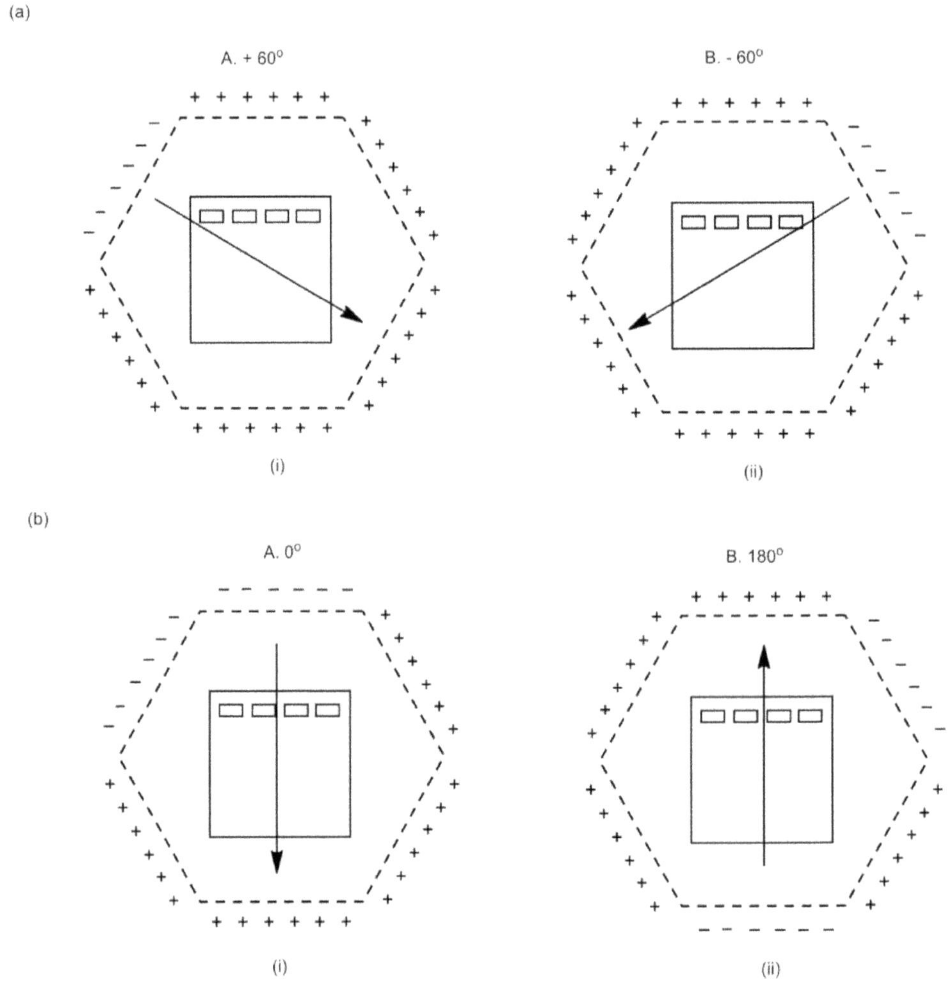

molecule, and its orientation towards to the pore. If molecules are placed perpendicular to their direction of migration, then it will require a very open gel structure to pass through. Similarly longer DNA molecule shall form different angles with respect to the pores, making it difficult to pass through the pores of the gel. Thus the mobility of the DNA molecule shall be limited when it is placed parallel to its direction of migration in the gel. Under such circumstances the DNA molecule shall move in a zigzag fashion which is called "reptation". It is essential that the DNA molecule has to maintain its conformation during such movements. Thermodynamically such highly ordered state of the DNA is unfavorable. Therefore the DNA will tend to revert to its "relax" form when conditions permit. The relaxed DNA molecule requires time to reorient itself for further reptation. The time required for reorientation is directly proportional to the length of the DNA molecule. In PFGE and FIGE techniques the electrophoretic separation of megabase size DNA molecules is done by considering this difference in reorientation time.

Several devices and systems for PFGE are available. It is important to optimize the gel running conditions and the size range to be resolved for each sample. Parameters which need to be optimized are: overall pulse lengths, pulse voltages, the ratio between forward and lateral pulse length, and the ramp rate between voltages. Since a general protocol does not work, it is usually recommended to consult the instruction manuals provided with the PFGE units for its optimum utilization.

In the original system known as Orthogonal Field Alternating Gel Electrophoresis (OFAGE), the electricity was applied alternately between a homogeneous and a non-homogeneous field. In this technique, straight migration lanes are not formed and therefore largely abandoned. In the Transverse Alternating Field Electrophoresis (TAFE), gel is mounted vertically in a rotating electrophoresis system and the electrodes are placed at either side of the gel. Reorientation of the electric field is done physically by moving the gel along the fixed electrodes. In the modified version the gel remains stationary while the orientation of the electric field is done through multiple electrodes controlled independently (Figure 11a).

Field Invasion Gel Electrophoresis (FIGE) is a variant of PFGE system, in which the electric field is oriented by 180^0 (Figure 11b). The polarity of the field is revised periodically along the axis of migration and the duration of the pulses are unequal. In this system, the strength and duration of the voltage pulses must be different, to make net progress of DNA through the gel. In FIGE, the voltage pulses timings must be matched to the DNA reorientation times. If a short duration of reversing pulse is applied, reorientation of the larger DNA molecules will not take place, while smaller molecules will migrate backwards after reorientation. When the forward field is resumed, the larger molecules will start reptation, while the smaller molecules will rapidly reorient, but shall move in the reverse direction to make up the lost distance. This will ensure that the longer DNA molecules will migrate at a faster rate than short molecules. To ensure good resolution over a wide range of sizes, it is essential to use progressively longer series of pulses. FIGE can be carried out in standard horizontal gel electrophoresis equipment and therefore easy to operate. But has the limitation of resolving size up to 2mb (about 750 kbp), whereas through PFGE it is possible to resolve molecules up to 5mb size. It can be performed with the conventional agarose gel electrophoresis equipment with a pulse controller, which can be used to change the polarity periodically.

The advanced version of PFGE electrodes are arranged hexagonally with an automatic program simulator. In this equipment electrodes are placed at their intermediate potentials, giving homogenous electric fields necessary for straight lanes. Since this instrument allow manipulating the pulse time, electrical field strength and pulse angle, it is possible to regulate the migration rate of DNA and resolution of the separation in agarose gel.

The duration for which the alternating electric field is applied to the DNA molecules is called pulse time in PFGE. Resolution of DNA molecule will be optimal when pulse time and re-orientation time are compatible. The rate of migration of DNA increases as the voltage increases. While selecting the field strength (voltage), it is important to come to compromise between the time for running the gel and resolution to be achieved. Field angle also affect separation of DNA molecules, with the decrease in field angle mobility of large molecules (>1 mbp) increases. Software are available by which all the three parameters can be optimized, for any size of DNA molecules.

The large size DNA molecules imposes certain constrains during sample preparation and handling. Large DNA molecules are easily cleaved during isolation and impart very high solution viscosity. To overcome these problems, preparation of DNA sample is carried out utilizing a gel medium. In this procedure the sample is first suspended in agarose solution and poured into molds. All subsequent operations (cell lysis, removal of protein, restriction digestion etc.) are carried out in the gel suspension molds by diffusing the reagents. The processed suspensions are then poured into agarose gel wells for electrophoresis.

In PFGE a constantly changing electric field is applied. In the originally version, alternating voltage electric fields oriented at 90° to each other was applied to the gels. In current versions electric fields are applied at 120° angles. More complex versions 3 or more angled voltages are used. The basic principle in all these systems is to force the larger DNA molecules to continuously reconfigure itself and to move in a new direction.

AUTORADIOGRAPHY

Autoradiography is a technique to visualize molecules or fragments of molecules that have been radioactively labeled by using X-ray (or occasionally photographic) film or emulsion. The relative intensities and positions of the radiolabeled molecule can be recorded permanently in the film. Biomolecules are usually labeled with ^{3}H, ^{32}P or ^{35}S, and can be detected by exposing to a film. Tritium (^{3}H) cannot be detected by autoradiography unless the sample has 1mCi of it in the band. The isotopes commonly used and their required amounts for autoradiographic detection are shown in Table 1.

Table 1. Autoradiography detection limits

Isotope	CPM necessary for detection	Energy per emission (MEV)
^{3}H	$>10^{7}$	0.0055
^{14}C	2000	0.050
^{35}S	1000	0.167
^{32}P	100	0.70
^{125}I	100	0.0355

In direct autoradiography, the sample present in a gel or filter is placed in contact with a sheet of X-ray film, and left to form a latent image. But this process will work if emitted radiations can reach and absorbed by the film. Usually the whole process is done using commercial cassettes, where the sample and film can be sandwiched between two perplex, glass or aluminum sheets, clipped or taped together. The assembly is then left at room temperature for several hours or days (depending upon the level of radioactivity in the sample) to allow adequate exposure of the film. The film is developed and fixed as per the instructions of the manufacturers. In the case of non-dried gel, (e.g. sequencing gel) it is essential to expose the films at -70^{0} C.

Based on the experimental type, autoradiography is classified as *in vivo* autoradiography and *in vitro* autoradiography. In *in-vivo* autoradiography the receptors are labeled in intact living tissue by systematic administration of the radioligand, and then the tissue is removed, processed and visualized. Whereas, in the case of *in-vitro* autoradiography, slide-mounted tissue sections are incubated with radioligand so that the receptors are labeled under controlled conditions.

Radioactive Exposure of Film

Radiolabelled-nucleotides emit β-rays which can penetrate the emulsion on the film to a depth proportional to their energy. While passing through the emulsion, the silver halide crystals present on them are activated. When the exposed film is developed, activated crystals appear as black silver grains. The activated silver halide crystals thus produced are unstable. A silver halide crystal needs at least five "hits" from the radioisotope to be stably activated and detected. However, if the autoradiography carried out at -70°C, the activated crystals get stabilized with increased sensitivity. Sensitivity can also be increased exposing the film to a microsecond burst of light called "preflashing" before the film is used. Only single "hit" per grain is required for "preflushed" film. Such films are not only very sensitivity in low signal intensity but also ensure linearity between the amount of radioactivity and the signal received by the film.

Fluorography

Fluorography can increase the sensitivity of autoradiography several fold by converting the radioactive emissions into light. Compared to β-particles, light can penetrate the film more efficiently and makes detection easy. When phosphor compound (e.g. AutoFlour) is dried and converted into a gel, it can absorb the energy from β- or γ-rays and re-emit it as light. Best result can be obtained when flash films are exposed at -70° C. The flashing particularly helps in detection of weak bands or spots. In certain cases there exist thresholds below which nothing can be detected. The threshold can be reduced by flashing.

Unlike most other isotopes used for autoradiography, ^{32}P produces high energy β-particles which penetrate through the film without activating the emulsion. In such situation phosphorescent intensifying screens are used to retain the energy within the emulsion. Such screens works most efficiently at low temperatures (-70°C). Exposure time required for different isotopes is presented in Table 2.

Digital Autoradiography

With the help of silicon detector it is now possible to detect presence of radioactively labeled molecule in various samples. Scanning of the autoradiogram is carried out by placing the detector very close (1mm) to the sample. The image captured through the digital detector can be analyzed faster by using suitable software.

Table 2. Fluorography exposure times

Isotope	dpm/band	beq/band	Exposure (hrs)
^{3}H	500	8.3	48-72
^{3}H	5000	83	24
$^{14}C/ ^{35}S$	300	5	24
$^{14}C/ ^{35}S$	1000	17	8-12

POLYMERASE CHAIN REACTION

Mullis et al. (1986) invented another novel and revolutionary technique called Polymerase Chain Reaction (PCR). Basically it is a technique to amplify specific DNA sequences by a direct enzymatic process. In the simplest form, a small amount of nucleotide sequence (about 20) at each end of the molecule should be known, and oligonucleatides complimentary to that sequence are synthesized. These synthesized oligonucleotides are used as primers for enzymatic amplification. The procedure is as follows (Figure 12).

1. A reaction mixture containing a sample of DNA that include the region to be amplified, the primers in large molar excess, deoxynucleoside triphosphate (dNTPs) and a heat-stable DNA polymerase (Taq DNA polymerase) should be prepared. The Taq DNA polymerase is isolated from the thermophilic bacterium *Thermus acquaticus*, which can be grown in the laboratory at 75^{0}C and above. Thus the enzyme produced by the bacterium does not get denatured by repeated heating and cooling that is required in the amplification process of PCR.
2. The reaction mixture is then heated to 94^{0}C, to denature the dsDNA molecule.

Figure 12. Amplification of DNA through PCR. The primers are synthetically prepared oligonucleotides having sequences complementary to the DNA on either side of the segment of DNA to be amplified.

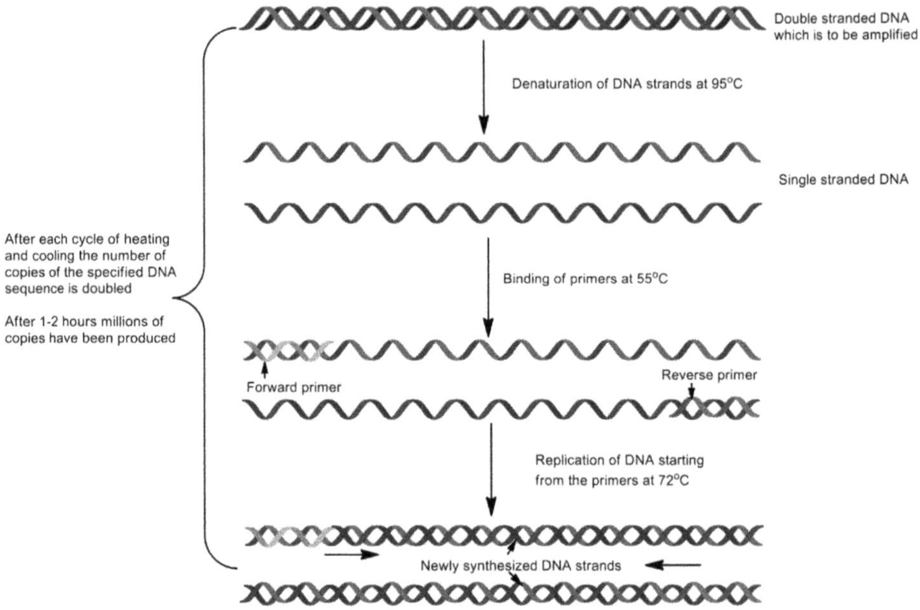

3. The mixture is then cooled to 50-60°C. This will induce rejoining of the complimentary sequences of single stranded molecules. Although it is possible for the sample DNA to self-anneal, but this is rendered less likely in the presence of short oligonucleotide primers, which anneal to the DNA molecules at specific positions.
4. The temperature is raised to 74°C, so that Taq DNA polymerase can start synthesizing new strands of DNA (using the dNTPs provided), complimentary to the template molecule, after getting attached to one end of each primer. This will result into four strands of DNA.
5. The temperature is then increased back to 94°C, which will again denature the dsDNA molecules to single strands.
6. The cycle of treatments should be repeated as described above, so that the process is repeated. At the end of second cycle there will be four dsDNA molecules.
7. By repeating the cycle 25 times, over 50 million new dsDNA molecules can be generated, each one a copy of the region starting molecule delineated by the annealing sites of the two primers.

Because of the repetitive nature of the process, it has become easy for automation of the process and several companies have developed PCR machines or Thermal Cyclers, which can carry out a series of programmed cycles of heating and cooling.

It is important to note that not all the molecules generated will be of defined length. If the process is applied to genomic DNA ('full-length' molecule), half the molecules will be full-length, after the first cycle, and the other half will start with the primer and have an undefined end. The length of the molecule will be determined by how far the polymerase reaction progressed during DNA synthesis ('intermediate' molecule). In the next cycle, each full length molecule will generate one full-length and one intermediate molecule, and each intermediate molecule will generate one intermediate molecule and one full defined target molecule (beginning and ending at the primer site). With the advancement of each cycle, the number of full-length molecules remain constant, whereas the number of intermediate molecules increases arithmetically, and the number of target molecules increases geometrically. After many cycles, gel electrophoresis of PCR product will indicate, molecules of a single size, corresponding to a single band (representing the target molecule) in the gel. A full set of PCR cycles normally takes a few hours. Primers can be obtained from commercial suppliers.

Limitations of Taq Polymerase

Taq polymerase can polymerise about 50-60 nucleotides per second. The enzyme has 5'-3' DNA polymerase and 5'-3' exonuclease activities. However, it has a number of properties which may be disadvantageous, such as:

1. Taq polymerase lacks 3'-5' exonuclease (proof reading) activity. Consequently about one nucleotide out of 10^4 incorporated may be incorrect. If this happens in an early cycle, a large fraction of PCR products will have the altered sequence.
2. Taq polymerase may dissociate from the template before it has synthesized a reasonably long piece of target DNA. Dissociation may takes place due to incorporation of an incorrect nucleotide (i.e. could not pair with the template base). Since the enzyme cannot correct the error, it cannot elongate the strand being synthesized, and thus dissociates.
3. The half-life of Taq polymerase is about 40 min at 95^0C. Thus there will be significant loss of activity over the 25-30 cycles used in a typical

PCR experiment. Therefore, it may be necessary to add fresh enzyme during the course of an experiment.

4. Taq polymerase may incorporate an extra adenine residue at the 3' end of the newly synthesized molecule, which has no complementary base in the template. This extra residue may help in cloning the product of PCR.

Other Thermophilic Polymerases

A number of thermophilic polymerases are available from other *Thermus*, *Thermococcus*, and *Pyrococcus* species. These include Tf1 and Tth enzymes from *Thermus flavus* and *Thermus thermophilus*, respectively. These enzymes do not have 3'-5' exonuclease (proof reading) activity. The enzymes Tli and VentR from *Thermococcus litoralis*, Pfu from *Pyrococcus furiosus* and DeepVentR from *Pyrococcus* sp. GB-D are more thermo-stable then Taq polymerase and have 3'-5' exonuclease (proof reading) activities.

Primer Design

Several computer programs are available to suggest suitable primer sequence for specific situations. However, following guidelines may help to design a specific primer.

1. Length of the primer: Designing of the primers is an important step towards success in PCR experiment. For designing appropriate sequences for the primer, it is important to know the sequences present at the flanking regions of the target region on the template molecule. Each primer must be complimentary (not identical) to its template strands in order to hybridization to occur. If the primers are too short they might hybridize to non-target sites and give undesired amplification products. However, short primers may offer sufficient specificity when amplifying using a simple template such as a small plasmid. But for eukaryotic genomic DNA long primers may be required. Let us take an example, imagine that total genomic DNA from rice is used for a PCR experiment utilizing two primers of eight nucleotide in length (in PCR jargon these are called "8-mers"). The expected attachment sites for these primers should be once every $4^8 = 65536$ bp, which means there should be about 46000 possible sites in a DNA molecule with 300000

Kbp of nucleotide sequences that make up the rice genome. This implies that by using a pair of 8-mer primers it would not be possible to obtain specific single amplification product from genomic DNA of rice. With 17-mer primers the expected frequency attachment sites will be once every 4^{17} = 17179869184 bp. This value is five times more than the total rice genomic DNA. Therefore, 17-mer primer should have only one site for hybridization in the genomic DNA of rice and produce one specified amplification product.

This is not advisable to make primers as long as possible, as the length of the primer adversely affect the hybridization rate: longer the primers slower the rate of hybridization. Thus complete hybridization to the template molecules cannot occur within the time allowed in the reaction cycle, which leads to inefficiency in PCR reaction. In practice, primers with 20-30 nucleotides give satisfactory results.

2. Melting temperature: DNA-DNA hybridization is a temperature dependent phenomenon. At too high or two low temperature, there may be no hybridization or mismatched hybridization respectively. The ideal temperature can be determined by knowing the melting temperature or T_m of the primer-template hybrid. The T_m is the melting temperature of DNA at which the DNA strands are half denatured, meaning half double stranded half single stranded. A temperature 1-2^0 C below T_m should be ideal for normal primer-template hybridization. T_m can be obtained experimentally or through the following formula:

$$T_m = [4 \text{ x } (C+G)] + [2 \text{ x } (T+A)]^0 \text{ C}$$

Where, C+G is the number of C and G nucleotides in the primer sequence and T+A is the number of T and A nucleotides.

The two primers may be designed having relatively similar T_m value, so that appropriate reaction is possible between the two primers and the template. The similarity of melting temperature will mean that the primers have a similar nucleotide composition.

3. Allowing mismatches: The 3' end of the primer should be correctly base-paired to the template so that the polymerase can bind properly for the extension of strand. It is always better to have C or G as the 3' terminal

nucleotide, as it would make the binding more stable compared to A or T at the 3' end. Some mismatches in the body of the primer sequence may be allowed at the 5' end, which will not anneal to the template. This may be necessary to incorporate restriction endonuclease recognition sites (not present in the template) to facilitate subsequent cloning and manipulation of the PCR products.

4. Internal repeat structure: The primers should not have any internal repeat sequence, so that the molecule does not fold back on itself, and not available to bind to the template.
5. Primer-primer annealing: It also important to see that the two primers do not anneal to each other. Annealing of primers lead to formation of a primer dimer, which may behave as an efficient template (as they are small) for amplification in the subsequent round of PCR.

In specific situations it is also possible to design a primer when the DNA sequence of the primer annealing sites is not known with certainty. Such situations are described below:

1. Primers based on amino acid sequence. There may be a situation, when the amino acid sequences of some or all of a purified protein are known and that information can be used to make primers to amplify the coding sequence. Since the genetic code is degenerate, it is not possible to predict the coding sequence with certainty, except for methionine and tryptophan as they have only one codon.
2. Primers based on related amino acid or nucleotide sequence. There may be another situation where it is required to amplify members of a gene family (multiple copies of a gene in one organism or homologous copies of the same gene in different organisms). Since members of a gene family are not absolutely conserved, they will differ in their base sequence. Therefore, it will not be possible to determine the sequence of the primer annealing sites with complete accuracy.

Many organisms preferentially use some codons for a given amino acids to others. So it may be possible to guess the codon likely to be used for a given amino acid, if information on codon preference for other genes is available. If it is not possible to determine which particular nucleotide to include in a particular site, then more than one nucleotide should be included at that position. This is called mixed site. Alternatively, a nucleotide having broader pairing capabilities e.g. inosine should be included. By using a mixed site, it can be made sure that a fraction of the primer molecules will have the correct sequence and they will have better chance to anneal at the end of the template.

Applications of PCR

PCR has a number of applications. These include, cloning, DNA sequencing, diagnostics, forensic, population genetics and archaeology and evolution. Details of some of these applications are described in Chapter 6.

Precautions and Drawback of PCR

Size: Inclusion of proof-reading enzyme will increase the size of the PCR product up to 10 kbp or more, as the incorrectly incorporated nucleotides can be removed thereby avoiding chain termination. The use of enzyme mixtures to generate large products is called long-range PCR. In such case, it may be necessary to adjust the time for DNA synthesis in the cycle times. If the template is heavily degraded, then it may not be possible to obtain large PCR products. Reports on ancient DNA suggest that an inverse correlation exist between the age of the template and the size of template and the size of the amplified molecule.

Amplifying the wrong sequence: Incorrect annealing may be avoided by using long primers, as this will ensure annealing at the specific site. Raising the temperature and adjusting the concentration of magnesium ions (which stabilize primer-template binding) can be used to increase the specificity of primer binding.

Modifications in PCR Technology

Improving specificity: One of the major problems is the annealing of primers at the wrong location(s), which will generate wrong PCR product(s). Although it is possible to reduce this problem by raising the temperature or by increasing the magnesium ion concentration, several other approaches have been suggested to overcome the problem.

1. **Hot-Start PCR:** In this process, DNA polymerase is not added to the reaction mixture until they reach the DNA melting temperature of the first cycle, so that the action of DNA polymerase cannot start before complete denaturation of the molecule. This is the basis of hot-start PCR.

For small number of samples this may work satisfactorily. But for large samples it would be more convenient if the polymerase activity can be made unavailable till the appropriate temperature (T_m) is reached. This can be achieved by incorporating the polymerase or the magnesium (required for polymerase to function) salt into wax beads. The wax beds will melt at high temperature releasing the enzyme or magnesium salt to start the reaction. Another approach is to inactivate the polymerase by combining with an antibody, which will denature at high temperature allowing polymerase to function.

2. **Touch-Down PCR:** To ensure stable binding the annealing temperature is usually kept several degrees lower than the maximum temperature, so that the primers can remain bound to the template. At this low temperature a small amount of mismatching between primers and template may take place, which may allow binding to incorrect sites and generating undesirable products. This can be reduced by touch-down PCR. In this procedure, initially a high temperature is used during annealing. The annealing temperature is reduced in the following rounds. At some point the correct temperature will be reached where annealing between primer and template will be ideal matched and no incorrect pairing will take place. The later cycles may be under less stringent conditions. But since the early cycles are be carried out under most stringent temperature conditions, the desired products will be produced most abundantly.
3. **Nested PCR:** In nested PCR two PCRs are carried out successively. The first PCR is carried out in the conventional manner. The products of the first PCR are then used as the template for the second PCR with the same primers used in the first PCR. It is expected that first PCR may generate some non-specific products along with the desired products. In the second PCR it is unlikely that the non-specific products will contain annealing sites for both the primers. Therefore, only the desired products from the first PCR cycle will be the suitable templates for the second cycle and the undesirable products will be eliminated.

Jumping PCR

When degraded DNA is amplified through PCR, it may not give the product which represents the entire distance between the two primer sites. In the first round of synthesis, the primer would be extended to the end of the fragmented molecule, but not to the second primer site. In the second round of synthesis,

Figure 13. Steps involved in jumping PCR (For details see text)

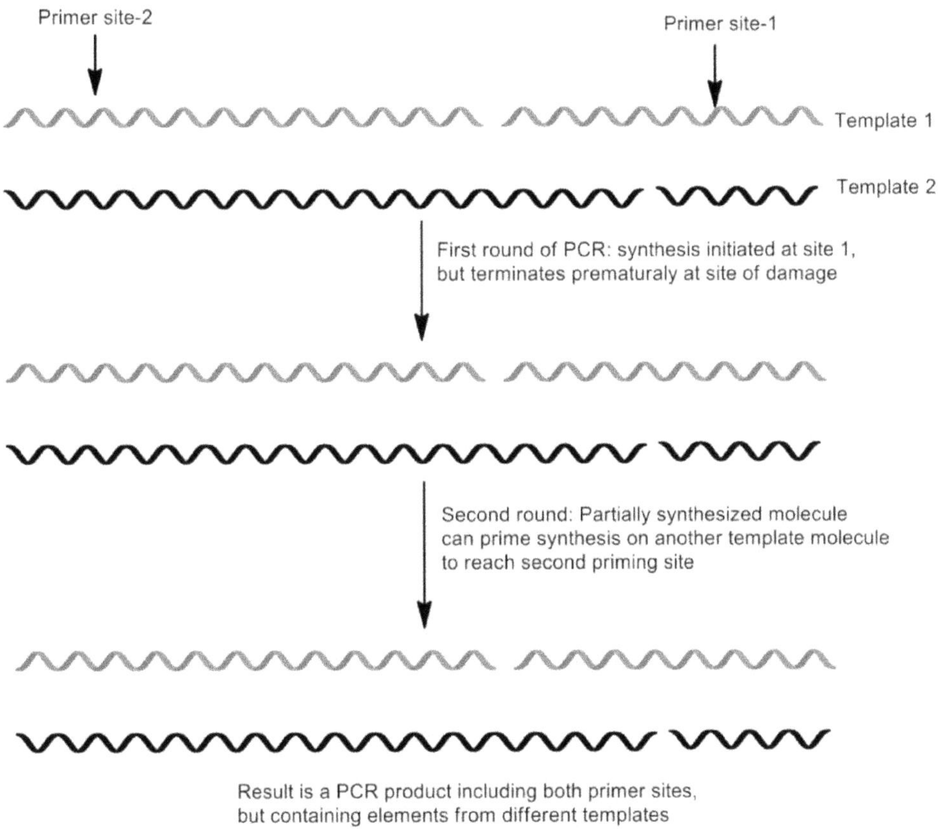

the truncated amplification product may anneal to a different DNA fragment that contains the remaining region intact. This would then allow synthesis of the full PCR product. This is called jumping PCR (Figure 13). This can be advantageous when amplifying badly degraded DNA molecule. But in the case of an individual, which is heterologous at two sites within an amplified region, jumping PCR could generate recombinant molecule for these loci. Such results may be disadvantageous.

Inverse PCR (IPCR)

In the conventional PCR reaction, amplification of the sequence between the primers is carried out. However, it is also possible to amplify the sequences outside the primers, which is known as inverse PCR (IPCR). The sample DNA is first cut with an enzyme outside the region whose sequence is already

Figure 14. Steps involved in inverse PCR (For details see text)

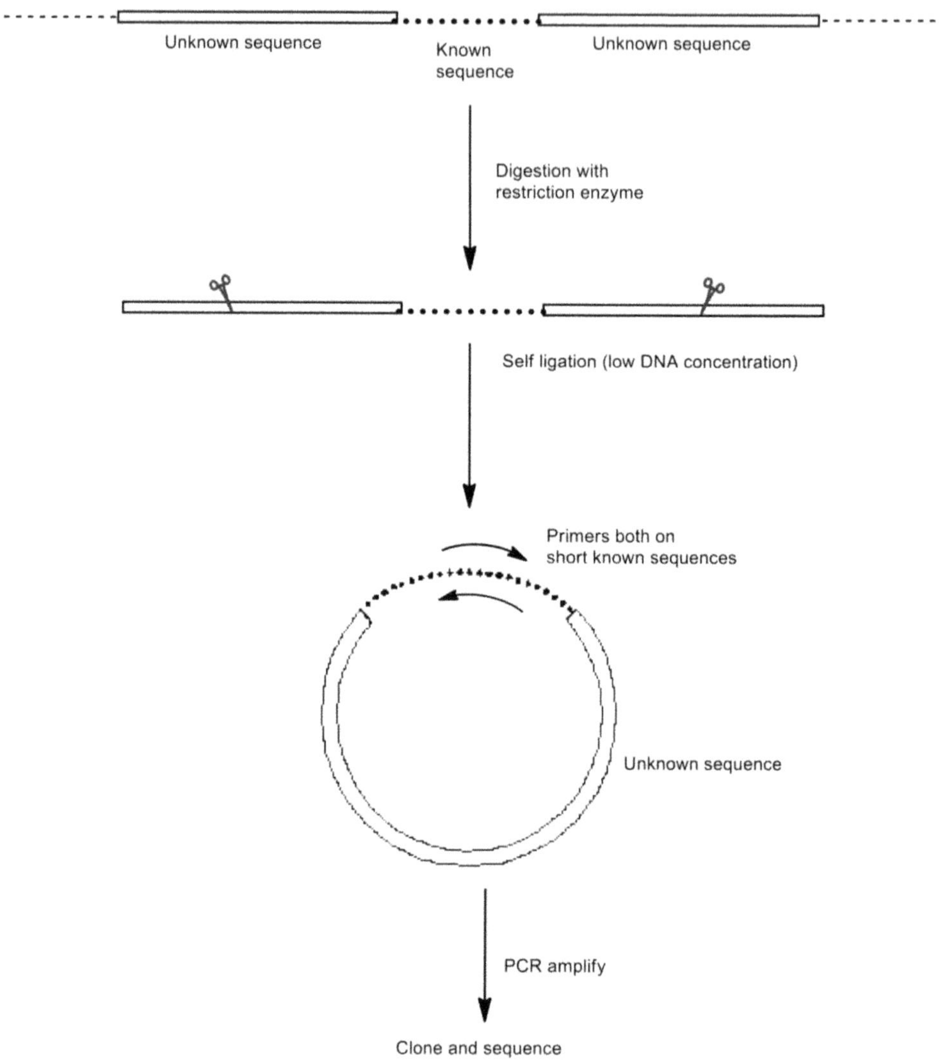

known. The resulting linear molecules are then circularized by ligation. Then the circular DNA is cut within the known sequence through restriction digestion. This treatment will make the known sequences turned inside out (Figure 14). Thereafter primers complimentary to the known sequence can be used to amplify the region of interest.

Reverse Transcriptase PCR (RT-PCR)

Sometimes to know about their abundance in a sample, it may be necessary to amplify mRNA molecules before cloning. This can be done by the use of reverse transcriptase enzyme and a single primer, so that a single strand cDNA is made prior to PCR reaction. The primer for reverse transcription should be specific to the particular mRNA.

In-Situ PCR

PCR can also be done *in situ* using pre-metabolized tissue, such as thin sections on a microscopic slide. To facilitate such reaction the PCR machine is combined with an adopter to accommodate the slide. The PCR products are normally detected by *in-situ* hybridization, through which it is possible to locate the target nucleic acid within the tissue. Sometime it is combined with RT-PCR to identify the location of a particular transcript.

Quantitative PCR (qPCR) or Real-Time PCR

Quantitative estimate of a particular nucleic acid molecule can be made by PCR and RT-PCR. There are two approaches to quantify the molecule. In the first approach, the PCR product can be quantified through gel electrophoresis, and comparing with a standard or control. In this method the end product of PCR has to be used and therefore called end-point measurement. In the second method, quantification can be done while the PCR is in progress (i.e in real time). This is normally done in two ways. In the first, a fluorescent, dsDNA-binding dye (such as SYBR green) is added in the PCR mixture. As the dsDNA product accumulates, the amount of fluorescent from the dye increases, and this can be detected through a fluorescence detector attached to the PCR machine. Since this method detects dsDNA, it gives a measurement of the amount of dsDNA produced at a given time regardless of whether it is from the correct region or not Heid et al. 1996, Wong and Medrano 2005).

A specifically synthesized oligonucleotide probe is used in the second method (Figure 15). This probe is designed to anneal within the region to be amplified. It carries a reporter fluorescent dye at one end and a quencher at the other end of the molecule. When the quencher and the reporter are in close proximity (i.e. attached to the same oligonucleotide), then the quencher stops the reporter from fluorescing. During PCR, the probe will anneal to single strand DNA within the target region. When the polymerase comes in contact with the annealed probe, the 5'-3' exonuclease activity of the polymerase shall degrade the probe, separating the reporter from the quencher. These will results into accumulation of reporter during the course of the PCR.

Figure 15. Steps involved in real-time PCR. The figure shows the reaction monitored with a probe carrying reporter and quencher. Hydrolysis of the probe by the polymerase liberates the reporter, which fluoresces.

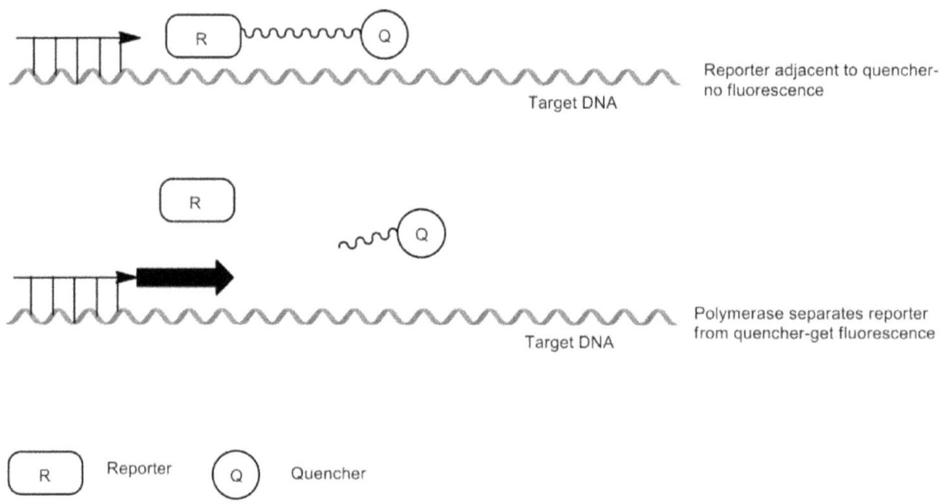

In a modified method of real-time PCR, two oligonucleotide probes are used. One of the probes is tagged with a molecule that absorbs light and the second with a molecule that can accept energy from the first and then re-emit energy at a different wavelength. When the two probes are annealed to the adjacent sites of their target DNA, then energy transfer is possible, otherwise little or no energy transfer can take place. Thus, quantitative assessment of the target DNA can be done by measuring the amount of fluorescence from the second tag.

Asymmetric PCR

By reducing the amount of one of the two primers, it is possible to regulate preferential amplification of one of the strands of a DNA molecule. This will result into single strand DNA molecule, which has several applications including DNA sequencing. The second amplification primer can be used as a primer for the sequencing reaction. Preferential amplification of one strand by adopting the technique described above is known as asymmetric PCR.

Anchored PCR

When sequence of one segment a nucleic acid region of interest is known

Figure 16. Steps involved in anchored PCR. The double stranded DNA molecule with known sequence is anchor that provides one of the two priming sites for PCR. The primers are shown by (i) and (ii).

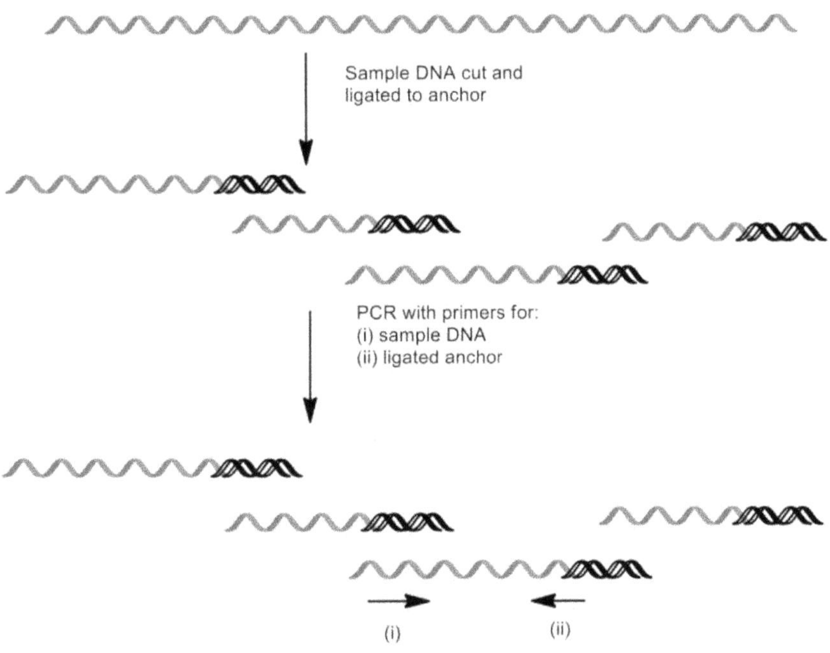

(thus having one priming site), it can be amplified through a process known as anchored PCR. Basically the region to be amplified is attached to a known sequence and then that site is used as a second priming site (Figure 16).

Emulsion PCR or Droplet PCR

Usually PCR reactions are carried out in plastic micro-tubes. These micro-tubes are not suitable when it is required to amplify DNA of much smaller size. To overcome the problem, such reactions are carried out inside lipid droplets. It is much easier to increase or decrease the temperature in small droplets very quickly. Since these lipid droplets contain a single template molecule, all the products within a droplet results from the amplification of a same template.

Droplet Digital PCR (ddPCR) has been developed on the basis of the technology called water-oil emulsion droplet. About 20,000 droplets are generated from each sample, and then each droplet is used for amplification of the template molecules through PCR. The droplets are comparable to the wells in the microplate in which the PCR is performed, but in a much smaller format. By creating thousands of droplets it becomes possible generate thousands of data points from a single sample, and makes statistical analysis more effective. This technique can work efficiently with small size compared to other commercially available digital PCR system. This helps to reduce cost and preserve precious sample. After completion of the PCR process each droplet is analyzed through a flow cytometer and compared with the PCR-positive fraction in the original sample.

The benefit of ddPCR include absolute quantification, unparallel precision, increased signal-to-noise ratio, reduced error rates, simplified quantification, reduced cost, superior partitioning of the sample, and lower equipment costs.

Degenerate Sequences and Non-Standard Bases

There may be occasions when exact DNA sequence for a PCR target is not known. Such situations arise, for example, when using sequence from one species to amplify a homologous sequence or a similar gene in another species. In such circumstances, it will be required to make primers that are more flexible in their specificity so that they can amplify a product. This is where the degenerate primer comes in. There are a few simple guidelines that can be used to increase the chances of getting an appropriate amplicon, as outlined below.

First, in the case of amplifying a gene homolog, it is important to select a region where the sequence of the amino acids is conserved in as many species as it is possible to find in the potential target. While there is no hard and fast rule, it is best to have such a conserved region run for at least six to eight amino acids. This will provide more choices as to which part to be used. It is also good if as many amino acids as possible, in the run, have less codons (one or two). Table 3 shows the codon specifications for the twenty amino acids.

As an example, the following amino acid sequence may be considered; GYPVVTCQWD.

Using the standard nucleotide coding system: A,C,G,T; R(G or A); Y (T or C); K (G or T); M (A or C); S (G or C); W (A or T); B (G,T,C); D (G,A,T); H (A,C,T); V (G,C,A); N (all), all possible DNA sequences encoding this peptide can be specified as:

Gly Tyr Pro Val Val Thr Cys Gln Trp Asp

GGN TAY CCN GTN GTN ACN TGY CAR TGG GAY

The total number of possible degenerate primers for this peptide is 16,384. There are ways to make this more manageable. We can begin by specifying

Table 3. Distribution of amino acids by codon specificity

One Codon	Two Codons	Three Codons	Four Codons	Six Codons
Met (M)	Cys (C)	Ile (I)	Ala (A)	Leu (L)
Trp (W)	Asp (D)		Gly (G)	Arg (R)
	Glu (E)		Pro (P)	Ser (S)
	Phe (F)		Thr (T)	
	His (H)		Val (V)	
	Lys (K)			
	Asn (N)			
	Gln (Q)			
	Tyr (Y)			

a fixed 3' end. The last six nucleotides are TGGGAY. If we remove Y and the last five nucleotides are confined as TGGGA, then the next step would be to take help of a Codon Usage Table for the species whose DNA is being targeted. Preference for certain codons by some species has been observed. The best source for such data is the Codon Usage Database (http://www.kazusa.or.jp/codon/).

There are thousands of species listed (some of which contain more complete information than others) by their genus and species names. It is essential

Figure 17. Universal base. (a) 3-nitropyrrole, and (b) 5-nitroindole

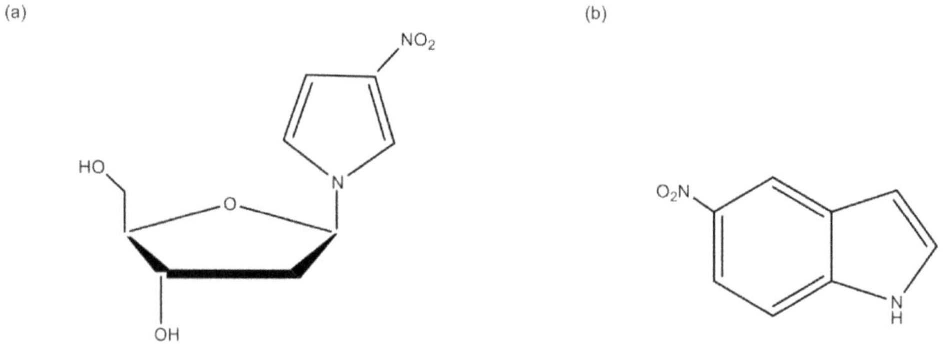

to know the basic information about the organism to use the database. An alternative means of designing degenerate primers that can be used along with or instead of the mixed base sites shown above is the universal base approach. Universal bases are analogue compounds that can be used to replace any of the four bases of DNA which does not destabilize base-pair interactions. Two universal bases commonly used are 3-nitropyrrole and 5-nitroindole (Figure 17). The first universal base was 2'- deoxyinosine. It is still used extensively, but does display a slight bias towards hybridization of nucleotides with dI:dC being favored over other pairings. In the past few years, truly universal base analogues have been engineered that have no pairing bias and do not alter stability. A new tool recently added to the array of non-standard bases in oligonucleotides that can be used for a variety of applications is the locked nucleic acid or LNA.

Loop-Mediated Isothermal Amplification (LAMP)

In the conventional PCR, repeated heating and cooling is required for amplification of nucleic acids. To overcome the problem, a process has been developed which allows templates to be amplified at a constant temperature (around 65^0 C). This has been achieved by using a DNA polymerase with strand-displacing activity. Therefore, there is no need of heating the molecules to high temperature for denaturation. This technique has practical utility such as detection of pathogens in the field, without the use of PCR machine.

DNA FINGERPRINTING

DNA fingerprinting is a technique for determining the likelihood of similarities between genetic materials derived from different individuals or groups. It has been established that in human 99 percent of DNA is identical between individuals. The one percent by which they differ enables scientist to distinguish between different individuals. This is also true for most of the eukaryotic species.

The genetic information of any living organism is carried by its DNA. The DNA carries all the functional genes of the organism, which are expressed to perform different tasks assigned to them. However, our entire DNA does not contain useful information. A large amount of DNA is called "junk" or "non-coding" DNA which does not code for any proteins. Like coding sequence, changes in the base sequence do often occur within these regions of junk DNA. When the change occurs in a functional gene, it may affect the organism adversely leading to its death, thereby remove that altered gene from the population. However, when the changes occur in the junk DNA, they are normally retained by the organism, as they do not contribute for the survival of the organism.

As a consequence, random variations accumulate in the non-coding regions of DNA which could be such that in every 200 bases there exist a variation. These variations are exploited through DNA fingerprinting to create a visible pattern to assess similarity or differences between two or more samples. After restriction digestion, highly variable fragments of DNA can be separated through electrophoresis or by PCR amplification. Each DNA sample will produce a unique banding pattern (based on their number and size) in the gel. Higher the genetic similarity between two individuals, higher will be the similarity in their banding profile and vice versa. Thus the probability of similarity or otherwise between two or more samples can be easily ascertained. Steps involved in DNA fingerprinting technique is shown in Figure 18.

The application of DNA fingerprinting can be explain through an example. In human, DNA sequences in the non-coding regions are frequently repeated. Such variations are called Variable Number Tandem Repeats (VNTRs). Within a population different person can have different number of repeats. Thus such variations can be used to construct genetic fingerprints of individual person. Suppose a person 'X' is having only 6 repeats in the VNTR region, and person 'Y' is having 11. When the genomic DNA from both are digested with a restriction enzyme say, *Eco*Rl (assuming that *Eco*R1 can cuts at either end

Figure 18. Steps involved in DNA fingerprinting technique. The process begins with a cell sample from which DNA is extracted. The DNA is cut into fragments using restriction enzymes. The fragments are then separated into bands by electrophoresis through an agarose gel. The DNA band pattern is transferred into a nylon membrane. The radioactive DNA probe is introduced. The DNA probe binds to specific DNA sequences on the nylon membrane. The excess probe material is washed away leaving the unique DNA band pattern. The radioactive DNA pattern is transferred to X-ray film by direct exposure. When developed, the resultant visible pattern is the DNA fingerprint.

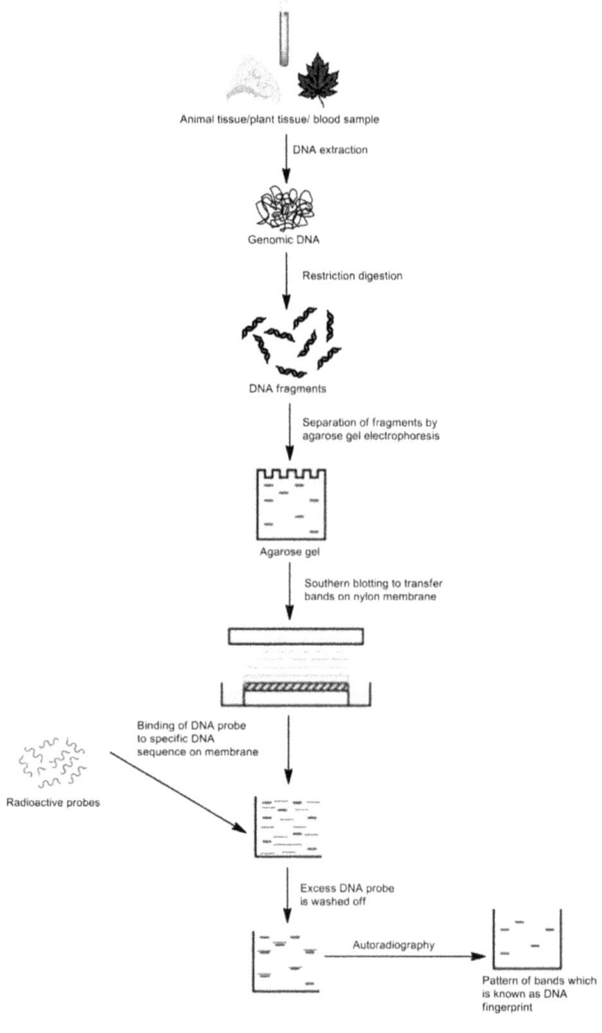

Figure 19. Paternity test through DNA fingerprinting. There is a match between a fragment (band) of mother and the child. There is also a match between fragments of possible father A and the child. But there is no match of fragments between the child and possible father B.

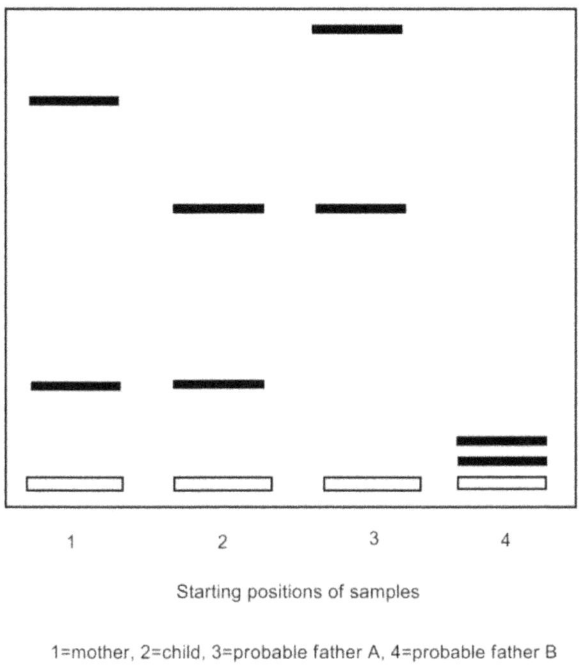

of the repeated sequence), the DNA fragment produced by Y will be nearly double compared to the fragments from X. Thus there will be difference in their banding sites in the gel (Figure 19). By analyzing the entire VNTRs sites of the genomic DNA of an individual, a "DNA fingerprint" unique to every individual can be constructed. The "DNA fingerprints" thus generated can be used for all future applications.

DNA fingerprinting can be used to prove heredity of an individual, as base pair sequences carried by the parents are transmitted to their offsprings. Pattern of migration and claims of ethnicity can also be studied by DNA fingerprinting. However knowledge about traditional sociological methodologies is essential for such analysis.

Perhaps best known use of DNA fingerprinting is in forensic science. DNA samples obtained at a crime scene can be used prepare DNA fingerprints and then compare with the DNA fingerprints of any suspect(s), to verify their similarities or otherwise. Since DNA fingerprints databases are available for only known offenders, it is not possible to identify the probable offender from the general public. With the advances in technology such a possibility cannot be overruled in the future.

Genetic fingerprinting can contribute significantly in medical science. Genetic basis of inherited diseases can be traced through DNA fingerprinting. Identification of the genes involved in disease susceptibility (genetic disorder) and the inherent physiological mechanism of disease development can be immensely helpful for developing therapies. Presence of inherited abnormalities in the parents and the fetuses can be screen through DNA fingerprinting, and necessary precautionary measure and therapeutic treatments can be prescribed.

DNA SEQUENCING

DNA sequencing is a technique by which the precise order of nucleotides in a piece of DNA can be determined. Although DNA sequencing methods were available for quite some time, rapid and efficient sequencing methods have been developed on during late 1970s.

Two sequencing techniques for DNA were invented almost simultaneously. They are known as Maxam-Gilbert (chemical cleavage) method and Sanger (or dideoxy) method. In Maxam-Gilbert method nucleotide-specific cleavage are induced chemically and is suitable for sequencing short oligonucleotides, less than 50 base-pairs. In Sanger method chain termination is done by dideoxy nucleotide and elongation through PCR. This method is technically easier to apply and therefore popular. Advancement and automation of PCR technology has made the technique easy to apply for long DNA strands.

The sequencing reactions in Sanger method are similar to the replicating reactions of DNA through PCR. The reaction mix is composed of: free nucleotides, template DNA, a primer, and an enzyme (usually a variant of Taq polymerase). The primer is a small single stranded DNA fragment (about 20-30 nucleotides long) that hybridizes to the template DNA. The dsDNA is first separated by heating, and then DNA polymerase elongates the primer that gets attached to its intended location in the ssDNA molecule. This leads to formation of a new strand of DNA. If we start with a million identical piece of template DNA, we can get a million new copies of one of its stands.

For DNA sequencing instead of normal nucleotide a dideoxyribonucleotide is used. Dideoxynucleotide is similar to the normal nucleotide, except it has no 3'- hydroxyl group (Figure 20). Therefore, if it is incorporated to the end of a DNA strand, it will stop elongation.

Figure 20. Structure of: (a) Deoxyribonucleotide and (b) Dideoxyribonucleotide. The reaction is carried out in the presence of a dideoxyribonucleotide. This DNA is like regular molecule, but without the 3' hydroxyl group. Thus, if added to the end of a DNA molecule, it cannot continue to grow. The reaction mixture most of the nucleotides are regular ones and only a fraction of them are dideoxynucleotides.

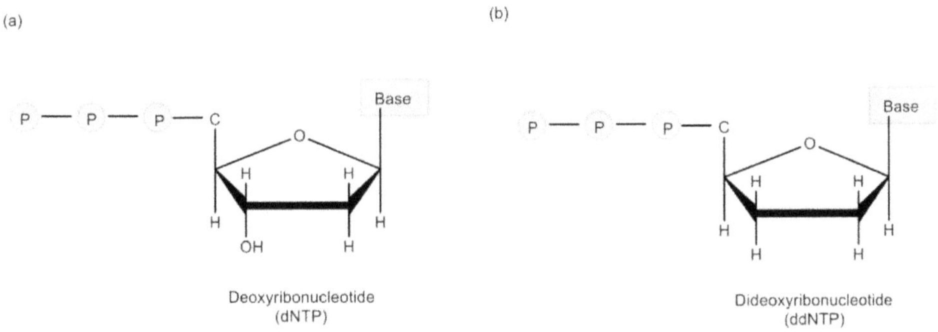

Let us take an example, suppose in the reaction mixture most of the thymine (T) nucleotides are normal ones and a certain percentage (small fraction) of them are dideoxynucleotides. While replicating a DNA strand, whenever a 'T' is required to be incorporated in the new strand, a normal nucleotide shall be picked-up for incorporation in most of the time. After adding a 'T', more nucleotides shall be added as the strand grows. However, in about 5% of the insertions, a dideoxy-T will be incorporated. In such situation the DNA strand will stop elongation. The fragment will eventually break away.

Over time, all of the newly synthesized copies will be terminated by a 'T'. However, termination point for each new strand will be random. If millions of new strands are initiated it will be possible to obtain strands terminated at every possible 'T' that are present in the sample DNA molecule. All the newly synthesized strands will start from the same point, but will terminate at a different point along the DNA molecule. Out of say, one billions new terminated strands, many millions will be found to have terminated at each possible 'T' position. To identify the sites of all the 'T's in the newly synthesized strand, the size of each terminated DNA fragments have to be determined (Figure 21). This can be done with the help of gel electrophoresis.

Now let us imagine that the reaction mixture contain certain percentage (small fraction) of all the four of the dideoxy nucleotides (A, T, G, and C) and with different fluorescent colors attached to each one of them, along with the normal nucleotides. After running the gel electrophoresis for the

Figure 21. DNA sequencing. The growing DNA strands stops at all possible 'T' along its length. Synthesis of all of the strands started from one position, but culminated at different positions according to the incorporation of dideoxynucleotide "T". Out of the billions of terminated fragments, many millions will be identical in terms of its terminated 'T' positions. Each fragment having similar specific "T" termination can be identified and measured through gel electrophoresis.

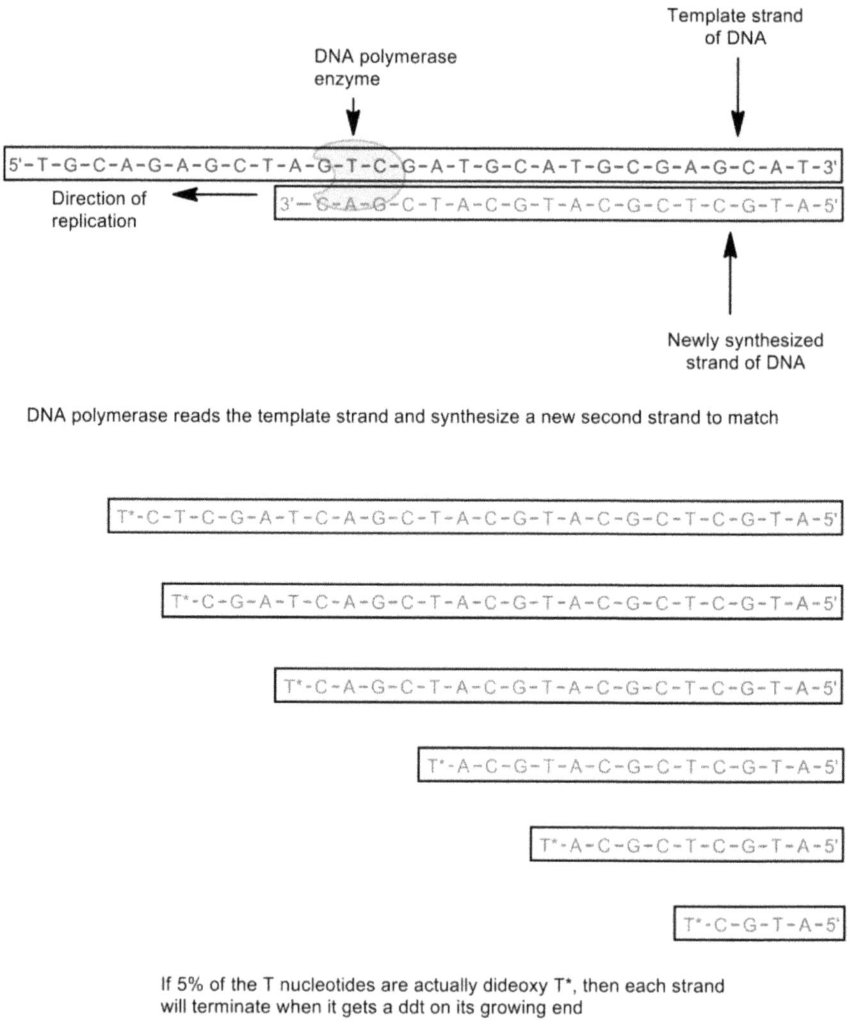

fragments obtained from the reaction mixture, it will produce a kind of gel as shown in Figure 22. The sequence of the DNA can be determined on the basis of the color code. Only thing that need to be done is to read the colors from bottom to top of the gel.

Figure 22. DNA sequencing. Reaction mixture contains certain percentage (small fraction) of all the four of the dideoxynucleotides (A, T, G, and C) and with different fluorescent colors attached to each one of them, along with the normal nucleotides. (a) After running the gel electrophoresis for the fragments obtained from the reaction mixture, it will produce a kind of gel as shown above. (b) From the color code it is possible to determine the sequence of the DNA. Only thing that need to be done is to read the colors from bottom to top of the gel.

(a)

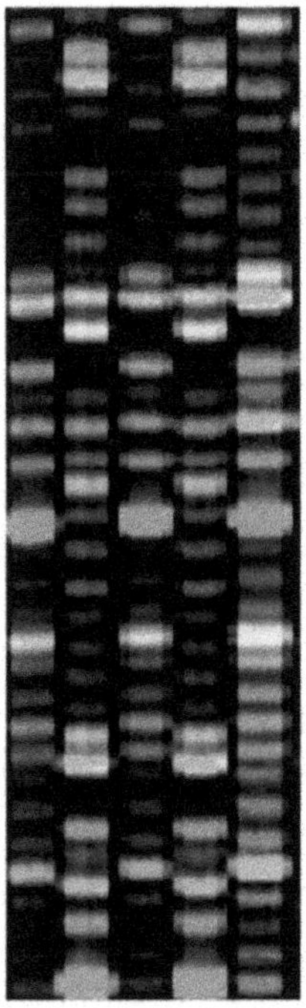

(b)

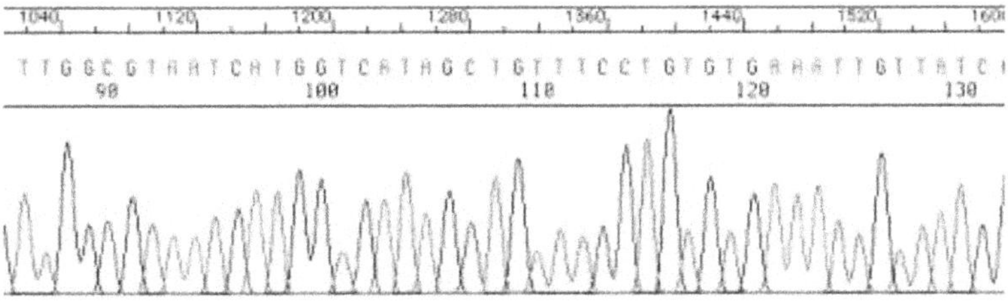

Automated DNA sequencing is done with the help of a machine that can run different steps of electrophoresis and can detect different colors produced by the nucleotides. The machine uses 'capillary electrophoresis', in which DNA fragments are passed through a tiny glass-fiber capillary, which comes out in different size-order through the far end. An ultraviolet laser is passed through the liquid emerging from the far end of the capillaries. The pulses of fluorescent colors emitted by the nucleotides are recorded. Normally 96 samples can be analyzed at a time in a sequencer. An image of the fragments of sequencing gel is shown in Figure 22a. Each nucleotide is represented by four different colors red, green, blue and yellow. A chromatogram showing the peak of each color is generated for each fragment with the help of a laser beam (Figure 22b), from which template DNA sequence can be determined.

NEXT-GENERATION DNA SEQUENCING

The scenario of DNA sequencing is changing rapidly due to new innovative and commercially viable approaches, called next generation sequencing (NGS) technologies, also known as high-throughput sequencing. The instruments became commercially available in 2004, and the techniques became popular due to several advantages. Apart from the fact that these techniques are much less expensive and takes fraction of the time required compared to conventional techniques, they can be used for a variety of biological samples to address new biological inquiry, like ancient genome, characterization of ecological diversity and identification of unknown etiologic agents etc. These technologies have revolutionized the study of genomics and molecular biology.

Three critical steps are shared by the next generation high throughput sequencing platforms: preparation of DNA sample, immobilization of the DNA molecules, and sequencing of the nucleotides (Figure 23). In general, during sample preparation, genomic DNA is fragmented randomly to critical sizes and then defined sequences, known as "adapters", are added to the ends of these fragments. Addition of adopters to the universal nucleic acid ends is called "sequencing library." Adopters are used to help the DNA fragments of the sequencing library to anchor to a solid surface, thereby immobilizing the fragments at a fixed site to carry out the sequencing analysis (Figure 24). Except PacBio RS platform, all the other high-throughput sequencing systems need sequencing library DNA amplification, to produce distinct and detectable sequencing features (Figure 25). Amplification is done *in situ* through PCR. Through this a clusters of clonal DNA copies are generated. Sequencing is

Figure 23. High-throughput sequencing workflow. Three steps are involved in high-throughput sequencing: preparation, immobilization, and sequencing. Genomic DNA is fragmented randomly and adapter sequences are added to the ends of the fragments. The fragments obtained from the library are immobilized on a solid support. Parallel cyclic sequencing reactions are performed to determine the nucleotide sequence.

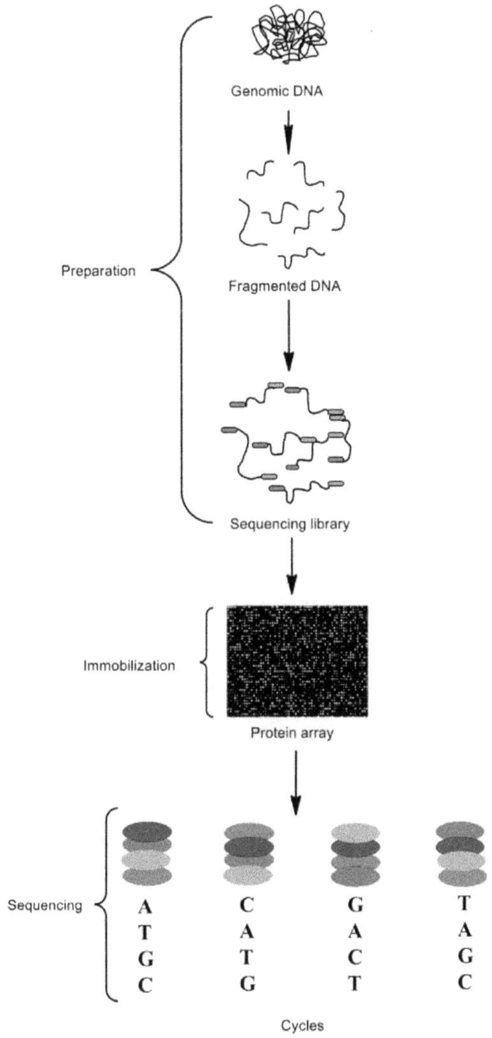

done either by using DNA polymerase synthesis for fluorescent nucleotides or by ligation of fluorescent oligonucleotides (Besser et al. 2018).

Figure 24. Sequencing library preparation. For adapter sequences addition and sequencing library preparation three approaches are available. (a) In GS FLX, Genome Analyzer, and SOLiD systems linear adapters are used. In both the ends of the genomic DNA fragments specific adapter sequences are added, (b) In CGA platform, circular adapters are used. Circular template DNA is produced by internalizing four distinct adaptor sequences, (c) In PacBio RS sequencing system bubble adapters are used. A circular molecule is generated when bubble adapters (hairpin forming) are added to dsDNA fragments.

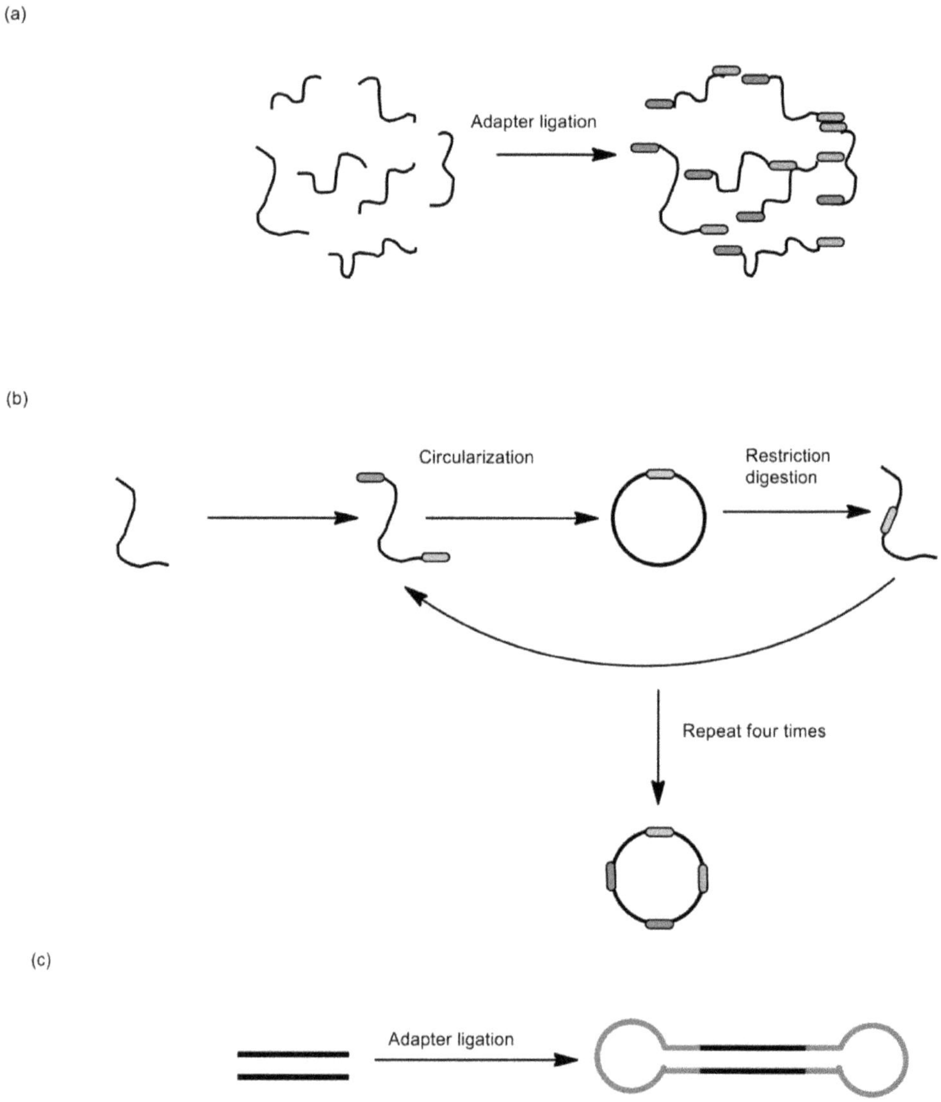

Figure 25. Sequencing features generation. Different approaches have been made to generate to detect sequencing features in high-throughput sequencing systems. (a) In GS FLX and SOLid system emulsion PCR is applied. Within an aqueous reaction bubble, emulsification is done for single enrichment bead and sequencing library fragment. Clonal copies of the template are then subjected to PCR. Beads containing the immobilized clonal DNA are put onto a Picotiter plate (in GS FLX) or on a glass slide (in SOLiD), (b) The in situ clusters of amplified sequencing library fragments is generated on a solid support through bridge-PCR, with the help of immobilized amplification primers, (c) Long stretches of DNA that fold into nanoballs is generated through rolling circle amplification, (d) In the PacBio RS system, the bubble adapted templates are attached to the biotinylated DNA polymerase. Complex of Polymerase/template is immobilized on the bottom of a zero mode wave guide (ZMW).

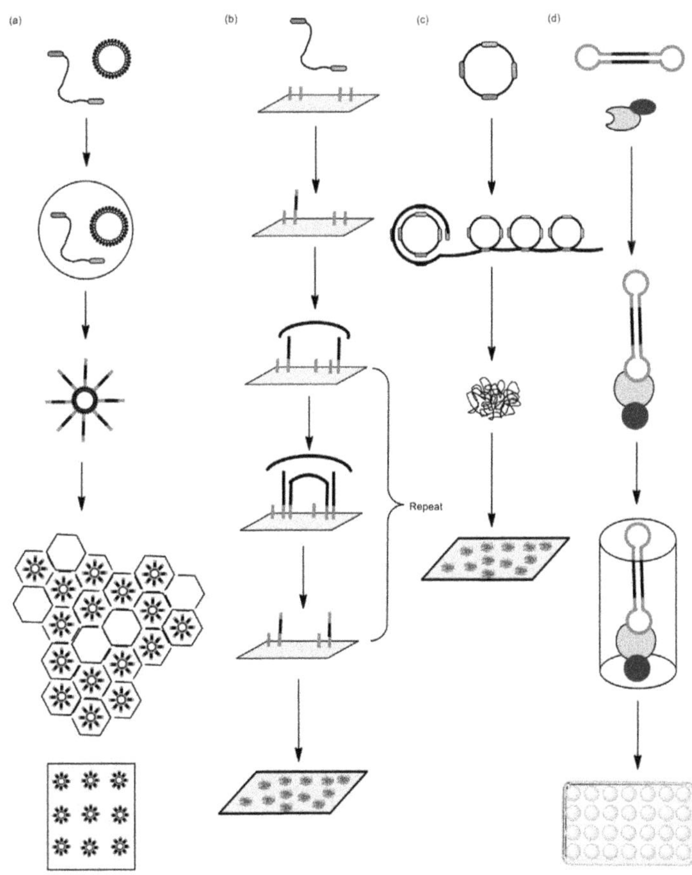

Varieties of fluidic and optic technologies have been integrated into the NSG platforms to carryout and record the sequencing reactions. The platforms are also fitted with micro-liter scale fluidic devices to assist immobilization and sequencing of DNA. These equipment are also capable of maintaining automated flow of reagents onto the immobilized DNA fragments for cyclic interrogation of nucleotide sequence (Glenn 2011). The NGS include the following technologies (Shendure and Ji 2008, Metzker 2010, Mondal and Das 2016).

Illumina (Solexa) Sequencing

In this technique, first the input samples must be cleaved into short sections, called the reads. The length of these sections (reads) will depend on the particular machinery used. Usually reads of 100-150 size are used. Slightly longer fragments are used to ligate to generic adapters. Then the reads are amplified through PCR, and spots are created with many copies of the same read. The DNA molecules are then converted to single strands for sequencing.

The illumine system utilizes a sequencing-by-synthesis approach. For this more than required amounts of nucleotides and DNA polymerase are added in the reaction mixture, so that they can freely get incorporated into the oligo-primed cluster fragments. A unique fluorescent label is incorporated into each nucleotide and their 3'-OH group is blocked chemically. This makes each incorporation event unique, and ensures incorporation of only one base at a time.

The next cycle, the 3'-OH block (terminator) is removed chemically, allowing incorporation of the next base, with the help of DNA polymerase. The fluorescent signal of the first nucleotide is also removed, so that it cannot contaminate the next image signal. The process should be repeated, so that only one nucleotide is added at a time and the image is captured in between.

Each read is provided a quality value with the help of a base-calling algorithm. Then the illumine data obtained from each run is evaluated by a quality checking pipeline, and poor quality sequences are removed. The base at each site in each image is then evaluated with the help of specific software and thereby the sequence is constructed.

Pyrosequencing

Pyrosequencing is an alternative method of DNA sequencing which detects the pyrophosphate release on nucleotide incorporation, rather than termination. Through this technique it is possible to sequence much longer reads than through illumine system. Roche-454 sequencing platform is the first pyrosequencing based NSG platform to achieve commercial applications in 2004.

In this method, after incorporation of each nucleotide by DNA polymerase, a pyrophosphate is released. A series of downstream reactions are initiated by the released pyrophosphate, which ultimately leads to production of light by luciferase, the firefly enzyme. The number of nucleotide incorporate shall produce equivalent amount of light which can be recorded.

In this method, the DNA or RNA is fragmented into shorter reads of 400 to 600 bp and genetic adapters are added to the ends. They are then mixed with a population of agarose or resin beads whose surface carry oligonucleotides complimentary to adopter sequence which are 454-specific present in the fragment library. The double stranded DNA fragments are made single stranded. Thus each bead is linked to a single fragment. With the help of adapter-specific primers, the fragments are amplified through PCR. Since each bead is placed in a different well, each well will contain all the PCR amplified copies of a single sequence. The DNA polymerase and sequencing buffers are also added to these wells. On the surface of each bead, about one million copies of each DNA fragments are produced. These amplified single molecules are then sequenced.

First, the beads are placed into the wells of a picotitre plate (PTT), each well containing a single bead. The PTT contains several thousand wells. Thus such fixed locations allow sequencing reactions to be monitored individually. Enzymes that catalyze various steps of pyrosequencing reactions are then added to the PTT. The slide is then flooded with one of the four NTPs. When the nucleotide is incorporated into the complementary DNA chains, the enzymes contained in each picotitre wells, shall transform the chemicals produced during nucleotide incorporation into light. The four nucleotides (A,T,G,C) are flowed sequentially through four different washes covering the entire PTT. The wells are washed automatically after carrying out every reaction, so that no residue of the previous reaction is left. After each reaction the light emitted by each bead are recorded by CCD camera. The number of nucleotide incorporated is equivalent to the light signal strength at a particular site. For example, if two Ts are incorporated sequentially at a particular site, the signal strength generated at that site shall be double compared to a site where only one T was incorporated. A graph is generated for each sequence read from the light signals. The graph represents the signal density for each signal wash. The sequence can be identified with the help of software.

Ion Torrrent: Proton/ PGM Sequencing

Unlike illumina and 454 sequencing techniques, Ion torrent sequencing does not make use of optical signals. Instead, in this technique the release of H^+ ion after addition of dNTP to a DNA polymer is exploited.

First the sample DNA is fragmented to about 200 bp, and then sequencing adapters are added. Thereafter, one molecule of the fragmented DNA is attached to a bead, and amplified on the bead by emulsion PCR. PCR amplification is done by placing each bead into a single well of a slide along with the PCR reagents. Following amplification the emulsion is disintegrated by chemical treatment and centrifugation. Enrichment of the amplified beads is done by glycerol gradient. The unamplified beads are pelleted at the bottom.

Like 454 sequencing method, the slides are flooded with a single dNTP, along with buffers and polymerase, one NTP at a time. The pH is detected in each of the wells, as each H^+ ion released will decrease the pH. The changes in pH can be used to identify the base, and how many thereof, was added to the sequence read.

This technique does not require optics, and depends on relatively cheaper components and disposable chips. Without the optics, that the system is also free from its dependence on slow image scans and thus the sequencing reactions are relatively fast. Absences of chemiluminescence (fluorescence) reaction keeps the nucleotides unmodified, and can be reused thereby make the system comparatively cheaper.

SOLiD Sequencing

In Support Oligonucleotide Ligation Detection (SOLiD) platform, DNA library is immobilized by binding to a solid support through emulsion PCR and cycle sequenced-by-ligation chemistry. This process include ligation, detection, and cleavage steps which are repeated several times during reaction for extending the complimentary strand to a length determined by the number of cycles.

In this process DNA sample is fragmented to a size ranging from 400 to 850 bp, end repaired and DNA adapters such as “P1” and “P2” is ligated to the ends of the fragments. Immobilization of the sequencing DNA onto "P1" coated paramagnetic beads is carried out by emulsion PCR. High-density, semi-ordered colony arrays are generated by functionalizing the 3' ends of the templates and immobilizing the modified beads to a glass slide.

SOLiD technology applies partially degenerate, fluorescently labeled, DNA octamers with dinucleotide compliment sequence recognition core. The oligonucleotides used for detection are hybridized to the template and the sequences that are perfectly annealed are ligated to the primer. After imaging, non-extended strands are capped and flurophores are cleaved. A fresh cycle starts from 5-bases upstream of the priming site. After completing seven sequencing cycles, the primer of the first sequencing cycle comes off and a second primer is hybridized to the template at the n-1 site. For sequencing five sequencing primers (n, n-1, n-2, n-3, and n-4,) are applied. To improve the accuracy of sequencing, 35- base insert is sequenced twice.

Complete Genome Sequencing

In Complete Genome Analysis (CGA), diverse technologies are brought together to create a comprehensive system for large scale sequencing of complete genomes. This system integrates a sequencing platform that is the combination of technology advancements in libraries, arrays, sequencing assay, instruments and software. The technology is based on preparing of circular DNA libraries and rolling circle amplification (RCA) to create nanoballs of DNA that are attached to a solid support.

DNA is fragmented to 200 to 500 bp, and the ends of the fragment are end-repaired and dephosphorylated. Through nick translation fragments are bound to common adapters. Then uracils are incorporated into the product through PCR and uracil containing primers. Overhangs are created by removing the uracils from the products. The products are then digested and methylated with *Acu*I and circularized with the help of T4 DNA ligase in the presence of splint oligonucleotide. Residual linear DNA molecules are degraded by treatment with exonuclease. By repeating the process it is possible to produce circular sequencing molecules having four unique adapters. Before the final circularization step, a single-stranded template is purified first by separating the strands and then by treating with exonuclease. At the end of the reaction it will produce two genomic DNA inserts having 26 bases and two other genomic DNA inserts having 13 bases adjacent to the adapter sequences.

The single-strand DNA library is amplified through rolling circle amplification (RCA) method. RCA creates long DNA strands from circular DNA library templates having short palindrome sequences. These palindrome sequences promote intramolecular coiling within the long linear molecules, and thereby induce formation of the DNA nanoballs (DNBs). A nanoball is a three dimensional, condensed, spherical sequencing object formed by long strand of repetitive fragments of amplified DNA.

CGA Platform uses a strategy called combinational probe anchor ligation (cPAL) for sequencing. Initially an anchor molecule and one of the unique adapters establishes linkage by hybridization. Then in the first position of the probe, four degenerated 9-mer oligonucleotides are labeled with specific fluorophores that corresponds to four specific nucleotides (A,T,G,C). For determination of sequence the correct matching probe has to hybridize to a template, which in turn has to be ligated to the anchor. The ligated anchor-probe molecules are denatured after imaging. The process is repeated five times using new sets of fluorescently labeled 9-mer probes that contain known bases at four positions.

The sequencing is continued after five cycles by resetting the reaction mixture, along with an anchor and degenerated region of the bases. A second round of five cycles of sequencing is done by ligation using fluorescently labeled 9-mer probes. By using unique anchors, the cycle of sequencing 10 bases can be repeated, up to eight times, which will resolve 62 to 70 base long reads from one DNA nanoball (DNB).

Pacific Bioscience RS II (PacBio RS II) Sequencing

PacBio RS II sequencing uses proprietary SMRT (Single Molecule Real Time) technology, which can detect incorporation of nucleotide during elongation of the replicated DNA strand from non-amplified single strand template, in real-time. This technology uses phospholinked nucleotides, in which phosphate chain of the nucleotide is labeled fluorescently rather than the base. Thus, incorporation of nucleotides is detected on the basis of the release of the associated fluorophore dissipated due to breakage of the phosphate chain.

Incorporation of nucleotide in nanoscale space which is detected in real time is called Zero Mode Waveguide (ZMW). ZMWs are nanofabricated on a glass surface, and the volume of the wells, coated with nanometre-sized aluminum layer, is in zeptolitre scale. For polymerase immobilization, SMRT cells are prepared by surface coating with streptavidin. To initiate the sequencing reaction, it is essential to incubate a biotinylated Phi29 DNA polymerase with primed SMRT cell DNA templates. By using a biotin-streptavidin reaction, the products obtained from above reaction are immobilized to the SMRT cell. The tethered Phi29 polymerase is a highly progressive strand-displacing enzyme capable of performing rolling circle amplification (RCA).

Once the sequencing reaction starts, nucleotides with individually phospho-linked fluorophores are incorporated to the growing chain, by the tethered polymerase. It is important to note that each fluorophore corresponds to a specific base (A,T,G,C). During base incorporation event, the fluorescent nucleotide is brought into the active site of the polymerase and near to the ZMW glass surface. A high resolution camera placed at the bottom of the ZMW records the fluorescence of the nucleotide being incorporated. A phosphate-coupled flurophore is released from the nucleotide during incorporation in the growing DNA chain, while dissociation diminishes the fluorescent signal. Incorporation of successive nucleotides in the growing chain is recorded in a movie-like format. Light pulse collected from the elongating nucleotides at the 150 k ZMW are monitored and analyzed in parallel using an optimized set of algorithms for the analysis. For genome analysis, DNA is broken randomly and subjected to end-repairing. Thereafter, 3' adenine is added to the fragmented DNA, to facilitate ligation of an adopter with a T overhang. The adapter forms an intra-molecular hairpin structure and is composed of a single DNA oligonucleotide. Although SMRT bell DNA template is a linear molecule, the bubble adapters create a circular molecule. Comparison of different features of the different NGS platforms is presented in Table 4 (Quail et al. 2012).

THIRD GENERATION SEQUENCING

High-throughput sequencing his revolutionized research in plant evolutionary biology. Development of third-generation sequencing technologies (Bethune et al. 2019), such as MinION, has provided the researchers to target long fragments of chloroplast DNA to improve genome assembly. Accordingly, it has now possible to develop a new method to capture and sequence long DNA fragments in plants, which promises to significantly improve assembly of plastid genomes and advance biodiversity research.

The new sequencing technologies like the MinION can play a pivotal role in accelerating biodiversity discovery and understanding its functioning and

Table 4. Comparison of different features of next generation sequencing technologies

Platform	Sequencing library	Support	Feature generation	Sequencing reaction	Detection method
GS FLX (Pyrosequencing)	Linear adapter	Picotitre plate	Emulsion PCR	Synthesis	Pyrosequencing
Genome Analyzer (Illumina sequencing)	Linear adapters	Flow cell	Bridge PCR	Synthesis	Fluorophore labelled reversible terminator nucleotides
SOLiD	Linear adapters	Flow cell	Emulsion PCR	Ligation	Fluorophore labelled oligonucleotide probes
CGA	Circular adapters	DNA nanoball arrays	Rolling circle amplification (RCA)	Ligation	Fluorophore labelled oligonucleotide probes
PacBio RS	Bubble adapters	Zero mode waveguide (ZMW)	Single molecule	Real-time Synthesis	Phospholinked fluorophore labelled nucleotides

evolution, especially in the tropics. This is because sequencing is becoming more portable and easy to do (Bethune et al. 2019).

Targeted sequencing combined with TGS technology could improve genome assembly, or the computational stitching together of many sequenced DNA fragments into larger continuous blocks. The short reads produced by older sequencing technologies, typically 100-400 base pairs long, make bioinformatic assembly of certain genomic regions like repetitive sequences very difficult. Third-generation sequencing technologies such as the portable MinION produce longer reads that could help produce these assemblies. However, these sequencers generally have lower data output than older sequencing technologies. Therefore, to sequence regions of interest efficiently, these regions must be enriched through methods like targeted capture and sequencing (Bethune et al. 2019).

GENOTYPE-BY-SEQUENCING (GBS)

A highly multiplex system developed for construction of reduced representational libraries for the illumine NSG platform, in the Buckler lab is known as genotype-by-sequencing (GBS) system. Through this, large number of SNPs can be generated for genetic analysis and genotyping. The

Figure 26. Procedure for genotyping-by-sequencing (GBS) for plant breeding

Plant tissue (leaves)

↓

Isolation, quantification and normalization of DNA from leaf tissue

↓

Digestion of DNA with restriction enzymes

↓

Ligation of adaptors including bar coding region in adaptor 1 in random *Pst*I-*Mse*I restricted DNA fragments

↓

Representation of different amplified DNA fragments with different bar codes from different biological samples/lines, which represent the GSB library

↓

Analysis of sequences from library on a NSG sequencer

↓

Bioinformatic analysis of NGS sequencing data

system is preferred due to low cost, minimum sample handling, reduced PCR and purification steps, absence of size fractionation, absence of reference sequence limits, easy barcoding and easy scaling-up. The steps involved in the GBS technology and some potential applications are shown in Figure 26.

GBS offers a simplified library production procedure which can encompass large number of individuals/lines. GBS combined with genome-independent imputation can be used to construct genetic maps in pseudo-testcross progeny. A GBS protocol using two enzymes (*Pst*I and *Msp*I), has been used to reduce greater degree of complexity and production of uniform library sequence in wheat and barley (He et al. 2014).

With Ion PGM system, two different GBS strategies have been adopted: (i) restriction enzyme digestion, in which no specific SNPs have been identified, and can be used to discover new markers for MAS. The complexity of genome is reduced by digesting the DNA with selected enzymes followed by ligation of adopters, and (ii) multiplex enriching PCR, in which a set of SNPs are defined for a particular section of the genome. In this approach, specific PCR primers are used to amplify the area of interest.

The efficiency of the GBS can be increased by incorporating a multiplex sequencing strategy that uses a barcoding system, which is usually inexpensive. Barcodes are introduced in one of the adapter sequences, upstream of RE cut site in genomic DNA, which eliminates the second illumine sequencing read. GBS is less complicated than RAD method. It requires single-well digestion of genomic DNA, need reduced sampling handling, need fewer DNA purification steps, and fragments are not size selected.

The low cost GBS method has now been applied in a variety of crops for discovering and genotyping SNPs. GBS is suitable for characterization of germplasm, population studies, plant genetics, and breeding diverse crops. Construction of GBS libraries is comparatively simple, quick, specific, and highly reproducible (He et al. 2014).

CAPILLARY ELECTROPHORESIS

In advanced automated sequencing, instead of gel electrophoresis, Capillary electrophoresis (CE) is used. Like conventional slab gel electrophoresis, capillary electrophoresis is used for separation of DNA fragments. In CE the DNA fragments are passed through a narrow glass fibre capillary during electrophoresis and they go out in ordered sizes. An ultraviolet laser source, built into the system, throws beams through the liquid coming out from the end of the capillaries, and records the fluorescent pulse that emerges. In commonly used sequencers, 96 samples can be analysed through as many capillaries (lanes). The four colours (blue, red, yellow and green) represent four different nucleotides.

CE is faster, highly sensitive, provide better resolution and easier to handle generated data compared to slab gel electrophoresis. The requirement of the sample DNA is in the nanogram range and it takes few minutes for separation. Detection is done either by fluorescent labelling or UV absorption. Resolution up to single-base within as fragment can be done up to several hundred base pairs.

Automated sequencing techniques can generally be accurately applied up to a maximum of 700-800 bp in length. However, by using methods such as Primer Walking and Shotgun sequencing (step-wise methods), it is possible to sequence larger genes including whole genome.

Figure 27. DNA sequencing by shotgun method. The Shotgun sequencing entails randomly cutting the DNA segment of interest into more appropriate (manageable) sized fragments, sequencing each fragment, and arranging the pieces based on overlapping sequences. This technique has been made easier by the application of computer software for arranging the overlapping pieces

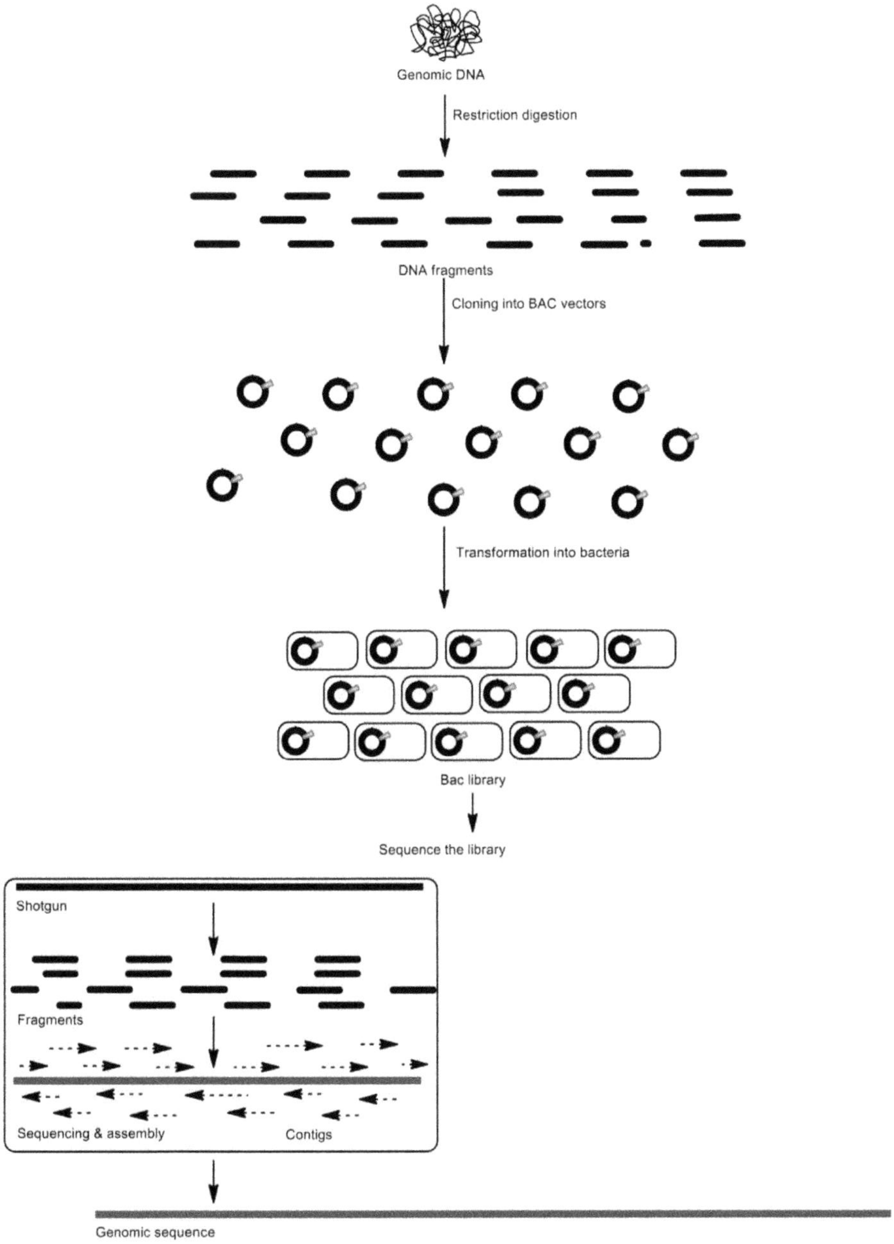

In Primer Walking, first Sanger method is used to sequence a small workable portion of a larger gene. Then reliable segment of the sequence is used to generate new primers, and used to sequence segments of the gene present outside the original segments. In Shotgun sequencing the DNA segment of interest is cut into appropriate (manageable) sizes, and each fragment is sequenced. The sequenced fragments are arranged on the basis of overlapping sequences, with the help of software (Figure 27).

CONCLUSION

Methods and analyses in all areas of molecular biology include the preparation and analysis of DNA, RNA and proteins. There have been many developments over the past three decades that have led to the efficient manipulation and analysis of nucleic acid and proteins. Many of these have resulted from the isolation and characterization of numerous DNA-manipulating enzymes, such as DNA polymerase, DNA ligase, and reverse transcriptase. However, perhaps the most important was the isolation and application of a number of enzymes that enabled the reproducible digestion of DNA. These enzymes, termed *restriction endonucleases* or *restriction enzymes*, provided a turning point for not only the analysis of DNA but also the development of recombinant DNA technology.

Since its introduction in the 1980s, PCR has become a standard tool in molecular biology research. One advantage of PCR is its extreme sensitivity which makes possible the detection and analysis of low abundance DNAs. This is especially helpful when limited amounts of starting material are available, or when few copies of the target sequence are present. Applications of PCR include the cloning of known and novel genomic DNA and cDNA sequences, DNA sequencing, construction of mutant or chimeric DNAs, and quantification of mRNA and DNA.

The molecular biology techniques have now become cost effective, efficient and reliable, although improvement in the overall efficiency of the techniques is a continuous process. Combination of various molecular biology techniques are being used to resolve our understanding about the structure and function of the macromolecules in agriculture, medicine, environmental science, forensic and many more.

REFERENCES

Alwine, J. C., Kemp, D. J., & Stark, G. R. (1977). Method for detection of specific RNAs in agarose gels by transfer to diazobenzyloxymethyl-paper and hybridization with DNA probes. *Proceedings of the National Academy of Sciences of the United States of America*, *74*(12), 5350–5354. doi:10.1073/pnas.74.12.5350 PMID:414220

Besser, J., Carleton, H. A., Garner-Smidt, P., Lindsey, R. L., & Trees, E. (2018). Next-generation sequencing technologies and their application to study and control of bacterial infections. *Clinical Microbiology and Infection*, *24*(4), 335–341. doi:10.1016/j.cmi.2017.10.013 PMID:29074157

Bethune, K., Mariac, C., Couderc, M., Scarcelli, N., Santoni, S., Ardisson, M., Martin, J.-F., Montúfar, R., Klein, V., Sabot, F., Vigouroux, Y., & Couvreu.r, T. L. P. (2019). Long-fragment targeted capture for long-read sequencing of plastomes. *Applications in Plant Sciences*, *7*(5), e1243–e1248. doi:10.1002/aps3.1243 PMID:31139509

Burnette, W. N. (1981). "Western Blotting": Electrophoretic Transfer of Proteins From Sodium Dodecyl Sulfate—Polyacrylamide Gels to Unmodified Nitrocellulose and Radiographic Detection With Antibody and Radioiodinated Protein A. *Analytical Biochemistry*, *112*(2), 195–203. doi:10.1016/0003-2697(81)90281-5 PMID:6266278

Glenn, T. C. (2011). Field guide to next-generation DNA sequencing. *Molecular Ecology Resources*, *11*(5), 759–769. doi:10.1111/j.1755-0998.2011.03024.x PMID:21592312

He, J., Zhao, X., Laroche, A., Lu, Z. X., Liu, H., & Li, Z. (2014). Genotype by sequencing (GBS), an ultimate marker-assisted selection (MAS) tool to accelerate plant breeding. *Frontiers in Plant Science*, *5*, 1–8. doi:10.3389/fpls.2014.00484 PMID:25324846

Heid, C. A., Stevens, J., Livak, K. J., & Williams, P. M. (1996). Real time quantitative PCR. *Genome Research*, *6*(10), 986–994. doi:10.1101/gr.6.10.986 PMID:8908518

Lee, J., Daugharthy, E., Scheiman, J., Kalhor, R., Ferrante, T. C., Terry, R., Turczyk, B. M., Yang, J. L., Lee, H. S., Aach, J., Zhang, K., & Church, G. M. (2015). Fluores.cent *in situ* sequencing (FISSEQ) of RNA for gene expression profiling in intact cells and tissues N*ature. Protocols*, *10*(3), 442–458. doi:10.1038/nprot.2014.191 PMID:25675209

Metzker, M. L. (2010). Sequencing technologies- the next generation. *Nature Reviews. Genetics*, *11*(1), 31–46. doi:10.1038/nrg2626 PMID:19997069

Mondal, T. K., & Das, A. (2016). Next generation sequencing and its utilization for managing the plant genetic resources. In P. C. Deka (Ed.), *Biotechnological tools for genetic resources* (pp. 1–18). Daya Publishing House.

Mullis, K., Faloona, F., Scharf, S., Saiki, R., Horn, G., & Erlich, H. (1986). Specific enzymatic amplification of DNA in vitro: the polymerase chain reaction. *Cold Spring Harbor Symposium Quantum Biology, 51*, 236-273. 10.1101/SQB.1986.051.01.032

Quail, M. A., Smith, M., Coupland, P., Otto, T. D., Harris, S. R., Connor, T. R., Bertoni, A., Swerdlow, H. P., & Gu, Y. (2012). A tale. of three next generation sequencing platforms: Comparison of Ion Torrent, Pacific Biosciences and Illumina MiSeq sequencers. *BMC Genomics*, *13*(1), 341–346. doi:10.1186/1471-2164-13-341 PMID:22827831

Shendure, J., & Ji, H. (2008). Next-generation DNA sequencing. *Nature Biotechnology*, *26*(10), 1135–1145. doi:10.1038/nbt1486 PMID:18846087

Southern, E. M. (1975). Detection of specific sequences among DNA fragments separated by gel electrophoresis. *Journal of Molecular Biology*, *98*(3), 503–517. doi:10.1016/S0022-2836(75)80083-0 PMID:1195397

Towbin, H., Staehelin, T., & Gordon, J. (1979). Electrophoretic Transfer of Proteins From Polyacrylamide Gels to Nitrocellulose Sheets: Procedure and Some Applications. *Proceedings of the National Academy of Sciences of the United States of America*, *76*(9), 4350–4354. doi:10.1073/pnas.76.9.4350 PMID:388439

Wong, M. L., & Medrano, J. F. (2005). Real-time PCR for mRNA quantitation. *Biotechnique*, *39*(1), 1–11. doi:10.2144/05391RV01 PMID:16060372

ADDITIONAL READING

Ahmed, A. A., Mukhopadhya, A., & Bahar, B. (2016). Application of high throughput molecular techniques for breeding of farm animals against major diseases. In P. C. Deka (Ed.), *Biotechnological tools for genetic resources* (pp. 309–345). Daya Publishing House.

Ashley, N. E., Jessica, S., & David, M. S. (2012). Application of next-generation sequencing in plant biology. *American Journal of Botany*, *99*(2), 175–185. doi:10.3732/ajb.1200020 PMID:22312116

Bernardo, A., Wang, S., Ahmed, P. S., & Bai, G. (2015). Using next generation sequencing for multiplexed trait-linked markers in wheat. *Public Library of Science (PLoS). ONE*, *10*, e143890. doi:10.1371/journal.pone.0143890

Bhagyawant, S. S., & Shrivastava, N. (2019). *Recent advances in plant molecular biology*. Himalaya Publishing House.

Celik, O. (2018). *New age molecular techniques in plants. Peptide Synthesis Services*. doi:10.5772/intechopen.79360

Davey, J. W., Hohenlohe, P., Etter, P. D., Boone, J. Q., Catchen, J. M., & Blaxter, M. L. (2011). Genome-wide genetic marker discovery and genotyping using next generation sequencing. *Nature Reviews. Genetics*, *12*(7), 499–510. doi:10.1038/nrg3012 PMID:21681211

During, K. (1993). Non-radioactive detection methods for nucleic acids separated by electrophoresis. *Journal of Chromatography. A*, *618*(1-2), 105–131. doi:10.1016/0378-4347(93)80030-8 PMID:8227252

Elshire, R. J., Glaubitz, J. C., Sun, Q., Poland, J. A., Kawamoto, K., Buckler, E. S., & Mitchell, S. E. (2011). A robust, simple genotyping-by-sequencing (GBS) approach for high diversity species. *Public Library of Science (PLoS). ONE*, *6*(5), e19379. doi:10.1371/journal.pone.0019379 PMID:21573248

Ganal, M. W., Altmann, T., & Roder, M. S. (2009). SNP identification in crop plants. *Current Opinion in Plant Biology*, *12*(2), 211–217. doi:10.1016/j.pbi.2008.12.009 PMID:19186095

Garg, R., & Jain, M. (2011). Pyrosequencing data reveals tissue specifuc expression of lineage-specifuc transcripts in chickpea. *Plant Signaling & Behavior*, *6*(11), 32–44. doi:10.4161/psb.6.11.17879 PMID:22057340

Gharizadeh, B., Herman, Z. S., Eason, R. G., Jejelowo, O., & Pourmand, M. (2006). Large-scale Pyrosequencing of synthetic DNA: A comparison with results from Sanger dideoxy sequencing. *Electrophoresis, 27*(15), 3042–3047. doi:10.1002/elps.200500834 PMID:16800029

Hardin, S. H. (2008). Real-time DNA sequencing. In J. M. Weinheim (Ed.), *Next generation genome sequencing: towards personalized medicine* (pp. 97–102). Wiley-VCH Verlag and Co. doi:10.1002/9783527625130.ch8

Huang, X. Q., Feng, Q. Q., & Zhao, Q. (2009). High-throughput genotyping by whole-genome resequencing. *Genome Research, 19*(6), 1068–1076. doi:10.1101/gr.089516.108 PMID:19420380

Khan, M. S., Khan, I. A., & Barth, D. (Eds.). (2012). *Applied molecular biotechnology: the nest generation of genetic engineering*. CRC Press.

Kircher, M., & Kelso, J. (2010). High-throughput DNA sequencing-concepts and limitations. *Biotechnology Essays, 32*, 524–536. PMID:20486139

Kricka, R. E. (1992). *Nonradioactive DNA probe techniques*. Academic Press.

Kubik, K. B., & Sugisaka, G. (2002). From molecular biology to nanotechnology and nanomedicine. *Biosystematics, 65*(2-3), 123–138. doi:10.1016/S0303-2647(02)00010-2 PMID:12069723

Kumar, S., Banks, T. W., & Cloutier, S. (2012). SNP discovery through next-generation sequencing and its applications. International Journal of Plant Genomics. *Article ID, 831460*. Advance online publication. doi:10.1155/2012/831460 PMID:23227038

Maliga, P., Klessig, D. F., Cashmore, A. R., Gruissem, W., & Verner, S. E. (1995). *Methods in plant molecular biology. A laboratory course manual.* Cold spring Harbour Laboratory Press.

Miesfield, R. L. (1999). *Applied molecular genetics*. Willy-Liss.

Myllykangas, S., Buenrostro, J., & Ji, H. P. (2012). Overview of sequencing technology platforms. In E. Rodriguez (Ed.), Bioinformatics for high throughput sequencing (pp. 11-24). New York: Springer Science+Business Media. doi:10.1007/978-1-4614-0782-9_2

Nawaz, M. A., Baloch, F. S., Rehman, H. M., Shahid, M. Q., Yildiz, M., ... Chung, G. (2016). Development of a competent and trouble free DNA isolation protocol for downstream genetic analysis in glycine species. *Turkish Journal of Agricultural Food Science and Technology*, *4*(8), 700–705. doi:10.24925/turjaf.v4i8.700-705.788

Rapley, R. (2009). Basic molecular biology techniques. In J. M. Walker & R. Rapley (Eds.), *Molecular Biology and Biotechnology* (5th ed.). Royal Society of Chemistry. doi:10.1039/9781849730211-00001

Russel, D. W., & Sambrook, J. (2001). *Molecular cloning: a laboratory manual*. Cold Spring Harbor Laboratory Press.

Shajahan, A. (2011). *Laboratory manual of basic techniques in plant molecular biology*. Trichi: ZAZYM Publication.

Sung, S., Hug, F., & Chen, Z. J. (2012). International plant molecular biology: A bright future for green science. *Genome Biology*, *13*(11), 323–338. doi:10.1186/gb-2012-13-11-323 PMID:23164288

Thudi, M., Li, Y., Jackson, S. A., & Varshney, R. K. (2012). Current state-of-art of sequencing technologies for plant genome research. *Briefings in Functional Genomics*, *11*(1), 3–11. doi:10.1093/bfgp/elr045 PMID:22345601

Wilhelm, B. T., & Landry, J. R. (2009). RNA-seq quantitative measurement of expression through massively parallel RNA-sequencing. *Methods (San Diego, Calif.)*, *48*(3), 249–257. doi:10.1016/j.ymeth.2009.03.016 PMID:19336255

APPENDIX

1. Describe the principles of Fluorescence *in situ* Hybridization (FISH) technique. What are the applications of FISH technique?
2. What is Flow Cytometry? Describe its applications.
3. Why all DNA molecules move in the same direction when electrophoresed? Is this also generally true in the case of proteins? Explain.
4. In SDS-electrophoresis of proteins, what features of the technique cause protein molecules to move in a single direction?
5. A mixture of different proteins is subjected to electrophoresis in three polyacrylamide gels, each having a different pH value. In each gel, six bands are seen. Can one reasonably conclude that there are only six proteins in the mixture? Explain.
6. What is Enzyme Linked Immunosorbent Assay (ELISA) technique? Describe its applications.
7. Describe the principles of pulse field gel electrophoresis.
8. Explain whether it is possible to determine the size of a PCR product through gel electrophoresis?
9. During Southern blotting DNA fragments are first separated through gel electrophoresis, and then transferred to a membrane filter. Before it is transferred, the gel is soaked in an alkaline solution to denature the double stranded DNA, and then neutralized. Explain why it is required to denature the double stranded DNA?
10. What is electroelution? How electroelution is performed?
11. What information and materials are required to amplify a segment of DNA through PCR?
12. Both PCR and RT-PCR can be used to quantify DNA and RNA in any experiment. If we assume that the efficiency of PCR process is 100 percent in every step, how many copies of a template would be amplified after 20 cycles of a PCR reaction if the number of starting template molecules were: i) 10, ii) 100, iii) 1000, and iv) 10,000.
13. Unlike *Taq* DNA polymerase, which polymerase enzymes lacks proof-reading activity? Some other DNA polymerases like *Vent*, have proof reading activity. What advantages are there for using DNA polymerases having proof reading activities for PCR?
14. What modifications are required to be made to the PCR to use this method for site-specific mutagenesis?

15. What is DNA fingerprinting? How this method could be used to establish parentage? How this method could be used in forensic science laboratories?
16. How are dideoxynucleotides (ddNTPs) used in the chain termination method of DNA sequencing?
17. In a typical PCR experiment, what phenomenon take place during variations in temperature range from (i) 90 – 95^{0}C, (ii) 50 – 70^{0}C, and (iii) 70 – 75^{0}C?
18. What specific properties, the DNA polymerases should have to be amenable for using in PCR reaction?
19. What is autoradiography? How autoradiography is different from *in-situ* hybridization? What are its applications?
20. How the qPCR is different from normal PCR? What are the advantages of qPCR?

About the Author

Pradip Chandra Deka is currently the Vice Chancellor at Sir Padampat Singhania University, Udaipur, Rajasthan, India. He obtained his Bachelor's degree from Assam Agricultural University, Jorhat, Assam, India and thereafter Master's and Doctorate degree in Genetics and Plant Breeding from Banaras Hindu University, Varanasi, India. Prof. Deka was the recipient of ICAR Senior Fellowship for pursuing doctorate degree. Prof. Deka was awarded WINROCK International Fellowship for pursuing post-doctorate research in Molecular Genetics at the University of Florida, Gainesville, USA and the University of California, Berkeley, USA. Prof. Deka was the Professor & Head, Department of Agricultural Biotechnology, Assam Agricultural University, Jorhat, India and also the Dean, Faculty of Agriculture at the same University. He served as Visiting Professor at the Institute of Plant Genomics, ETH, Switzerland and at Seoul National University, Seoul, South Korea. Earlier Prof. Deka was the Vice Chancellor, Tezpur University, Tezpur, Assam, India. Prof. Deka's research interest is on Plant Molecular Genetics and Plant Tissue Culture. He has published more than 200 research papers in reputed national and international journals and authored five books on Molecular Genetics and Biotechnology.

Index

F

G

H

I

L

M

N

www.ingramcontent.com/pod-product-compliance
Ingram Content Group UK Ltd.
Pitfield, Milton Keynes, MK11 3LW, UK
UKHW051028041026
15470UKWH00015B/1723

9 781799 84312